Salvation from the Sky

U.S. Navy, Royal Australian Air Force,
and Royal New Zealand Air Force
Heroic Air-Sea Rescue in the
Pacific in World War II

Imagine being a fighter or bomber pilot. You and your crew have been in the heat of battle when, suddenly, your plane catches fire or your engine conks out. You have to bail out or ditch in the water below. Who will save you? In World War II, survivors of Allied aircraft who found themselves in such straits, looked skyward in desperate hope, particularly those within range of Japanese shore guns, or adrift in enemy waters. Their prayers were answered when large, ungainly PBY Catalina or PBM Mariner seaplanes, whose engines thundered in noisy disproportion to the speed they generated, alighted on the water nearby. In the face of gunfire from enemy shore batteries, every second spent as a helpless, fixed target invited disaster for the pilots and aircrews of these plucky planes. Nevertheless, they willingly risked their lives to bring the survivors of downed aircraft, and sunken vessels, back from the shadow of death on slow, sure wings. Air-sea rescue operations were often hazardous, even in the absence of enemy threat. Seemingly calm whitecaps viewed from the air, might well be rolling swells twenty feet high, forcing pilots to put down on moving slopes of water. Gigantic bounces in heavy seas often resulted in damage that prevented their taking flight again. In this companion book to *Eyes of the Fleet* and *Ingram's Fourth Fleet*, readers take flight with the heroic aircrews of rescue aircraft scouring ocean waters for their fellow Allied servicemen. *Salvation from the Sky* also visits four future American presidents—John F. Kennedy, Richard Nixon, Gerald Ford, and George H. W. Bush—who were then serving in the Pacific Theater.

One hundred seventy-nine photographs; maps and diagrams; appendices; a bibliography; and an index to full-names, places, and subjects add value to this work.

Salvation from the Sky

U.S. Navy, Royal Australian Air Force,
and Royal New Zealand Air Force
Heroic Air-Sea Rescue in the
Pacific in World War II

Cdr. David D. Bruhn, USN (Retired)
and
Stephen Ekholm

HERITAGE BOOKS
2020

HERITAGE BOOKS
AN IMPRINT OF HERITAGE BOOKS, INC.

Books, CDs, and more—Worldwide

For our listing of thousands of titles see our website at
www.HeritageBooks.com

Published 2020 by
HERITAGE BOOKS, INC.
Publishing Division
5810 Ruatan Street
Berwyn Heights, Md. 20740

Copyright © 2020 Cdr. David D. Bruhn, USN (Retired)
and Stephen Ekholm

All rights reserved. No part of this book may be reproduced or transmitted in any form or by any means, electronic or mechanical, including photocopying, recording or by any information storage and retrieval system without written permission from the author, except for the inclusion of brief quotations in a review.

International Standard Book Number
Paperbound: 978-1-55613-032-8

Heritage Books by Cdr. David D. Bruhn, USN (Retired)

Battle Stars for the "Cactus Navy":
America's Fishing Vessels and Yachts in World War II

Enemy Waters:
Royal Navy, Royal Canadian Navy, Royal Norwegian Navy,
U.S. Navy, and Other Allied Mine Forces Battling the
Germans and Italians in World War II
Cdr. David D. Bruhn, USN (Retired) and Lt. Cdr. Rob Hoole, RN (Retired)

Eyes of the Fleet:
The U.S. Navy's Seaplane Tenders and Patrol Aircraft in World War II

Gators Offshore and Upriver:
The U.S. Navy's Amphibious Ships and Underwater Demolition Teams,
and Royal Australian Navy Clearance Divers in Vietnam

Home Waters:
Royal Navy, Royal Canadian Navy, and U.S. Navy
Mine Forces Battling U-Boats in World War I
Cdr. David D. Bruhn, USN (Retired) and Lt. Cdr. Rob Hoole, RN (Retired)

Ingram's Fourth Fleet:
U.S. and Royal Navy Operations Against German Runners,
Raiders, and Submarines in the South Atlantic in World War II

MacArthur and Halsey's "Pacific Island Hoppers":
The Forgotten Fleet of World War II

Nightraiders:
U.S. Navy, Royal Navy, Royal Australian Navy, and
Royal Netherlands Navy Mine Forces Battling the
Japanese in the Pacific in World War II
Cdr. David D. Bruhn, USN (Retired) and Lt. Cdr. Rob Hoole, RN (Retired)

On the Gunline:
U.S. Navy and Royal Australian Navy Warships off Vietnam, 1965–1973
Cdr. David D. Bruhn, USN (Retired) and
STGCS Richard S. Mathews, USN (Retired)

Salvation from the Sky: U.S. Navy, Royal Australian Air Force, and
Royal New Zealand Air Force Heroic Air-Sea Rescue in the Pacific in World War II
Cdr. David D. Bruhn, USN (Retired) and Stephen Ekholm

Support for the Fleet:
U.S. Navy and Royal Australian Navy Service
Force Ships That Served in Vietnam, 1965–1973

We Are Sinking, Send Help!:
The U.S. Navy's Tugs and Salvage Ships in the African,
European, and Mediterranean Theaters in World War II

Wooden Ships and Iron Men:
The U.S. Navy's Ocean Minesweepers, 1953–1994

Wooden Ships and Iron Men:
The U.S. Navy's Coastal and Motor Minesweepers, 1941–1953

Wooden Ships and Iron Men:
The U.S. Navy's Coastal and Inshore Minesweepers,
and the Minecraft that Served in Vietnam, 1953–1976

Contents

Foreword by Dwight R. Messimer	xiii
Foreword by Commodore Hector Donohue AM RAN (Ret)	xvii
Foreword by George H. S. Duddy	xxi
Acknowledgements	xxv
Preface	xxix
1. Rescue of USS *Indianapolis* Survivors	1
2. *Coos Bay* Leaves Builder's Yard and Prepares for War	11
3. Duties of Tenders at Espiritu Santo	25
4. RNZAF No. 6 Squadron	33
5. Operations from Halavo Seaplane Base	47
6. JFK Rescues Marine Paratroopers on Choiseul Island	57
7. Duty at Rendova Harbor	63
8. *Coos Bay* Steps Forward to Blanche Harbor, Treasury	69
9. Capture of Green and Emirau Islands	81
10. VP-34 "Black Cat" Catalina Pilot earns Medal of Honor	87
11. Duty at Green Island	99
12. RNZAF No. 6 Squadron at Halavo	105
13. Santo/Sydney/Noumea Interlude	115
14. Martin PBM Mariner Patrol Bomber	119
15. Duty at Saipan, Mariana Islands	123
16. Activities of Two Other Future U.S. Presidents	131
17. Liberation of the Philippines	141
18. Philippines Campaign Air-Sea Rescue	155
19. Air-Sea Rescues from Madang, New Guinea	171
20. Two RAAF Bombers and a Catalina lost in Dutch East Indies	179
21. No. 113 Air-Sea Rescue Flight	185
22. Assault of Iwo Jima	191
23. Invasion of Okinawa	197
24. The British Pacific Fleet	219
25. Air-Sea Rescue, Ryukyus	231
26. Mission to Rekata Bay, and War's End	249
27. Post-war RAAF ASR Operations	255
Postscript	261
Appendices	
A. RAAF and RNZAF Honours and Awards	267
B. RNZAF No. 6 Squadron Rescue Missions	269
C. *Barnegat*-class Seaplane Tender Battle Stars	271

D. McCampbell MOH, and Rushing Navy Cross Citations 273
E. Solicitation of Goldfish Club Members 275
F. VH-3 Instructions to Downed Airmen 277
G. Mills and Hastie DFC Award Citations 279
Bibliography/Notes 281
Index 309
About the Authors 335

Photos and Illustrations

Foreword-1: Lt. Robert Hampton Gray, RCNVR xxiv
Acknowledgements-1: Richard DeRosset xxv
Acknowledgements-2: Hector Donohue xxvi
Acknowledgements-3: George Duddy xxvi
Acknowledgements-4: John M. MacFarlane xxvii
Acknowledgements-5: Dwight Messimer xxvii
Preface-1: PBY-5 Catalina of VP-52, circa 1944 xxix
Preface-2: Lt. J. Thorvaldson with rubber boat xxx
Preface-3: Oil painting by John Hamilton xxxi
Preface-4: Diorama by Norman Bel Geddes xxxi
Preface-5: USS *YP-72* (ex-purse seiner *Cavalcade*) xxxii
Preface-6: Lifeboat fitted to a RAF Warwick aircraft xxxv
Preface-7: Douglas R4D-6 and PBM Mariner aircraft xxxvii
Preface-8: RNZAF No. 6 Squadron group photograph xl
Preface-9: Supermarine Walrus amphibian aircraft xlii
Preface-10: Flight Lieutenant Ian James Lock Wood xliii
Preface-11: Large seaplane tender USS *Salisbury Sound* xlvii
Preface-12: Small seaplane tender USS *Avocet* xlix
Preface-13: Destroyer seaplane tender USS *Osmond Ingram* l
Preface-14: Large seaplane tender USS *Wright* li
Preface-15: Painting *Salvation from the Sky* by Richard DeRosset lx
1-1: PV-1 Ventura patrol bombers 3
1-2: A PBY-5A on the water off Morotai Island 3
1-3: Heavy cruiser USS *Indianapolis* preparing to leave Tinian 6
1-4: Destroyer escort USS *Cecil J. Doyle* 9
1-5: Lt. Robert A. Marks and his aircrew in front of a PBY 10
2-1: Seaplane tender *Coos Bay* off Houghton, Washington 11
2-2: Battleship *Arizona* with Seagull scout biplane on deck 16
2-3: Fleet tug *Kewaydin* with a target hove in astern 18
2-4: Destroyer Base, San Diego 19
2-5: Fritiof L. Ekholm on the bridge wing of *Coos Bay* 20
2-6: ADM Chester W. Nimitz being toured in a jeep 21

2-7:	Recruiting poster of sailors on the beach at Waikiki	22
2-8:	Pearl Harbor Naval Air Station on Ford Island	23
3-1:	Allied shipping in the Segond Channel	25
3-2:	Maj. Gregory Boyington, commanding officer of VMF-214	31
4-1:	RNZAF No. 6 Squadron PBY-5 Catalina and crew	33
4-2:	Oil painting of USS *Curtiss* by RADM John W. Schmidt	34
4-3:	Rear Admirals John S. McCain and Aubrey W. Fitch	35
4-4:	Vought OS2U Kingfisher on a seaplane ramp	36
4-5:	SS *Cape San Juan* listing to starboard, and down by the bow	37
4-6:	U.S. Navy World War II recruiting poster	38
4-7:	A Martin PBM-3R Mariner transport aircraft taking off	40
4-8:	Cockpit of a PBM-3R Mariner	41
4-9:	ADM William F. Halsey Jr, USN	41
4-10:	RNZAF PBY-5 Catalina (No. 4017 XXT)	42
5-1:	Japanese Destroyer *Kikuzuki* under salvage in Halavo Bay	47
5-2:	7th Australian Infantry Battalion camp on Stirling Island	53
5-3:	Landing craft circle off Bougainville Island	54
6-1:	Painting *Salvation from the Sea* depicting JFK's USS *PT-59*	57
7-1:	USS *PT-164* with demolished bow in Rendova Harbor	63
7-2:	Douglas SBD Dauntless scout-bomber aircraft	65
7-3:	USAAF twin-engine B-25 Mitchell medium bomber	68
8-1:	Munda Airfield, New Georgia Island, Solomons	75
8-2:	Lockheed PV-1 Ventura patrol bombers	78
9-1:	Tank landing ship USS *LST-446* beached at Nissan Island	81
9-2:	Invasion forces off Emirau Island	84
9-3:	ADM William F. Halsey Jr., commander Third Fleet	85
10-1:	LT Nathan G. Gordon with VADM Thomas C. Kinkaid	87
10-2:	Aerial view of Langemak Bay, New Guinea	90
10-3:	U.S. Navy carrier aircraft strike at Kavieng, New Ireland	91
11-1:	LCDR Richard M. Nixon, USNR	99
12-1:	Supermarine Seagull amphibious flying boat	106
12-2:	RNZAF No. 6 Flying Boat Squadron PBY-5	108
12-3:	Minesweeper USS *YMS*-89 in San Francisco Bay	114
13-1:	Garden Island in Sydney Harbour, Australia	115
13-2:	Paddington Town Hall, Australia	116
13-3:	USS *Coos Bay* leaving Sydney	116
13-4:	Pontoon Pier at the Seaplane Base, Noumea	118
14-1:	Martin PBM-3D Mariner patrol bomber	119
14-2:	PBM Mariner starboard waist gunner	120
14-3:	PBM Mariner takes off with JATO assistance	122
15-1:	Japanese ships in Tanapag Harbor, Saipan	123
15-2:	PB2Y-3 Coronado at Naval Air Station Honolulu	124

15-3:	LTJG Alexander Vraciu aboard USS *Lexington*	126
15-4:	USS *Ellet* going astern	129
16-1:	George H. W. Bush sitting in the cockpit of an aircraft	132
16-2:	Seaplane base and town on Chichi Jima Island	133
16-3:	TBM Avenger taking off from USS *San Jacinto*	134
16-4:	TBM-1C Avenger torpedo bomber	135
16-5:	Launching of the submarine USS *Finback*	136
16-6:	LT Gerald Ford taking a sextant reading	139
17-1:	Destroyer USS *Buchanan* in Kossol Passage	143
17-2:	Japanese shipping under attack by aircraft off Manila	144
17-3:	Smoke rising from the beachhead on Leyte	145
17-4:	General MacArthur and Philippine president Osmena	146
17-5:	CDR David McCampbell in the cockpit of his Hellcat	147
17-6:	Crewmembers of the sinking Japanese carrier *Zuikaku*	149
17-7:	American survivors of the Battle off Samar	152
17-8:	Personal flag of RADM Jesse B. Oldendorf, USN	154
18-1:	Destroyer USS *Cooper* under way	157
18-2:	Fillipino guerillas pose with captured Japanese items	159
18-3:	A PBY Catalina skims the waters of San Pedro Bay	160
18-4:	USS *Hollandia* arriving at Naval Air Station Alameda	162
19-1:	Madang Harbour, New Guinea	171
19-2:	RAAF Northern Command Headquarters, at Madang	172
19-3:	Flight Lieutenant Wood and Warrant Officer Dawson	173
19-4:	RAAF Crash Launch of No. 111 Air-Sea Rescue Flight	174
19-5:	*Endeavor*, an Avro York transport aircraft	175
19-6:	Vice-Regal crest on the *Endeavor*	176
19-7:	Walrus amphibian aircraft of a RAAF Air-Sea Rescue Flight	176
19-8:	Two native police boys on the Karawari River	178
19-9:	2nd Lt John Carter, and those involved in his rescue	178
20-1:	PBY Catalina on water near Flores Island	179
20-2:	RAAF B-24 bomber crashes in the sea near Flores Island	181
20-3:	Two aircrew of a Fairey Fulmar aircraft cling to a dinghy	182
20-4:	Japanese Irving 11 Navy reconnaissance aircraft	182
20-5:	Waist blisters of two PBY Catalina seaplanes	183
20-6:	Japanese light cruiser *Isuzu* at Yokohama	183
20-7:	Flight Lieutenant Robin Morton Corrie at Darwin	184
21-1:	Flight Lieutenant Wally Mills in aircraft cockpit	185
21-2:	Japanese Rufe II Navy fighter seaplane	187
21-3:	Japanese airstrip at Liang, Ambon Island	188
21-4:	No. 31 Squadron RAAF members standing by aircraft	189
22-1:	Landing craft staging for assault landings on Iwo Jima	192
22-2:	Commodore Dixwell Ketcham, USN, and his staff	193

22-3:	Shore bombardment of Iwo Jima	195
23-1:	Japanese battleship *Yamato* blowing up following air strikes	204
23-2:	PB4Y-2 Privateer patrol planes over Miami, Florida	207
23-3:	Vought OS2U Kingfisher floatplane on the water	208
23-4:	Japanese Rufe and Jack, and Tojo fighter aircraft	213
24-1:	Aircraft carrier HMS *Formidable* at sea	219
24-2:	Royal Navy Fleet Air Arm Avengers, Seafires, and Fireflies	222
24-3:	ADM Arthur W. Radford, USN	224
24-4:	Japanese escort ship *Etorofu* at sea	226
25-1:	Group photograph of Rescue Squadron VH-6 personnel	231
25-2:	FG-1D Corsair aboard the aircraft carrier USS *Yorktown*	236
25-3:	Japanese Judy Navy dive bomber	237
25-4:	Japanese Tojo 2 Army fighter	237
25-5:	Japanese Battleship *Hyuga* off Kure	239
25-6:	Transfer of survivors from USS *Pomfret* to VH-6 plane	244
26-1:	Flight Officers Carr-Smith, Scott, and Sgt Andy Bettie	249
26-2:	A RAAF Hudson bomber landing on an airfield	250
26-3:	A Boeing 314 "Clipper" in flight	253
26-4:	Anzac Club in New York City	253
26-5:	RAAF De Havilland DH82 Tiger Moth biplane	254
27-1:	Royal Australian Air Force Base Rathmines	255
27-2:	Flight Lieutenant Wally Mills in flight from Morotai	256
27-3:	Squadron Leader Ronald J. Rankin AFC RAAF	257
27-4:	A mural of pin-up girls adorning the wall of RAAF mess	258
27-5:	Manokwari, Dutch New Guinea	259
27-6:	Squadron Leader Robert J. Rankin and Maj Stuart Peach	260
Postscript-1:	U.S. Navy sailors celebrating VJ Day at sea	261
Postscript-2:	Empty beer bottles along Mindil Beach, Darwin	262
Postscript-3:	Fleet Admiral Nimitz, USN, arrives at Tokyo Bay	263
Postscript-4:	Nimitz signs the Instrument of Surrender	264
Postscript-5:	Insignia worn by Nimitz during the ceremony	264
Postscript-6:	Painting *Coming into Golden Gate Bridge*	265

Maps and Diagrams

Preface-1:	Hawaiian Islands	xxxii
Preface-2:	Political control of the Western Pacific islands	xxxix
Preface-3:	Movement of Allied Forces toward Japan	liv, lv
Preface-4:	New Britain and New Ireland Islands	lix
1-1:	Palau Islands in the Western Caroline Islands	2
3-1:	Central and Eastern Solomon Islands	26
5-1:	Guadalcanal, Savo, and Florida Islands	48

5-2: Tanambogo, Gavutu, and Gamoi Islands	49
5-3: Central Solomon Islands	52
5-4: Treasury Islands, Solomon Islands	53
7-1: New Georgia and Rendova Islands	64
7-2: Bougainville Island and Shortland Islands	65
7-3: New Ireland and New Britain Islands	67
8-1: Central Solomon Islands	69
8-2: Treasury Islands, Solomon Islands	70
9-1: Papua, New Guinea	82
9-2: Nissan Island, Green Islands	83
10-1: Finschhafen area, northeastern coast of New Guinea	89
12-1: New Britain and New Ireland islands	110
12-2: Tonga Islands	113
13-1: French Colony of New Caledonia	117
14-1: Views of the Martin PBM-3D Mariner patrol bomber	121
15-1: Central Pacific island chains, and Lower Mariana Islands	125
17-1: Philippine Islands	142
18-1: Bismarck Archipelago	163
18-2: Central east coast of Palawan Island, Philippines	168
20-1: Lesser Sunda Islands	180
21-1: Ambon, Moluccas ("Spice Islands")	188
23-1: Builder's plans for IJN Battleship *Yamato*	202
25-1: Okinawa, Ryukyu Islands	232
25-2: Inland Sea of Japan	238
25-3: Locations of rescues of 273 airmen	248

Foreword

The first Navy air-sea rescue occurred on 21 August 1918 when Ens. Charles H. Hammann landed his Macchi M-5 seaplane fighter outside of the Pola Harbor entrance (now Pula, Croatia) to rescue his flight leader, Ens. George H. Ludlow who had been shot down. Ensign Hammann made a difficult landing on choppy water within range of the harbor defenses while the enemy were still in the area. The rescue had many of the features that were typical of the rescues made by Army and Navy flying-boats in the Pacific thirty-five years later: close proximity to the enemy, a rough sea, and a difficult take-off in an over-loaded plane.

Despite the example that Ensign Hammann set, dedicated air-sea rescue units were not organized by any nation until long after World War I because there was no perceived need for them during the interwar period. Even after Pearl Harbor was bombed, there seemed to be little need for the Navy and the Army Air Force (AAF) to establish specialized Air-Sea Rescue units. In fact, until the summer of 1944, the U.S. Navy did not have dedicated air-sea rescue squadrons. Instead, aircraft in the patrol squadrons were detailed for rescue work on an ad hoc basis as the situation demanded.

Only the United States Coast Guard specialized in air-sea Search and Rescue, but it had a limited rescue capability that was confined to coastal waters. Traffic which was not coastal depended on commercial shipping for assistance, which at the time seemed to be adequate.

By the spring of 1943, it had become evident to both the Army Air Force (AAF) and the U.S. Navy that the over-water distances in the Pacific Theater created a need for a specialized air-sea rescue organization. In that year, the Joint Chiefs of Staff looked for a way to achieve coordination of effort between the Navy and the Army Air Force. There was also a need to develop a close liaison with Allied services. There was common agreement that an integration of existing Navy and Army Air-Sea Rescue organizations would lead to standardization of equipment and would establish common rescue procedures among the Armed Services. Though the Army and the Navy agreed on the need for improvement, there was disagreement on how to achieve it.

The Army preferred to retain its own rescue service, but favored the establishment of a liaison committee to provide coordination of training, efforts, procedures, and equipment. On February 15, 1944, the committee became the Air-Sea Rescue Agency (ASRA), but it was not an operations command.

Coast Guard

The Coast Guard operated as part of the Navy during the war, and its air units were employed in anti-submarine warfare (ASW) and air-sea rescue work (ASR). The Coast Guard had been given ASW duties within the Sea Frontiers beginning in early 1942, but by mid-1943, the submarine offensive off the coasts of the United States had diminished significantly and the Coast Guard air stations gradually shifted the focus to ASR missions. The ASR operations were based on the British system and Coast Guard officers received training at the ASR school Blackpool, England. As air traffic and merchant marine operations increased in scope, incidents requiring ASR occurred more often so that by the end of 1943 the Coast Guard air stations on the West Coast operated primarily as ASR units.

The Coast Guard air station at San Diego was the home of the West Coast air rescue units. Between Jan 1, 1944 and December 1, 1945, 124 aircraft went down in the San Diego area. Of the 201 pilots and crews involved, 137 were saved, 59 were killed by collision or impact with the water and five were lost because of lack of or improper use of equipment.

Army Air Force

The Army was the first of the services to form dedicated air-sea rescue squadrons. In the fall of 1943, the AAF drafted plans to organize seven Emergency Rescue Squadrons (ERS), but only six were organized before the war ended. The new rescue squadrons were to be operational by the spring of 1944 to support the Pacific Air Forces, and provide rescue support for the Air Transport Command. Initial PBY training was provided by the Navy at NAS Pensacola.

The AAF procured Canso PBY-5As which the Army called OA-10s, that were manufactured under license by Canadian Vickers in Montreal. Delays in forming the squadrons occurred and only the 1st and 2nd ERS were operational by the summer of 1944. The 3rd became operational in the Southwest Pacific by the end of the year, the 4th, 5th, and 6th were not operational until mid-1945.

Until the arrival of the specialized ERS units, the Army assigned aircraft to air-sea rescue missions on an ad hoc basis. The practice put a strain on the tactical squadrons whether they were fighter or bomber squadrons. Until the specialized ERS units arrived in 1944, any airplane assigned to the mission was a land plane. Almost all aircraft types were used, but the workhorses of the Army's rescue program were Liberators, Flying Fortresses, and Super-fortresses. The Army's planes shared the assets of long range and adequate self-defense, but none of them could land on water.

A postwar analysis of the Army's use of the PBY said:

> The rescue record of the Catalinas was a spotty one, ranging from some of the most spectacular successes of the war to discouraging failures. Its range and load capacity were satisfactory, its cruising speed ideally slow, but the Catalina had trouble in landing in rough seas and in taxiing with a heavy load aboard. During one period in 1944, CBI's [China Burma India Theater] Eastern Air Command reported that half the Catalinas sent on rescue missions cracked up on landing, leaving two planes in trouble instead of one. The Catalina, awkward in flight and lightly armed, was quite vulnerable to enemy attack.[1]

Navy

The Patrol squadrons' mission was to locate the enemy at sea and report. Up to, and until several weeks after, the attack on Pearl Harbor, only PatWing-10, VP-101 and VP-102 faced the Japanese. Both Squadrons, equipped with PBY-4s, were based at Sangley Point and Olongapo. The squadrons were tasked with flying patrols, attacking Japanese ships, and aerial photography missions. From 27 April to 3 May 1942, the wing flew Army nurses, high-ranking officers, and pilots to Australia.

Demands for air-sea rescue picked up considerably during and after the Battle of Midway, 4-7 June 1942. As the U.S forces pushed deeper into the South Pacific, the patrol squadrons were called upon more frequently for air-sea rescues while at the same time, more collateral missions were added, such as mine-laying, defensive patrols around surface forces at sea, and diversionary, harassing attacks against enemy bases and islands.

By 1943, the ad hoc approach to ASR was putting a strain on the resources of the patrol and bomber squadrons that resulted in the creation of six specialty squadrons in 1944 and 1945 whose sole mission was air-sea rescue. The arrival of the specialty squadrons in the South Pacific did not end the practice of tapping the patrol squadrons for air-sea rescue work, but the new squadrons did alleviate the pressure. The six special units were employed in areas where the intensity of operations made calls upon their services frequent; elsewhere patrol planes continued to fly air-sea rescue missions on an ad hoc basis.

Not all the rescues involved an open-sea landing. If the sea was too rough, the planes remained aloft, circling the people to be rescued while calling-in surface vessels. The plane remained on station until the surface vessel arrived. If the men in the water were under enemy fire or were in imminent danger of being captured, the pilot landed regardless of the sea state.

> In the Central Pacific the problem of making an open-sea landing was acute. Only the most skillful and experienced seaplane pilots could land and take off again in the enormous swells, the job required as much seamanship as airmanship, and it became standard practice to avoid open-sea landings unless conditions were favorable and there was no other rescue agent available.[2]

A violent landing in fifteen-foot swells could cause severe structural damage. VH-3 had five of their PBMs declared a loss after suffering major damage during rescue operations. Squadrons that were equipped with the Martin PBM Mariner had the advantage of aircraft equipped with jet-assisted takeoff rockets (JATO), that made taking off in heavy seas, with a heavy load, or under fire, possible. A JATO-equipped PBM could take off in under 10 seconds versus about a minute without. But even with JATO, takeoffs in heavy seas often caused damage to the plane. PBM pilots would land in close to shore where the sea was calmer, and then taxi many miles to the rescue site. They did that because JATO allowed them to take off in seas they could not land in.

Even when the sea state was good enough to land, damage to the hull could occur. The PBYs were particularly susceptible to "popping rivets" on any hard landing. All PBY navigators had a hefty supply of wooden pencils on hand for the crew to plug the rivet holes. The standard wooden pencil was about the diameter of a rivet hole. Crewmen shoved the pencil in the hole, broke it off, and used the remaining piece to plug another hole. The "field expediency" worked well enough to withstand a takeoff but probably not the next landing. Plugging rivet holes with a pencil was a common practice on all PBYs.

By the end of the war, air-sea rescue had improved to the point where chances of rescue were good. Throughout the war nearly 5000 USAAF aircrew members were rescued, testifying to the improved conditions in air-sea rescue.[3]

Dwight R. Messimer, author, *In the Hands of Fate: The Story of Patrol Wing Ten, 8 December-1941-11 May 1942*, Naval Institute Press

Foreword

In his latest book, *Salvation from the Sky*, David Bruhn has again tackled a subject which has not received much attention from contemporary military historians. The details of the heroic rescues undertaken by these airmen have often been ignored by historians who tend to focus on offensive actions undertaken by forces the air-sea rescue units supported. David is to be commended for his description of the significant effort involved in rescuing survivors of Allied aircraft downed over water. The brave actions by the many aircrew involved in air-sea rescue during the final stages of the war against Japan in the Pacific, receive suitable recognition in *Salvation from the Sky*.

The Allied strategy for defeating Japan in the Pacific revolved around two loosely coordinated campaigns. The first was a thrust across the central Pacific by the US Navy under US Commander in Chief Pacific Fleet, Admiral Chester Nimitz. This began in 1942 and as well as the US victories in the Coral Sea and Midway, involved the invasion of the Solomon Islands and the progressive recapture of the occupied islands towards Japan. The second was an advance from Australia, through the island chain on the south west rim of the Pacific. This campaign started with a move by US and Australian forces under the Supreme Commander Allied Forces, Southwest Pacific, General Douglas MacArthur in south eastern New Guinea. The campaign continued along the north coast of New Guinea, then continued into the north-eastern islands of the Netherlands East Indies (NEI now Indonesia) and then to the Philippines which were reached in October 1944. The two campaigns coalesced at the Philippines with the US Third Fleet supporting MacArthur's forces.

As the Allied forces fought their way across the north coast of New Guinea, they were given air support by the USAAF Fifth Air Force, which included No. 9 RAAF Operational Group (No. 9 OG). By early 1944, Allied offensive operations in the South West Pacific Area swung to the north-west, by-passing the large base the Japanese had established at Rabaul. In order to ensure that the Japanese forces in Rabaul were contained, No. 9 OG remained in Papua, establishing a new headquarters at Milne Bay, effectively becoming the RAAF area command for Papua and south-eastern New Guinea. It was renamed Northern Command in April 1944 to better reflect its function.

The RAAF's mobile air operations were assumed by the No. 10 Operational Group (No. 10 OG). Formed in September 1943, No. 10 OG began combat operations in north-western New Guinea and the Allied campaign in Western (Dutch) New Guinea. By September 1944, during the Battle of Morotai, No. 10 OG had grown to encompass 12 flying squadrons and two airfield construction wings, as well as the requisite support units. As a result of this expansion in size, No. 10 OG was renamed the First Tactical Air Force in October 1944.

The RAAF started to develop its air-sea rescue capability by forming No. 1 Rescue and Communications Squadron in Port Moresby in October 1942. In the search and rescue role, the Squadron used Walrus, Dornier Do-24 and Catalina flying boats. Operating at low level, they proved versatile and quite effective in evading Japanese aircraft. Taking advantage of the low take-off and landing speeds, they were able to land in small clearings that other aircraft could not have utilised. The squadron's Walrus aircraft were also able to put down on narrow rivers where the larger Catalinas were unable to land. It was reformed as No. 8 Communications Unit in November 1943 at the major Allied base at Goodenough Island near Milne Bay off the eastern tip of Papua. In addition to communications duties, it provided rescue services north to New Britain. The Unit moved to Madang in November 1944.

From December 1944, a number of Air-Sea Rescue Flights (ASRF) were formed: 111 at Madang, 112 at Darwin, 113 initially at Morotai and then at Labuan. Some of these units were also equipped with motor launches that could be used to recover personnel when alighting on the water was not possible. In July 1945 two more Flights were formed: 114 at Cairns and 115 at Morotai (using four of 113's aircraft when they moved to Labuan).

ASRF carried a crew of seven, as opposed to nine for mining missions. In addition to rescues, ASRF aircraft flew SAR (Search and Rescue) patrols to cover strikes or transiting aircraft, support operations for amphibious landings, evacuation of wounded, transport of prisoners, insertion of commandos and extraction of escapees. In this last role ASRFs (mainly No. 113 Flight) cooperated with the Allied Intelligence Bureau.

By April 1945, Air-Sea Rescue had evolved to the point where it was part of the planning of an air operation. Air-Sea Rescue Catalinas provided cover for many of the Allied bombing missions flown in the region. In these operations, the Catalina held in a safe area within visual range of the target, observed the attack and provided assistance to any of the attacking aircraft in distress.

The Catalina flying boat, designated PBY by the US, was the primary aircraft used by the RAAF Air-Sea Rescue Flights. The RAAF made an early decision to buy the Catalina. The first batch was ordered in August 1940 and delivered between February and October 1941. The RAAF eventually acquired 168 Catalinas and formed four Squadrons, five Air-Sea Rescue Flights and some ancillary units. The Squadrons conducted a wide ranging and significant aerial mining campaign against the Japanese. The aircraft were retired soon after the end of the War and most were disposed into civilian service.

The Catalina flying boat was one of the durable and effective aircraft of the second World War, with the significant operational feature being its extreme range, with endurance of up to 31 hours flying time depending on the load carried. Although slow and ungainly, Catalinas distinguished themselves in World War II. They were used successfully in a wide variety of roles for which the aircraft was never intended. PBYs are remembered for their rescue role, in which they saved the lives of thousands of aircrew downed over water. Flying boats such as the Catalina placed a special demand on training air crews who not only learnt to fly the aircraft, but needed to learn manoeuvres in sea conditions which were usually associated with naval operations.

The Rathmines RAAF seaplane base at Lake Macquarie on the NSW central coast was a significant facility. It was a centre for training, housing the Operational Training Unit for Catalina crew, providing training to over 200 Catalina crews during the war. The Base was also important as a repair centre for the flying boats and was the location of a Flying Boat Repair Depot. New flying boats manufactured in the US were converted at Rathmines for RAAF operational duties.

The RAAF had made a significant contribution to the war in the Pacific. In August 1945 when the war against Japan ended, the RAAF had some 175,000 personnel operating 6,000 aircraft. A majority, almost 132,000, were serving in the Pacific. That theatre was also the focus for all but 20 of the RAAF's 75 flying squadrons. At this point the RAAF had evolved into the fourth largest air force by size behind the USA, USSR and UK.

However with the war over, Australia accelerated its demobilisation plans for the armed services, so that by the end of October 1946 the RAAF had dropped to 13,000 members. This process still had some way to go, with the post-war low for the RAAF reducing to a strength of just 8,000 at the end of 1948.

Salvation from the Sky covers the air-sea rescue activities from the time the Allied offensive against the Japanese began with the invasion of the Solomon Islands and the campaigns farther west in New Guinea and the Netherlands East Indies. It is not only a comprehensive treatment of the significant impact the small number of air-sea rescue aircrew made on saving lives during the final stages of the War in the Pacific, but is a very enjoyable read giving many examples of the acts of courage and sacrifice involved in the rescue and recovery of downed aircrew.

Commodore Hector Donohue AM RAN (Rtd)

Foreword

I was born in the small town of Fort St John in northern British Columbia, Canada, where my dad managed the Hudson's Bay Company store. Before he retired from the company and we moved to the coast, I recall as a small boy, the great construction activity and the presence of many army personnel and trucks around the town. Fort St. John was the second town on the recently opened Alaska Highway and the nearby large airport was part of the Northwest Staging Route for delivery of aircraft for the defense of Alaska and of the Soviet Union.

When I was five years old, I observed my dad decorating the old pump-up style gas pumps of our recently purchased country grocery store near Victoria, BC, with color crepe paper ribbon. He told me that it was because it was V-J day. I later learned about the significance of that day and of V-E day, and have often contemplated what our lives would have been like if they had not occurred.

Participating in the preparation of several of David Bruhn's books has contributed to this education. His subjects, usually ones that have either escaped the attention of naval historians, or been ignored, draw attention and honor to those often-unheralded individuals who made tremendous contributions in areas many of us are entirely unfamiliar with. In *Eyes of the Fleet* he introduced us to U.S. Navy seaplane tenders, and combat operations of the patrol aircraft they supported in World War II. This book focuses on the tremendous contributions made by their tended flocks of sea planes, as well as squadrons of Royal New Zealand and the Royal Australian Air forces of similar aircraft, to downed pilots in the Pacific through air-sea rescue.

Salvation from the Sky spans the war from the desperate days of 1942 when the Japanese were almost within striking range of Australia and New Zealand to the final days of it, when invasion of the Japanese homeland was imminent. From David's co-author Stephen Ekholm comes the story of the seaplane tender *Coos Bay*, a core thread that is followed and returned to often in this book. Hers is just one example of the histories of the many tenders that could have filled this role.

Naval Air Power proved its new, predominant role early in the war in Europe with the British attack on the Italian Fleet at Taranto, Italy, on 11 November 1940. Fairey Swordfish torpedo-bombers launched from the aircraft carrier HMS *Illustrious* sank one battleship and damaged two others, shattering greatly cherished beliefs about the dominance of battleships and naval gunnery. Japanese naval officers studied the attack carefully, particularly how the British had been able to employ airdropped torpedoes in the shallow harbor, without them striking the bottom, and used this information as a model for their subsequent 7 December 1941 carrier aircraft attack on Pearl Harbor. The sinking of the battleship HMS *Prince of Wales* and battlecruiser HMS *Repulse* in the Pacific off the eastern coast of Malaya by Japanese bombers on 10 December 1942 gave further proof of the ascendency of war planes.

Air combat, anti-aircraft fire and operational and weather problems produced huge numbers of aircraft casualties for the Allies. In some cases, uninjured pilots and aircrewmen were able to parachute from their damaged aircraft (bale out) or escape it before it sank following an ocean landing (ditch). Reaching the ocean surface was neither easy, nor a certainty, and those able to do so were left stranded in life jackets or life rafts, often in shark-infested waters which might also be under enemy control or patrolled by the enemy. It was intolerable that this situation could be allowed to exist given the huge sacrifices and the risks that these flyers faced in combat missions. Accordingly, the rescue measures described in this book were put into place.

Initially, air-sea rescue was performed on an as-required basis by seaplanes engaged in multiple types of missions, but as the numbers of combat aircraft and missions they flew dramatically increased, dedicated resources were allocated to rescue efforts. In addition to air-sea rescue by seaplanes, these measures also involved surface ships and submarines all under coordinated U.S. Navy and U.S. Army Air Force command. Even putting humanist values aside, the business case was easy. The benefits of avoiding the cost and time of recruiting replacement crew far outpaced that of the costs of recovery of existing ones. Further, air-sea rescue measures provided a huge morale booster. Aircrews flying combat missions day-after-day, had some reassurance that if shot down or forced down by mechanical failure, they had a good chance of being rescued. The many examples related in this book and the hundreds of rescues proved that the Pacific air-sea rescue practices were hugely successful.

I am grateful for having an opportunity in this foreword to highlight Canada's participation in the Pacific war. Many people have the perception that Canada's involvement in the Pacific was practically nonexistent compared to that in the European theatre. It may have been less but not insignificant given our country's then small population. Prior to Pearl Harbor, Canada had been engaged in the European conflict for two years with most of her expanding naval assets devoted to protecting merchant shipping braving the U-boat wolf pack-patrolled North Atlantic to bring life sustaining and military supplies to Britain.

Yet from 8 December 1941, the day after the attack on Pearl Harbor, and continuing to Christmas Day of that year, a 1,975-man Canadian force (Royal Rifles and Winnipeg Grenadiers battalions), part of the reinforcement forces, desperately defended Hong Kong against the Imperial Japanese Army's 38th Infantry Division while sustaining immense causalities. Those not killed spent the duration of the war in Japanese prison camps. At war's end only 1,225 of the original force returned to Canada.

Many do not realize that the long tentacles of Japanese imperialism extended to the North as well as South Pacific. Japanese submarines attacked Canada's and America's west coasts, sinking ships and so panicking local populations that the assets of Japanese immigrants were seized and their owners and their families were forcibly transported to detention camps in both countries. Further, Alaska's Aleutian chain was invaded. Canada, in spite of very limited resources, supplied troops, planes and naval vessels in combating this northern aggression and in patrolling coasts of both nations, even to the extent of basing squadrons of RCAF patrol aircraft in Alaska, where the Americans lacked resources to cover patrols and coastal defence.

Later, the 5,300-member 13th Canadian Infantry Brigade departed Nanaimo, BC, on 12 July 1943, aboard U.S. transports bound for Adak, Alaska to operate under U.S. command. This was a contingent comparable in size to Canada's better-known involvement in the Dieppe raid. Fortunately, this force suffered none of the combat losses resulting from the Dieppe disaster—the Japanese slipped away before they could be attacked. The Alaskan campaign is described in an excellent book, *War on Our Doorstep*, by Brendan Coyle.

Finally in 1944, after victory in Europe, Canada was able to participate actively in the British Pacific Fleet with our newly commissioned cruiser HMCS *Uganda*. Also, together with other Commonwealth nations, mainly New Zealand, she contributed through embedded air crew in the Royal Navy's Fleet Air Arm. Among those Canadians serving valiantly aboard the RN fleet aircraft carriers was Lt. Robert Hampton Gray, RCNVR, from Nelson, British Columbia. He was Canada's last winner of the Victoria Cross, one of the only two earned by the Fleet Air Arm during the war. He was killed, a few days before VJ Day, while sinking a Japanese escort vessel on 9 August 1945 at Onagawa Wan (Bay), Honshu, Japan.

Photo Foreword-1

Lt. Robert Hampton Gray, RCNVR.
Canadian government, photographer unknown

As a final note I salute the foundation aircraft of the air-sea rescue described in this book, the Consolidated PBY flying boat, widely known as the Catalina, but here in Canada in our amphibian version, as the Canso. Widely used and loved throughout the world both before and after the war it was a remarkable aircraft. On reflection, it gives me some pride in knowing that many of the planes described in this book were built by Canadians in a Boeing plant less than thirty miles from where I am writing this in White Rock, British Columbia. After the war this versatile machine was used all over the world for a wide variety of purposes. My dad's old employer, the ancient Hudson's Bay Company, founded 1670, even employed one in their fur trade operations in northern Canada and the Arctic.

The Canadian version was named for the strait in the province of Nova Scotia that separates Cape Breton Island from the mainland. An alternate anecdotal version of the naming is that when an inspection team from the RCAF first saw the strange yacht-like aircraft they remarked "it can never fly." The response by a Consolidated Aircraft Corporation engineer was "Can so."

George H. S. Duddy, P. Eng. Ret.

Acknowledgements

Maritime and aviation artist Richard DeRosset has created yet another stunning book cover. His painting *Salvation from the Sky* depicts a valiant PBY Catalina with fifty-six survivors of the USS *Indianapolis* aboard the overburdened seaplane, following sinking of the heavy cruiser by the Japanese submarine *I-58*. Those with the most severe injuries, including broken arms or legs, were placed on the wing, and covered with parachute material. As the destroyer escort USS *Cecil J. Doyle* searched with stabbing searchlight beam in darkness for the seaplane, she was aided by parachute flares dropped by aircraft. The PBY, critically damaged by wind and wave, not being designed for seaborne operations under such adverse conditions, could not lift off again.

Photo Acknowledgements-1

Richard DeRosset.

This book would not have been possible without much assistance and expertise lent by Hector Donohue, George Duddy, John MacFarlane, Dwight Messimer, Jenny Scott, and Stephen Ekholm.

Commodore Donohue, AM RAN (Ret.) has provided much Australian-related material, including excerpts from books and articles by him, and has penned forewords for several of my previous books. He has done so again, and provides valuable senior Royal Australian Naval officer perspective and assessment in his foreword to this one. Donohue began his career in the RAN in 1955 as a seaman officer and subsequently sub-specialized as a clearance diver and torpedo and anti-submarine officer. His service in the RAN included command of the destroyer escort HMAS *Yarra* and the guided missile frigate HMAS *Darwin*. Ashore, he held a number of senior positions in Defence policy and force development prior to retirement in mid-1991.

Photo Acknowledgements-2

Commodore Hector Donohue, AM RAN (Retired), 2020.

Canadian George Duddy is a retired Professional Engineer with a keen interest in maritime subjects, particularly those relating to the maritime history of western Canada and the Arctic. He has published several articles on these subjects on the Nauticapedia Project website: https://www.nauticapedia.ca/. He graciously provided Royal Canadian Navy material for previous books, and lent his critical eye and keen mind during technical review and editing of the manuscripts. He has "leaned in" on this project as well, and authored the chapter on the British Pacific Fleet. It details the contributions of Canadian flyers and aircrews aboard Royal Navy aircraft carriers, to Allied efforts in 1945 during Operation MERIDIAN attacks on Japanese-held oil refineries on Sumatra, the invasion of Okinawa, and Third Fleet operations against Japan.

Photo Acknowledgements-3

George Duddy, 2020.

Photo Acknowledgements-4

John M. MacFarlane

Canadian John M. MacFarlane assisted by having compiled, with Robbie Hughes, information about Canada's Naval Aviators in a book of the same title. Updated information about this subject is available at the website: http://www.nauticapedia.ca/Articles/NavalAviator.php. MacFarlane has been interpreting and preserving Canada's natural and cultural heritage resources since 1969. After a brief career in the navy he joined Parks Canada and stayed with them for twenty years. Afterward, he was the Director and Curator of the Maritime museum of British Columbia in Victoria. He is now the Curator of the website, The Nauticapedia, which shares a vast resource related to British Columbia's nautical history, and has two extensive searchable databases of ships and people. He has written many articles and books on this theme.

Photo Acknowledgements-5

Dwight Messimer, 2020.

Dwight Messimer is a former university lecturer, and acclaimed military historian and author. A specialist on the German Navy and U-boats, he has written nearly a dozen books on this subject and on naval aviation covering the period 1925-1942. One of his books, *In the Hands of Fate: Story of Patrol Wing Ten*, details the heroics of PBY Catalinas in offensive operations during World War II. Messimer's work has also appeared in many periodicals, including *The American Neptune*; *The Quarterly Journal of Military History*; *War, Revolution and Peace*; and *Naval History*. In addition to his review of *Salvation from the Sky*, Messimer was kind enough to pen a foreword for it, as he did previously for my books *Ingram's Fourth Fleet* and *Home Waters*.

New Zealander Jenny Scott was particularly kind in allowing me to use photographs from her collection, and providing material from her book *Dumbo Diary: RNZAF No. 6 Squadron 1943-1945*, chronicling the exploits of this Royal New Zealand Air Force squadron of seaplanes in World War II. As elaborated on in Chapter 26 of this book, her father, Flight Officer Alastair Scott, was a member of the squadron.

I am much indebted to co-author Stephen Ekholm who suggested the subject matter for this book. Stephen's father, Fritiof ("Fritz") L. Ekholm, enlisted in the U.S. Navy on 15 January 1942, and was discharged after the war on 9 October 1946. During his service, he rose from newly commissioned officer, serving as navigator, to second in command aboard the seaplane tender USS *Coos Bay*. Following the war, Fritz started a cannery in Alaska, and tragically died of polio in autumn 1949. Stephen, then an infant, never knew his father. Eager to learn about him, when of age to do so, he began attending annual reunions of former *Coos Bay* crewmembers.

Stephen was warmly welcomed by them and over decades of doing so, he amassed much information, but was most impressed by several central themes expressed by *Coos Bay* sailors over the years. These included their universal assertions that "she was a very good ship," their pride in being involved in the rescue of downed airmen, and their great affection for two executive officers they served under. William F. James and his successor, Fritz Ekholm, were greatly beloved and they, in turn, held the crew in high regard. Both men occasionally "took off their officer's hats" and went drinking with the crew to build esprit de corps.

Finally, a tip of the hat to Lynn Marie Tosello, the final editor of this text. Lynn has edited several books for me, and continues to amaze me with the worldliness, eloquence, and prose she lends to each one.

Preface

Air-sea rescue, which kept personnel losses to a minimum, preserved that element of military power most difficult to replace and bolstered the morale of all fighting men.

—U.S. *Naval Aviation in the Pacific*, issued by The Office of the Chief of Naval Operations, United States Navy, 1947.[1]

The excellence of the air sea rescue – dumbo – service provided by Fleet Air Wing ONE has done much to foster high morale in the carrier and land based squadrons.

—Admiral William F. Halsey Jr., USN, commander, Third Fleet, 13 September 1945.[2]

Photo Preface-1

PBY-5 Catalina of VP-52, circa 1944. The squadron was transferred to Woendi Lagoon on 15 July 1944, where it conducted anti-submarine patrols and Dumbo air-sea rescue missions for downed flyers in support of the bombing of Woleai, Truk, and Yap islands. National Archives photograph #80-G-223133

At the outset of World War II, operating procedure for the rescue of pilots and aircrews was practically non-existent. But there were a number of safety devices which permitted pilots to survive the unexpected failure of their planes. Standard equipment included parachutes, inflatable life jackets (popularly known as "Mae Wests"), and rubber life rafts equipped with emergency survival and signaling gear. However, for the first half of 1942, often little or nothing could be done to recover pilots who had survived crashes, engine failure, or being shot down.[3]

Photo Preface-2

Lt. J. Thorvaldson, pilot from a RAAF Kittyhawk, poses with the rubber boat on which he floated down the Markham River to Nadzab, New Guinea, after he made a forced landing following a single dogfight with four Zeros, 2 October 1943.
Australian War Memorial photograph 015902

A number of rescues were made as a result of individual initiative. Such occurred after the Battle of Midway, when PBY Catalinas picked up many pilots. This followed valuable contributions by the versatile seaplanes during the battle, including locating the enemy and carrying out a bombing attack on Japanese shipping.⁴

Photo Preface-3

Oil painting by John Hamilton of the discovery, by a PBY, of part of the Japanese Fleet on 3 June 1942, setting the course for an American victory in the battle. Naval History and Heritage command accession #80-142-P

Photo Preface-4

Diorama by Norman Bel Geddes, depicting the torpedo attack made by four PBY-5As on the Japanese Midway Occupation Force during the night of 3-4 June 1942. The oiler *Akebono Maru* was hit during this attack, but was able to continue on her mission. National Archives photograph #80-G-701846

Map Preface-1

Hawaiian Islands

At this point in the war, there was a scarcity of tenders to provide refueling and other services for the seaplanes assigned to the Midway Defense Forces. Accordingly, four ex-tuna clippers, acquired by the Navy from the Portuguese fishing community in San Diego for use as patrol vessels, were positioned at outlying islands to serve as support ships. The *YP-284* (ex-*Endeavor*) was allocated to Lisianski, *YP-290* (ex-*Picaroto*) to Layson, *YP-345* (ex-*Yankee*) to Gardner Pinnacles, and *YP-350* (ex-*Victoria*) to Necker. Also, two World War I vintage destroyers converted to seaplane tenders—*Thornton* (AVD-11) and *Ballard* (AVD-10)—were at French Frigate Shoals, located 761 miles to the southeast of Midway.⁵

Photo Preface-5

The wooden-hulled *YP-72* (ex-purse seiner *Cavalcade*) patrolled fog-shrouded waters, and supported PBY Catalinas of Patrol Wing 4 carrying out searches for Japanese naval forces during the Aleutian Islands portion of the Battle of Midway.
U.S. Navy Bureau of Ships photograph

CONSOLIDATED PBY-5 AND PBY-5A CATALINAS

The PBY-5 and 5A Catalina long-range maritime aircraft played an understated, but important role in World War II, providing Allied naval and air forces ground attack, anti-submarine warfare, reconnaissance, and air-sea rescue capabilities. In rare cases, these type aircraft were also employed as dive bombers. The PBY Catalina was used in the U.S. Navy and other services of the American military, and also in the air arms of Britain, Brazil, Australia, New Zealand, Canada, the Netherlands, and the Soviet Union. Additionally, squadrons of the RAF were manned exclusively by Norwegians who had escaped from their occupied land. (A Catalina of No. 333 squadron spotted Canadian Victoria Cross winner Hornell, whose story follows on page 9.) The locations at which the patrol aircraft were manufactured are identified in the table:

United States	Canada	Soviet Union
Buffalo, New York	Vickers of Canada,	Beriev's plants at Taganrog,
Norfolk, Virginia	Cartierville, Quebec	Rostov, and Oblast
New Orleans, Louisiana	Boeing of Canada,	
Philadelphia, Pennsylvania	Vancouver, BC	
San Diego, California[6]		

Interestingly, in 1939, engineers at the Consolidated plant in San Diego, California, carefully appraised the PBY-4 Catalina and decided that the flying boat was outdated. (Following introduction of the XP3Y-1 in response to an order from the Navy for a monoplane patrol aircraft, Consolidated produced improved PBY-2, PBY-3 and PBY-4 models between 1936 and 1939.) Two months later, the plane was put back in production with the long sleek fuselages on the assembly line essentially those of its predecessor aircraft designed in 1934.[7]

The twin-engine, high-wing flying boat could take off from land or water. The PBY-5 model, produced from September 1940 to July 1943, was characterized by high-power engines, waist gun blisters, and a large "wet wing." This term refers to the wing skin serving as the fuel tank, with no need for separate fuel tanks or bladders to be inserted into areas between ribs and spars—a substantial weight-saving feature, but one that requires that every seam and rivet be sealed or gasketed.[8]

Improvements offered by the PBY-5A (October 1941-January 1945), included hydraulically-operated retractable tricycle landing gear for amphibious operation, a new tail gun position, replacement of the bow single gun position with an "eyeball" turret with twin .30-caliber machine guns, improved armor, and self-sealing fuel tanks. Later models were equipped with radar and MAD (Magnetic Anomaly

Detection) gear to greatly enhance their capabilities to locate enemy submarines.⁹

PBY-5 and 5A Catalinas were powered by two 1,200 hp Pratt & Whitney R-1830-92 engines. Improvements and upgrades made to the PBY-5A resulted in increased aircraft weight, and associated slightly slower top speed and shorter range than the PBY-5. Crew size for the two aircraft varied between seven and nine personnel. The crew of a USN PBY-5A typically consisted of a pilot, co-pilot, radioman, navigator, flight mechanic, bow turret gunner, two waist gunners (one each for the port and starboard blisters), and a ventral gunner. Some positions could be combined. For example, Royal Australian Air Force Catalina crews utilized a Wireless Air Gunner position, incorporating both wireless (radio) operator and gunner duties.[10]

Aircraft characteristics and armament of the PBY-5A Catalina, which had a manufacturer's specified maximum speed of 196 mph, cruise speed of 124 mph, range of 2,520 miles, and maximum service ceiling of 15,800 feet, are provided in the table. The height listed is for an aircraft on wheels, and the maximum weight for a takeoff on wheels.

PBY-5A Catalina Characteristics

Dimensions	Armament
Length: 63 ft. 10 in.	Two .30-caliber machine guns in nose turret
Wingspan: 104 ft.	Two .50-caliber machine guns (one in each waist blister)
Height: 21 ft. 1 in.	One .30-caliber machine gun in ventral hatch at tail
Empty weight: 20,910 lbs Max weight: 35,420 lbs	2,000 pounds of bombs and two torpedoes, or four 325-lb depth charges[11]

"DUMBO" MONIKER

Organized rescue operations in the Pacific were first developed in the Solomon Islands Campaign. PBY Catalinas stripped of all heavy non-essential gear, and known with the adopted name "Dumbos," were dispatched to recover flyers who had been shot down. Initially, this was incidental to primary duties (such as long-range searches, anti-submarine patrols, and night bombing missions), assigned as a need for it arose. From this unstructured beginning, it evolved into a standard procedure of having a Dumbo circling near the scene of an air raid. Positions were reported as planes went down, and the Catalina, often protected by aircraft from the strike, recovered those in the water. Rescue crews became renowned for their bravery in landing in positions exposed to enemy shore battery fire.[12]

The nickname "Dumbo" was originally coined in reference to a Walt Disney character, the flying elephant (the protagonist of the 1941 Disney film of the same name), for land-based bomber aircraft that searched for downed airmen in the oceans. The planes assigned such duties got their name from their bulky appearance, because they had a large lifeboat attached beneath their fuselage, used for the purpose of the air-sea rescue. The aircraft included the American B-17, and British counterparts such as the Lancaster and Vickers Warwick. The appearance of a lifeboat under a bomber made the aircraft look unwieldy—like a flying elephant. The Dumbo aircraft searched for downed air crews in the water, and in some cases, aided the survivors of sunken ships. When such were sighted, the aircraft would fly in low and drop the boat.[13]

Photo Preface-6

An airborne lifeboat fitted to a Warwick aircraft of the RAF Air-Sea Rescue service, England, circa November 1944.
Australian War Memorial photograph SUK13115

"Dumbo" later became synonymous with ocean rescue by flying boats, although Cats (Catalinas) and other such aircraft did not carry ponderous lifeboats. Instead, they landed near survivors and recovered them, or dropped inflatable life rafts if alighting on the water was impossible. The difference between "Cat" and "Dumbo" set-ups of the Catalina, was that Dumbos were stripped of bombs or torpedoes they might otherwise carry, and retained only machine guns for self-protection.[14]

Dumbo operations were often difficult, and sometimes dangerous as well, owing to challenges presented by angry seas or combat action. When a PBY had to land on water to recover survivors and the enemy forces that had downed their aircraft or sunk their vessel were still present, both the survivors and the PBY were in peril. Catalinas often performed other missions as well, and this type plane carried out combat and scouting roles in all theaters. As an example, RCAF Flight Lieutenant Daniel E. Hornell posthumously won the Victoria Cross for attacking and sinking a German U-boat in a Canso (Canadian variant of PBY-5A Catalina) in June 1944.[15]

RCAF Squadron Leader Leonard Birchall piloting a Catalina flying boat had become known as the "Saviour of Ceylon" two years earlier, by foiling a Japanese attack on a Royal Navy force at Ceylon (today Sri Lanka). After identifying that a Japanese naval fleet was clearly on its way to attack Ceylon, and reporting this imminent danger, he was shot down by Japanese fighter aircraft from a carrier—a prelude to three years of imprisonment.[16]

ESTABLISHMENT OF RESCUE SQUADRONS

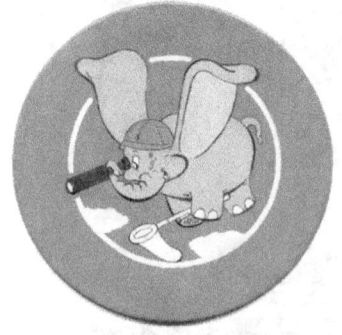 

Rescue Squadron VH-1 plaque, depicting the Disney-character Dumbo, and a sketch of VH-4's squadron insignia

Air-sea rescue on the open ocean, particularly in the Central Pacific, was challenging. Catalinas were employed extensively to search for survivors, to drop emergency gear, and to circle overhead until a ship could arrive. They also made rescues in passable conditions and in protected lagoons. Only the most skillful pilots could land and take off again in enormous swells. Because of the high risks of such operations, it became standard practice for PBYs to avoid open-sea rescues unless conditions were favorable, and there were no other means available.[17]

Meanwhile, the growing aircraft carrier strength in the Pacific, and associated increasing number of planes employed for strikes, made the problem more acute. In autumn 1943, it was decided to establish rescue squadrons, which would be specially trained and equipped for rescue work. Catalinas paved the way in the Solomons in 1942-1943, where they established a fine record. In 1944, USN rescue squadrons were organized to meet the war's growing demands.[18]

Commander Air Force, U.S. Pacific Fleet, proposed six VH squadrons, each comprised of six Douglas R4D-6 land planes for the evacuation of wounded, and six PBM-3R Mariner seaplanes for the rescue of survivors of downed aircraft or sunken vessels. However, once established, there was such a wide variance in duties performed by these two types of aircraft that difficulties arose in the administration, operation, and maintenance of their squadrons. On 12 December 1944, the land planes were formed into separate evacuation squadrons (VE). The original rescue squadrons retained six seaplanes apiece.[19]

Photo Preface-7

Left: Douglas R4D-6 aircraft of the Naval Air Transport Service at Alaskan airfield, circa 1945. Right: A PBM Mariner rescues LTJG J. M. Denison, shot down while operating from the escort carrier USS *Marcus Island* (CVE-77) in 1945.
National Archives photograph #80-G-K-5863, and USS *Marcus Island* (CVE-77) 1944-1945 cruise book photograph

The rescue squadrons equipped with PBM-3R Mariners (modified to carry newly developed rescue gear, and with pilots and crews given special training in the techniques of air-sea rescue) were able to carry out rescues under conditions impossible for Catalinas. During the Okinawa invasion, a 6-plane rescue squadron made 76 landings and rescued 183 survivors of all services.[20]

The commissioning dates of the squadrons, and the identities of their commanding officers follow.

U.S. Navy VH Rescue Squadrons

Sqdn	Comm	Commanding Officer(s)
VH-1	15 Apr 44	LCDR J. D. Adam, LCDR Russell R. Barrett Jr.
VH-2	1 Jul 44	LCDR Clarence A. Keller, LCDR Harold A. Wells
VH-3	1 Aug 44	LCDR William D. Bonvillian
VH-4	1 Sep 44	LT Forrest H. Norvell Jr.
VH-5	11 Sep 44	LCDR M. E. Brown
VH-6	20 Sep 44	LCDR Lealand O. Ebey[21]

VH Squadrons were employed only in areas where the intensity of operations made calls upon their services frequent; elsewhere in the Pacific, patrol plane squadrons continued to provide Dumbos as an adjunct to their other duties. These included the Royal New Zealand Air Force's No. 6 (Flying Boat) Squadron.[22]

ROYAL NEW ZEALAND AIR FORCE (RNZAF)

The Royal New Zealand Air Force was established on 1 April 1937, when the Air Force Act and the Air Department Act were passed. The first created the Air Force as a separate branch of the Armed Services, and the second instituted a new department of state for administration of aviation. The Air Department was responsible for the administration of both service and civil aviation.[23]

In April 1939 with war looming in Europe, a Pacific Defense Conference was held in Wellington, attended by representatives of the United Kingdom, Australian and New Zealand Governments. It was discussed that, should the Japanese attempt an invasion of Australia or New Zealand, it would be necessary for them to secure bases in the South Pacific and, accordingly, steps should be taken immediately to ensure the protection of potential bases against Japanese attack. At the conference, it was determined that New Zealand would take on this responsibility for territories formerly administered by Britain.[24]

The most important point in the South Pacific from New Zealand's perspective was the Fijian island chain. The harbor facilities at Suva on the southeast coast of the island of Viti Levu, and supply of fuel there made it one of the most important naval fueling bases in the South Pacific. Also, it had a communications station, which would become increasingly important as a center of air communications. Additionally, the Fijian islands produced plentiful food, which would enable the Japanese, if they were to successfully invade, to maintain a large force there, capable of easily attacking trans-Pacific shipping.[25]

To help prevent this possibility, the conference recommended several actions: New Zealand should immediately construct two aircraft landing facilities on Viti Levu, one near Suva and another on the island's northwestern coast; part of New Zealand's reserves of fuel, bombs, and ammunition should be held in Fiji; a survey should be made of Tonga to the east, to determine the feasibility of establishing landing facilities there for the Royal New Zealand Air Force; and finally, in time of war, New Zealand should carry out reconnaissance on a line along the New Hebrides-Fiji-Tonga island chains.[26]

Map Preface-2

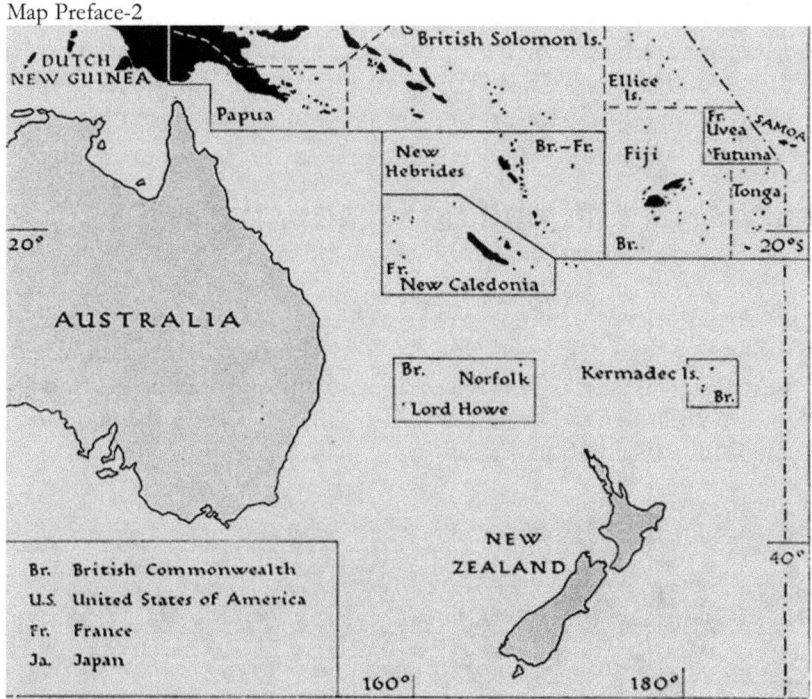

Political control of the Western Pacific islands in 1939 (The lines separating the various groups do not necessarily indicate extent of sovereignty.)

A few hours after the Japanese attacked the U.S. naval base at Pearl Harbor on 7 December, New Zealand declared war on Japan. (It was the 8th in New Zealand across the International Date Line). In addition to protection of shipping in New Zealand waters, the RNZAF also had its new responsibility to protect Fiji. Work began on an aerodrome at Nandi, near Lautoka on the west coast of Viti Levu, and an air strip at Nausori, fifteen miles from Suva on the east coast. At the same time, sites for an aerodrome and seaplane alighting area were surveyed at Tonga, followed by construction of the aerodrome.[27]

In 1942, to stem the tide of Japanese aggression in the South and Southwest Pacific, the Allies developed a chain of island bases stretching from Northern Australia through New Caledonia, the New Hebrides, Fiji and Tonga to Samoa. These were intended to serve as a protection for Australia and New Zealand, and also guard the vital supply line to Australia. Shipping bound from Bora Bora in French Polynesia to Australia had to pass through or close to the Cook Islands, then the Samoa, Tonga and Fiji groups, and finally, a thousand miles from the Australian coast, the New Hebrides group, and New Caledonia.[28]

These bases would later provide important supply and repair support as the Allied forces moved up through the Solomon Islands toward the Japanese home islands.[29]

RNZAF NO. 6 (FLYING BOAT) SQUADRON

Long-range patrols and reconnaissance flights were carried out from Fiji by PBY Catalinas of Royal New Zealand Air Force No. 6 Squadron. This squadron was commissioned at Fiji on 25 May 1943. Its first Catalina had been delivered at Lauthala Bay (to the east of Suva on the southern coast of Viti Levu Island), in April, by an American crew who flew it there from San Diego, California. The squadron's full allowance of PBYs was twenty-four aircraft.[30]

Photo Preface-8

Royal New Zealand Air Force No. 6 (Flying Boat) Squadron group photograph. Courtesy of Jenny Scott

No. 6 Flying Boat Squadron: PBY-5 Catalinas
Formed in Fiji on 25 May 1943
Disbanded on 9 September 1945

Commanding Officers

Wg Cr George Gatonby Stead	May-October 1943
Wg Cr I. A. Scott	October 1943-August 1944
Wg Cr John R. S. Agar	August 1944-June 1945
Wg Cr K. G. Smith	June-September 1945

Duty Assignments

Lauthala Bay, Viti Levu, Fiji Islands	March-October 1943
Flight at Nukualofa, Tongatapu Island	August-November 1943
Espiritu Santo, New Hebrides Islands	October-December 1943
Halavo Bay, Florida Island, Solomons	December 1943-August 1945[31]

While most of No. 6 Squadron operated from the Lauthala Bay seaplane base, carrying out ship escort, submarine search, and air-sea rescue missions, a Flight (detachment) of six aircraft and crews was sent to Tonga. Stationed at the U.S. naval base at Nukualofa, it was responsible for the protection of shipping in Tongan waters.[32]

The squadron left Fiji at the beginning of October, 1943, when it was posted to Espiritu Santo. Based aboard the seaplane tender USS *Wright* in Segond Channel, its PBYs searched for enemy ships and submarines to the west of "Santo." The flight based at Tonga rejoined the squadron in November.[33]

Just before Christmas in 1943, the entire unit moved forward to Halavo Bay on Florida Island, near Guadalcanal. On 9 February 1944, two aircraft and crews were sent to the Treasury Islands, to be based with the seaplane tender USS *Coos Bay*. The main body of No. 6 Squadron operated from Halvao Bay for the remainder of the war, with detached units at various times to Funafuti, Ellice Islands; Emirau, Bismarck Archipelago; Los Negros, Admiralty Islands; Green Island, Solomon Islands; and Jacquinot Bay, New Britain.[34]

ROYAL AUSTRALIAN AIR FORCE AIR-SEA RESCUE

The RAAF's first CSAR (Combat Search and Rescue) capability in World War II was provided by No. 1 Rescue Flight, located at Goodenough Island (in Milne Bay off the eastern tip of Papua), in November 1942, with detachments at various Allied bases around New Guinea. This unit (later renamed No. 1 Rescue and Communication Squadron, and eventually No. 8 Communications Unit) used Walrus, Dornier Do-24 and Catalina flying boats in the CSAR role.[35]

The Royal Australian Navy had previous experience with the Supermarine Walrus, a single-engine amphibian reconnaissance aircraft originally termed the Supermarine Seagull V, and known throughout the Royal Australian Navy as the "Pusser's Duck." The aircraft was designed to be catapulted from large Navy warships, for use in reconnaissance and gunnery spotting (observing the fall of gun rounds, and advising corrections by radio communications). Walruses obtained by the RAAF were used for coastal patrols and air-sea rescue tasks until they were phased out of service in 1946.[36]

From December 1944, the number of units specializing in CSAR greatly increased. Air-Sea Rescue Flights were formed at Darwin and Cairns, Australia; Madang, New Guinea; and Morotai Island, Netherlands East Indies. These Flights were equipped with PBY Catalina flying boats for search and rescue operations, often involving the recovery of aircrew stranded in enemy territory. Some of the units also had motor launches that could be used to recover personnel when alighting on the water was not possible. Catalinas also transported medical supplies to remote Army units as well as providing courier runs throughout the region.[37]

Photo Preface-9

A Supermarine Walrus amphibian aircraft of the Air-Sea Rescue Service on a landing ground in North Africa, circa 1942.
Australian War Memorial photograph MED0032

Summary information about the RAAF air-rescue flights follow: specifically, names of their commanding officers and locations at which the units were based. Several members of Flights 112 and 113 earned awards for valour, including the Distinguished Flying Cross,

Distinguished Flying Medal, Air Force Cross, and Mention in Despatches. The identities of these individuals, and Royal New Zealand Air Force counterparts similarly honoured are provided in Appendix A.

Distinguished Flying Cross

Distinguished Flying Medal

Air Force Cross

Mention in Despatches

Photo Preface-10

Madang, New Guinea, January 1945. Flight Lieutenant Ian James Lock Wood in the cockpit of a Catalina, of No. 111 Air-Sea Rescue Flight. While previously assigned to No. 8 Communications Unit, Wood had received an immediate award of the Distinguished Flying Cross for the dramatic rescue of six Americans off the coast near Boram in northeast New Guinea.
Australian War Memorial photograph OG2028

No. 111 Air-Sea Rescue Flight
(Formed at Madang on 13 December 1944 under Northern Command, which became Northern Area Command from December 1944; disbanded on 24 January 1947.)

Commanding Officers

FL Ian James Lock Wood (DFC)	13 December 1944
FL C. W. Miller (DFM)	6 January 1945
FL G. H. Priest	1 September 1945
FL B. Parker	1 June 1946

Duty Assignments

Madang, Papua New Guinea	13 December 1944
Port Moresby, Papua New Guinea	18 March 1946

112 Air-Sea Rescue Flight
(Formed at Darwin on 23 December 1944 under North Western Area Command. Final Unit History entry on 16 September 1947.)

Commanding Officers

SL Kenneth Arthur Crisp	23 December 1944
FL Robin Morton Corrie	20 March 1945
FL L. M. Cameron	1 September 1945
FL R. J. Rankin	21 November 1945
FL A. E. Delahunty	15 May 1946
SL F. S. Robey	1 April 1947
SL C. A. Voges	26 June 1947

Duty Assignments

Darwin, Australia	23 December 1944

No. 113 Air-Sea Rescue Flight
(Formed at Cairns on 10 January 1945; based at Morotai from 11 March through the duration of the war under First Tactical Air Force; final unit history entry on 31 January 1946.)

Commanding Officers

FL W. G. White	10 January 1945
FL Walter Raymond Mills	17 September 1945
FL E. M. Allison	29 November 1945

Duty Assignments

Cairns, Queensland, Australia	10 January 1945
Townsville, Queensland, Australia	20 February 1945
Morotai, Molucca Islands	11 March 1945
Labuan Island, Borneo	29 September 1945
Rathmines, New South Wales, Australia	28 January 1946

The most lauded aircrew was that of Flight Lieutenant Walter Mills of No. 113 Air-Rescue Flight, which had some interesting experiences, beyond those associated with typical air-sea activities. On one occasion, after delivering supplies to Kapit, Borneo, PBY Catalina A24-104 received tasking to support an operation by P-40 Warhawks. The Penghula (tribal chief) of the Kapit area was aboard the Catalina. (Presumably, to serve as an airborne guide to help them reach their destination.) By the time the seaplane reached the target area, the fighter/ground-attack aircraft had thoroughly strafed and burned the barracks at a Japanese base, resulting in many enemy casualties. The chieftain was wild with excitement and wanted to land in order to add more skulls to his collection. As the PBY was departing, Flight Lieutenant Walter Mills and the others aboard could see dozens of prahus (boats) full of headhunters converging on the area.[38]

After occupying Borneo, the Japanese had targeted the Kapit District, in southern Borneo. Bordered by the Rajang River, this area was home to many Dyak (who practiced headhunting). The ill-treatment spurred natives to join with the allies against a common enemy. Trained by American and Australian military leaders in guerrilla warfare in the jungle, the indigenous fighters in ensuing years captured or killed 1,500 Japanese and provided vital intelligence about enemy-held oil fields.[39]

No. 114 Air-Sea Rescue Flight
(Formed at Cairns on 18 July 1945 under North Eastern Area Command; ceased to function on 1 October 1947.)

Commanding Officers

SL G. M. Mason	18 July 1945
FL R. T. Clark	1 July 1946
FL R. J. Rankin	10 February 1947

Duty Assignments

Cairns, Queensland, Australia	18 July 1945
Garbutt, Queensland, Australia	September 1946

115 Air-Sea Rescue Flight
(Formed at Morotai on 16 July 1945 under 11 Group, replacing 113 which moved to Labuan; disbanded on 28 March 1946.)

Commanding Officers

FL Geoffrey Francis Gregerson	16 July 1945

Duty Assignments

Morotai, Molucca Islands	18 July 1945
Detachment at Balikpapan, Borneo	
Detachment at Balikpapan discontinued	mid-October 1945
1 plane detachment at Biak, Papua New Guinea	24 October 1945
Detachment at Biak discontinued	22 November 1945

Before leaving this preview of the Royal New Zealand Air Force, and No. 6 Squadron, some readers might find a comparison of USN, RNZAF, and RAAF officer ranks in World War II useful.

U.S. Navy	Royal Australian Air Force	Royal New Zealand Air Force
Fleet Admiral FADM (O-11)	Marshal of the Royal Australian Air Force MRAAF	No equivalent
Admiral ADM (O-10)	Air Chief Marshal ACM (O-10)	No equivalent
Vice Admiral VADM (O-9)	Air Marshal AM (O-9)	Air Marshal AM (OF-9)
Rear Admiral RADM (O-8)	Air Vice-Marshal AVM (O-8)	Air Vice-Marshal AVM (OF-8)
Commodore CDRE (O-7)	Air Commodore AC (O-7)	Air Commodore AC (OF-7)
Captain CAPT (O-6)	Group Captain GC (O-6)	Group Captain GC (OF-6)
Commander CDR (O-5)	Wing Commander Wg Cr (O-5)	Wing Commander Wg Cr (OF-5)
Lieutenant Commander LCDR (O-4)	Squadron Leader SL (O-4)	Squadron Leader SL (OF-4)
Lieutenant LT (O-3)	Flight Lieutenant FL (O-3)	Flight Lieutenant FL (OF-3)
Lieutenant (junior grade) LTJG (O-2)	Flying Officer FO (O-2)	Flying Officer FO (OF-2)
Ensign ENS (O-1)	Pilot Officer PO (O-1)	Pilot Officer PO (OF-1)

The ranks of Fleet Admiral and Commodore no longer exist in the U.S. Navy. The last fleet admirals were in World War II, and there were only four; Chester W. Nimitz, William D. Leahy, Ernest J. King, and William F. Halsey Jr. Commodore was a command rank in the Navy from 1862 to 1899. It was reestablished on 9 April 1943 for war service, and 147 officers held it as a temporary rank. After the war, the flag rank structure reverted to its prewar form. Navy captains promoted to flag rank jumped directly to two stars. There were no O-7 (one-star admirals. This changed in 1982, with the introduction of commodore admiral. This title was changed to rear admiral (lower half) after eleven months, and the latter convention continues today.

It is important to highlight that several variants in the abbreviations of military rank are used throughout the book. This is because, in addition to those identified on the previous page for the U.S. Navy, and Royal Australian and Royal New Zealand Air Forces, there are also references to Royal Navy, and U.S. Army Air Force and U.S. Marine Corps ranks. Those for the Royal Navy are only associated with a small portion of the book devoted to the British Pacific Fleet, while those of the USAAF and USMC pertain mostly to the identities of survivors.

AFLOAT SEAPLANE BASES

Photo Preface-11

Large seaplane tender *Salisbury Sound* (AV-13) tending P5M-2 Marlin anti-submarine patrol seaplanes at Tsugen Jima, Japan, in March 1957.
Painting by Richard DeRosset

The valuable Dumbo contribution of RNZAF No. 6 Squadron, which included the recovery of seventy-nine survivors (identified in Appendix B) of downed aircraft and sunken vessels, was but part of a much larger role of patrol aircraft and seaplane tenders in the war.

The U.S. Navy's sixty-seven seaplane tenders refueled, rearmed, and repaired the "eyes of the fleet," its scout planes and patrol planes. Later in the war, this included patrol bombers. The ships could function in any sizeable body of protected water where their tended aircraft could land and take off. (It is important to note that only a few tenders supported seaplanes wholly devoted to Dumbo operations, and generally only for certain periods of time. The squadrons assigned to most tenders carried out many missions, including long-range search, anti-submarine, and strike operations, and Dumbos, only as specifically tasked, or when a situation arose requiring air-sea rescue.)

Tenders were particularly valuable in advance areas which lacked facilities for land-based reconnaissance aircraft. The bulk of the Navy's seaplane tenders served in the Pacific Theater, due to the scarcity of existing airstrips or airfields across its oceanic vastness. Of the 109 battle stars collectively earned by tenders during the war, only six of them were outside the Asiatic-Pacific Theater.[40]

As General MacArthur drove up through Papua-New Guinea, Admiral Halsey through the Solomon Islands, and Admiral Spruance through the Central Pacific toward the Japanese home islands, tenders supported the assault forces. Their aircraft scouted for Japanese naval forces, and carried out attacks on enemy ships and shore targets as opportunities presented themselves. Once Naval construction battalion personnel ("Seabees") had built airstrips and supporting facilities in captured areas to host land-based fighter and attack aircraft, the seaplane tenders moved forward to new areas, repeating this cycle.[41]

These offensive operations are described in my book, *Eyes of the Fleet: The U.S. Navy's Seaplane Tenders and Patrol Aircraft in World War II*, a companion to this one.

As recounted in *Eyes of the Fleet*, cloaked by jungle foliage to escape detection, the unheralded seaplane tenders operated ahead of the Fleet, like the Navy's famed PT boats. The Japanese were keen to destroy the scouts and their floating bases, and seaplane tenders often lived a furtive existence, particularly early in the war. The PBY Catalinas and later PBM Mariners they tended combed the seaways for Japanese forces and carried out bombing, depth charge, and torpedo attacks on enemy ships and submarines. Their nighttime anti-shipping missions were dangerous and daytime combat operations even more so, when encounters with more maneuverable and heavily-armed fighters necessitated hiding in clouds to survive.

Seaplanes in forward areas where there were no aviation facilities ashore, were wholly dependent on tenders and the seadromes they maintained. In simple terms, a seadrome was a sheltered body of water, typically a bay or harbor, on which planes could land and take off, refuel, rearm, and make up to mooring buoys when not in flight. Because of their close association and intertwined dependency, the tenders' stories are inextricably linked to those of the aircraft they serviced. Among the Navy's oldest tenders were two classes of converted World War I vintage minesweepers and "four-piper, flush-deck" destroyers:

Ex *Lapwing*-class Minesweepers, 187 feet, 1,350 tons

Lapwing AVP-1 (ex AM-1)	*Avocet* AVP-4 (ex AM-19)	*Swan* AVP-7 (ex AM-34)
Heron AVP-2 (ex AM-10)	*Teal* AVP-5 (ex AM-23)	*Gannet* AVP-8 (ex AM-41)
Thrush AVP-3 (ex AM-18)	*Pelican* AVP-6 (ex AM-27)	*Sandpiper* AVP-9 (ex AM-51)[42]

Photo Preface-12

Small seaplane tender *Avocet* (AVP-4) carrying a Curtiss SOC Seagull scout biplane. U.S. Navy Photograph 1 Hanson Place Brooklyn, New York

| Preface

Photo Preface-13

Destroyer seaplane tender *Osmond Ingram* (AVD-9) under way off Norfolk Naval Shipyard, Portsmouth, Virginia, on 10 July 1943.
U.S. Naval History and Heritage Command Photograph # NH 42922

Former "Flush Deck" Destroyers, 314 feet, 1,215 tons

Childs AVD-1 (ex DD-241, ex AVP-14)	*Goldsborough* AVD-5 (ex DD-188, ex AVP-18, later APD-32)	*Osmond Ingram* AVD-9 (ex DD-255, later APD-35)	*Greene* AVD-13 (ex DD-266, later APD-36)
Williamson AVD-2 (ex DD-244, ex AVP-15)	*Hulbert* AVD-6 (ex DD-342, ex AVP-19)	*Ballard* AVD-10 (ex DD-267)	*McFarland* AVD-14 (ex DD-237)
George E. Badger AVD-3 (ex DD-196, ex AVP-16, later APD-33)	*William B. Preston* AVD-7 (ex DD-344, ex AVP-20)	*Thornton* AVD-11 (ex DD-270)	
Clemson AVD-4 (ex DD-186, ex AVP-17, later APD-31)[43]	*Belknap* AVD-8 (ex DD-251, later APD-34)	*Gillis* AVD-12 (ex DD-260)	

Beginning in 1943, as new, purpose-built *Barnegat*-class small seaplane tenders began to replace them, these ex-destroyers served in other roles including convoy escort; anti-submarine warfare; and local patrol, plane guard, and shakedown support for escort carriers. The newly constructed tenders included the *Coos Bay*, previously mentioned as supporting New Zealand's No. 6 flying boat squadron.[44]

Barnegat-class Small Seaplane Tenders, 311 feet, 2,750 tons

Barnegat (AVP-10)
Biscayne (AVP-11)
Casco (AVP-12)
Mackinac (AVP-13)
Humboldt (AVP-21)
Matagorda (AVP-22)
Absecon (AVP-23)
Chincoteague (AVP-24)
Coos Bay (AVP-25)
Half Moon (AVP-26)
Rockaway (AVP-29)[45]

San Pablo (AVP-30)
Unimak (AVP-31)
Yakutat (AVP-32)
Barataria (AVP-33)
Bering Strait (AVP-34)
Castle Rock (AVP-35)
Cook Inlet (AVP-36)
Corson (AVP-37)
Duxbury Bay (AVP-38)
Gardiners Bay (AVP-39)

Floyds Bay (AVP-40)
Greenwich Bay (AVP-41)
Onslow (AVP-48)
Orca (AVP-49)
Rehoboth (AVP-50)
San Carlos (AVP-51)
Shelikof (AVP-52)
Suisun (AVP-53)
Timbalier (AVP-54)
Valcour (AVP-55)

Fifteen large seaplane tenders served during World War II. Although designated seaplane tenders, they were often referred to as "large seaplane tenders" or "heavy seaplane tenders" because of their size. A converted merchant ship and a Navy collier were the first two of these tenders. The *Wright* had been laid down in the builder's yard as a Hog Island type "B" cargo vessel, and was later fitted out as a lighter-than-air aircraft tender and commissioned *Wright* (AZ-1) on 16 December 1921. She was reclassified a heavier-than-air aircraft tender (AV-1) on 2 December 1926.[46]

Photo Preface-14

USS *Wright* (AV-1) under way, location and date unknown.
Naval History and Heritage Command photograph # UA 475.19

Langley had been commissioned on 7 April 1913 as the collier *Jupiter*. Following conversion to the Navy's first aircraft carrier in 1920, she was renamed *Langley*. In 1937 the Navy converted her to a seaplane tender (AV-3). *Langley* was irreparably damaged by Japanese aircraft bombs south of Java in the Dutch East Indies, on 27 February 1942, and was sunk by the destroyer *Whipple* (DD-217) to avoid capture or salvage by the eneny.[47]

Large Seaplane Tenders No Name Class
Wright (AV-1) 448-feet, 12,142 tons | *Langley* (AV-3) 542-feet, 19,360 tons

Curtiss-class, 527-feet, 13,880 tons
Curtiss (AV-4) | *Albemarle* (AV-5)

Currituck-class: 541-feet, 14,300 tons
| *Currituck* (AV-7) | *Norton Sound* (AV-11) | *Pine Island* (AV-12) | *Salisbury Sound* (AV-13) |

Tangier-class: 492-feet, 8,950 tons
Tangier (AV-8) | *Pocomoke* (AV-9) | *Chandeleur* (AV-10)

Kenneth Whiting-class: 492-feet, 8,000 tons
| *Kenneth Whiting* (AV-14) | *Hamlin* (AV-15) | *St. George* (AV-16) | *Cumberland Sound* (AV-17)[48] |

Although predominantly in the Pacific, U.S. Navy seaplane tenders operated in every theater during World War II. The activities of tenders and patrol aircraft along the east coast of South American and in the narrows of the Atlantic between Brazil and west Africa, are described in my book *Ingram's Fourth Fleet*.

In the Pacific, seaplane tenders earned their first battle stars at Pearl Harbor, and during unsuccessful defenses of the Philippine Islands and the Java Sea. As combat spread across the Pacific, stars were won in the Aleutian Islands and far to the south-southwest in the Solomons.

Tenders also earned battle stars in the Santa Cruz, Gilbert, New Georgia, Bismarck, Western Caroline, and Marshall islands. More stars were garnered in Western New Guinea, and at Saipan, at Leyte, and Luzon in the Philippines and, as war drew nearer an end, at Iwo Jima and Okinawa. The chart on the following two pages shows the advance of Allied forces across the Pacific toward Japan, between 7 August 1942 and 1 March 1944, and from 1 March 1944 to 1 March 1945.[49]

COMBINED RESCUE OPERATIONS

Because fast-carrier operations during the latter stages of the war were generally deep in enemy waters, the only rescue facilities were seaplanes carried aboard battleships and carriers, the ships of the task force itself, and submarines. Lifeguard submarines were positioned at key points to rescue flyers forced down. In March 1944, the submarine USS *Tang* (SS-306) recovered twenty-two flyers off Truk, which resulted in her second war patrol being cut short. This was described by Motor Machinist's Mate Clayton Decker:

> We received a message to rescue a bunch of Navy carrier pilots who had been shot down during the battle of Truk Atoll. We pulled 22 flyers out of the water; they weren't hurt physically. We had a crew of 87 and now we had an extra 22 to accommodate—talk about "hot bunking!" Admiral [William F.] Halsey wanted all of those flyers back in service as fast as he could get them, so we went back to Pearl Harbor. We did not get a star in our combat pin for that patrol, but Admiral Halsey gave every member of the *Tang* crew the Navy Air Medal.[50]

In the vicinity of carrier task force operations, destroyers assigned to screening and picket duties performed rescues. Air strikes on the enemy fleet during the Battle of the Philippine Sea cost twenty planes shot down, and another fifty-five forced by lack of fuel to ditch on the water before reaching their carriers. Of approximately 180 personnel involved, all but 16 pilots and 22 aircrewmen were recovered, most by destroyers before dawn.[51]

Coordinated rescue operations reached their high point in connection with B-29 strikes against Japan. A chain of ships and submarines, each with a supporting aircraft circling overhead, was stationed along the route from the Marianas to the target. When a bomber went down, the assigned air-sea rescue plane (which on the most remote stations was usually a specially equipped B-29 "Super Dumbo" and on closer stations a Navy seaplane) searched for survivors and directed a ship or submarine to the scene.[52]

Map Preface-3a

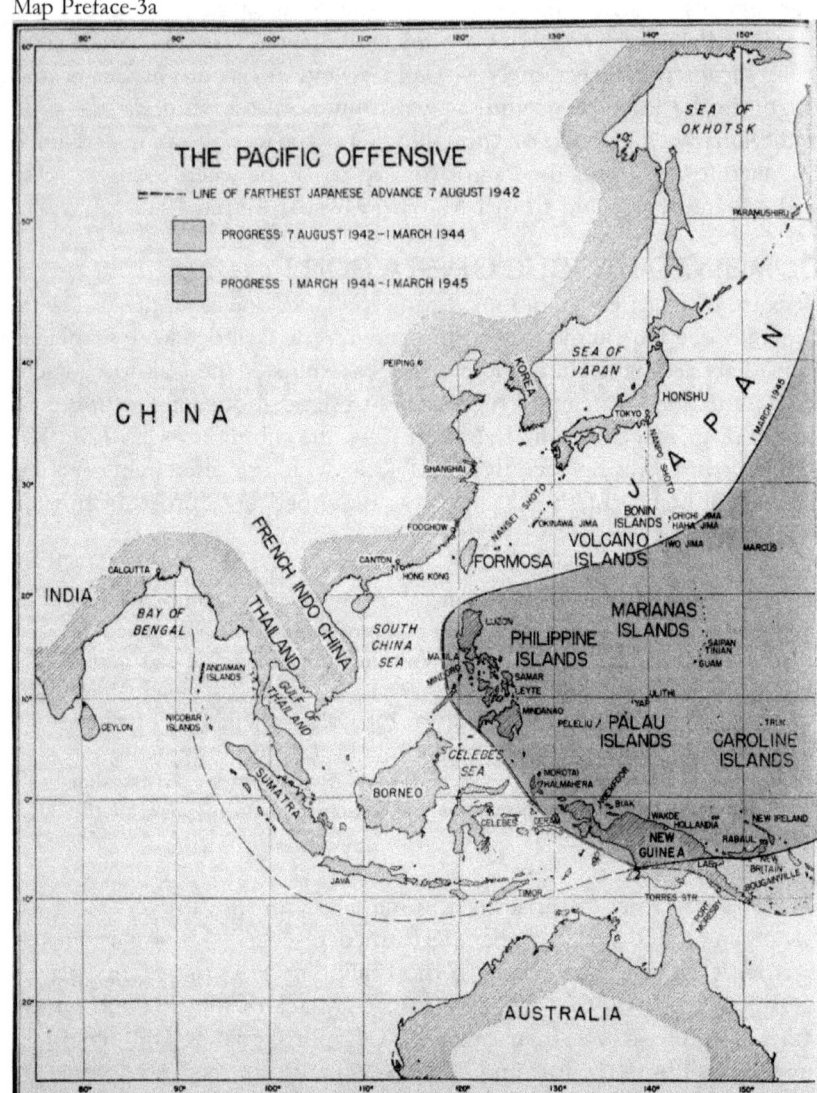

Movement of Allied Forces up through the South, South West, and Central Pacific toward the Japanese home islands
United States Navy at War, Second Official Report to the Secretary of the Navy, covering combat operations March 1, 1944, to March 1, 1945 by Fleet Admiral Ernest J. King

Map Preface-3b

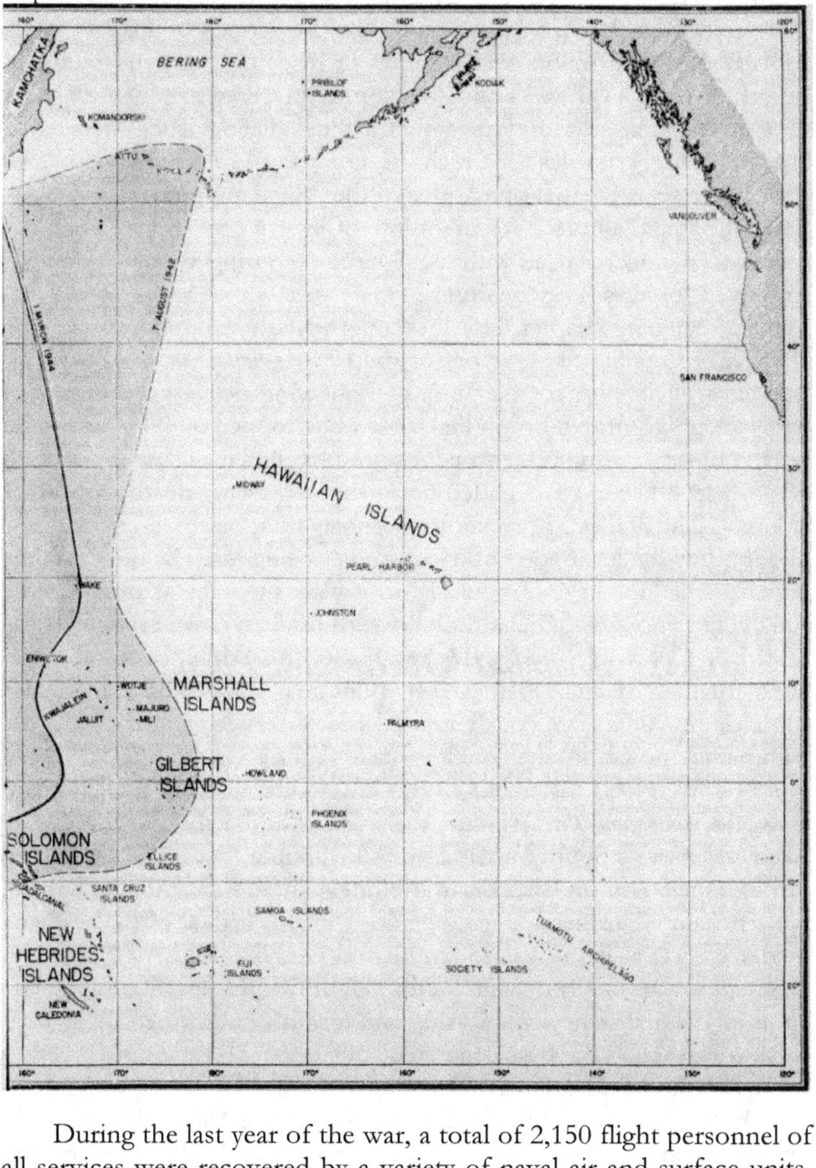

During the last year of the war, a total of 2,150 flight personnel of all services were recovered by a variety of naval air and surface units. The value of air-sea rescue transcended the lives saved, as it bolstered morale and spurred combat aircraft crews to even greater efforts against the enemy.[53]

PHILOSOPHY AND SCOPE OF THIS BOOK

The underlying theme of this book is heroic actions of PBY Catalina and PBM Mariner pilots and aircrews in rescuing survivors of Allied aircraft downed over water and, less commonly, crewmembers of ships sunk at sea. The slow and ungainly Catalinas distinguished themselves in a wide variety of roles for which the aircraft was never intended. *Eyes of the Fleet* details combat operations of the "Cats" (referred to as "Black Cats" if painted all black for nighttime offensive operations). *Salvation from the Sky* concerns itself with the "Dumbos," stripped down Catalinas employed for air-sea rescue duties.

The opening chapter describes a particularly daunting rescue by a U.S. Navy Catalina of survivors of the USS *Indianapolis* (CA-35). Her story is well known to the public. Following delivery to Tinian of components for the A-Bomb that would be dropped on Hiroshima, the heavy cruiser was sunk by torpedoes fired by the Japanese submarine *I-58*. Up to 800 survivors ended up in the sea, facing death from their wounds, dehydration, overexposure, exhaustion, or shark attack. The pilot of the Catalina, upon observing bodies being eaten by the fearsome predators, landed in enormous, 12-foot waves to save as many of the still living as possible. Richard DeRosset's brilliant cover art depicts the destroyer escort USS *Cecil J. Doyle* (DE-368) searching by spotlight at night for the Catalina. Damaged by pounding waves, and loaded with survivors including grievously injured laid atop her wing and covered with parachute fabric, she could not take flight.

Following this testament to the intrepidness of one Catalina and crew, *Salvation from the Sky* joins the war in the Pacific, on the eve of Admiral Halsey's South Pacific Forces beginning their drive northwest through the Solomon Islands toward the Japanese stronghold at Rabaul, New Britain. This beginning coincides with the arrival of the seaplane tender USS *Coos Bay* (AVP-25) and Royal New Zealand Air Force No. 6 Squadron at Espiritu Santo, New Hebrides, in late summer/early autumn 1943. The newly built ship and recently constituted flying boat squadron reached the base separately, but were soon working together as part of the Allied offensive.

As *Coos Bay* moved forward, helping facilitate Allied advancements up the Solomons, Catalinas operated from her and other tenders, and from shore bases. PBY-5 Catalinas, pure flying boats, required support from tenders anchored in the vicinity of their mooring buoys. Successor PBY-5As had retractable landing gear, allowing them to taxi up ramps and out of the water for maintenance and repair "on the hard." Once aviation facilities were constructed ashore in newly captured areas, seaplane tenders could "step farther forward" to provide needed support where there was none.

Late in the war, scores of tenders were in the Western Pacific, supporting Catalina and Mariner seaplanes carrying out a variety of missions. These included maritime patrol; anti-submarine warfare; night attack, and naval interdiction; and of course, search and rescue. As MacArthur's forces driving up through Papua-New Guinea (on his road "back to the Philippines"), and Halsey's and Spruance's through the Solomons and Central Pacific, began to converge in the final push toward Japan, Allied bombing attacks on enemy installations increased dramatically. So too did numbers of bombers lost to enemy action, and associated need for the rescue of downed flyers.

In late 1944 and through war's end, American and Australian bombers were also striking Japanese positions in New Guinea, and in parts of the Netherlands East Indies including, in the last action of the war, on Borneo (today Indonesia). The rescue of Australian and American flyers provided much work for Royal Australian Air Force Flights No. 111, 112, 113. Two additional flights—114 and 115—were formed too late to get into the war. Flying from bases in northern Australia and forward areas, Aussie Cats performed myriad important duties, including many heroic rescue operations.

With this introduction in our wake, it's time to take flight (vicariously) with the air-sea rescue aircraft of the U.S. Navy, and the Royal Australian, and Royal New Zealand Air Forces in World War II.

RAAF Ensign
1922–1948

USN Ensign
48-Star U.S. Flag
1912-1959

RNZAF Ensign
1939-present

But first, a testament to the love downed pilots and aircrews had for Dumbos, as evidenced by the poem, "Fifty Baker Twenty Eight." It was penned by a Marine fighter pilot while in sick bay aboard the *Coos Bay*, after being picked up by a PBY Catalina, responding to "50B28," the VHF radio call sign for a "Dumbo." A map following the poem may assist readers to understand the dire situation in which the aviator found himself, and why Dumbos were valued so greatly.

FIFTY BAKER TWENTY EIGHT

He was over Rabaul bombing
When some "flak" got in his way
And his engine coughed and sputtered
and then called it a day
He was gliding for the channel
and was cursing at his fate
When suddenly he remembered
Fifty Baker Twenty Eight

He opened up his R/T
and he broadcast loud and clear
"This plane of mine has had it,"
and the water's getting near
I'm fifteen east of Cape Gazelle
So please don't make me wait
Just send me out the "Dumbo"
Fifty Baker Twenty Eight

So that PBY came quickly
and its fighter escort too
Till they saw the PVs [patrol bombers] circling
as the PVs always do
They took one look and landed
and I'm happy to relate,
They got them all home safely
Fifty Baker Twenty Eight

Now remember this you fighters
and bombers, large and small,
If you ever get shot up
while bombing old Rabaul
Just head off down the channel
And get some other "crate"
To yell like hell for "Dumbo"
Fifty Baker Twenty Eight

Map Preface-4

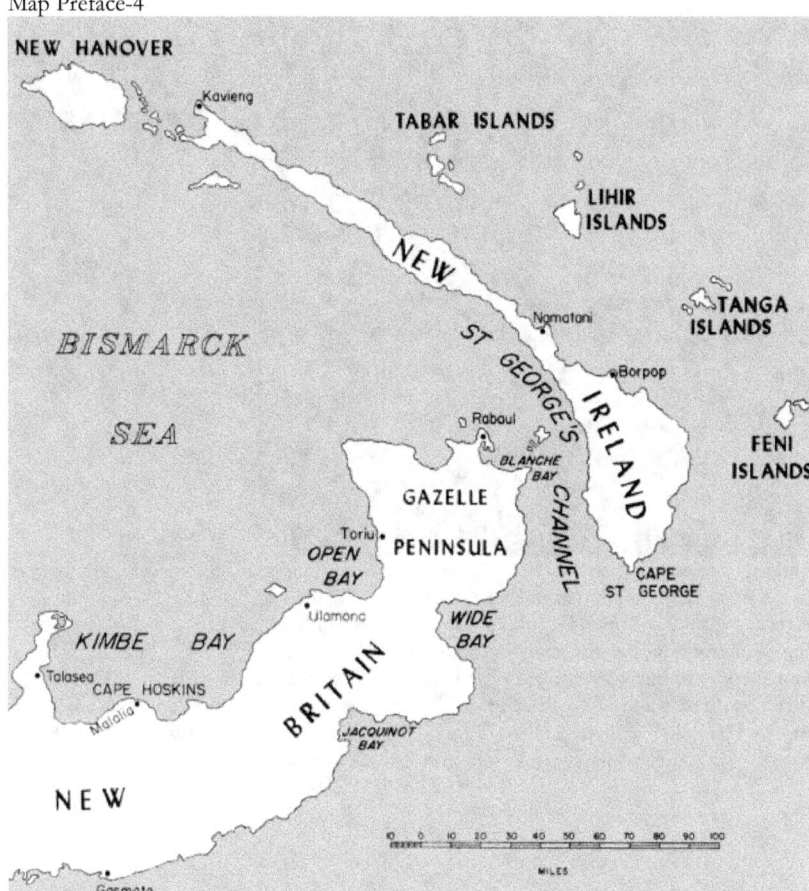

From the powerful Japanese naval and air base at Rabaul (on the northern tip of the island of New Britain), attacks from the air and sea were launched on Allied forces in the South and Southwest Pacific.
Henry I. Shaw Jr. and Douglas T. Kane, *History of U.S. Marine Corps Operations in World War II Volume II: Isolation of Rabaul* (Washington, DC: U.S. Marine Corps, 1963)

Photo Preface-15

Painting *Salvation from the Sky* by Richard DeRosset, depicts the destroyer escort USS *Cecil J. Doyle* searching in darkness for a PBY-5A Catalina seaplane alighted on the water. Earlier, Lt. Robert A. Marks had heroically landed in towering waves to rescue survivors of the heavy cruiser USS *Indianapolis*, sunk by the Japanese submarine *I-58*. Damaged by pounding seas, and heavily burdened by the weight of fifty-six survivors, the aircraft could not take off again. About twenty-five of the most grievously injured, many with broken arms and legs, are seen laid on the wing, covered with parachute-fabric to provide a measure of shelter from the night chill and wind-driven sea spray.

1

Rescue of USS *Indianapolis* Survivors

3 AUG 45
From: CO VPB 23
Action: CNO (DCNO) AIR
Information: COMMARIANAS, CTG 94.11, COMAIRPAC, COMAIRPACSUBCOMFWD, COMFAW 2, COMWESCAROLINES
PBY-5A in bound trip 6472 landed at scene of Indianapolis *sinking to take men aboard who might not have survived until ship arrived. Suffered damage on water during night due to heavy load of survivors and sea conditions. Unable to take off so abandoned and destroyed. Replacement required.*

—Message sent by commanding officer, Patrol Bombing Squadron VPB-23, regarding the loss of a Catalina patrol bomber while rescuing survivors of the cruiser USS *Indianapolis* (CA-35), sunk on 30 July 1945 by the Japanese submarine *I-58*.[1]

On 2 August 1945, Patrol Bombing Squadron VPB-23's duty officer received a phone call in late morning from his counterpart at VPB-152, informing him that one of its planes had sighted survivors approximately 275 miles from Peleliu at position 11°30'N, 133°30'E. The report ended with, "PV will stay on station until Dumbo arrives."[2]

The planes of both squadrons were operating from the U.S. Marine Corps air base on Peleliu, in the Palau island group, to provide air-sea rescue in support of U.S. Army Air Corps B-29 strikes on the Japanese home islands. Air-sea rescue missions were also provided in support of USMC air strikes on Japanese positions on islands bypassed by American forces as they leapfrogged across the Central Pacific.[3]

The Palau Islands lay in the eastern approaches to the southern Philippines. The group was made up of a cluster of volcanic islands, fragmented coral atolls, and islands of limestone composition, surrounded by reefs. Until its capture by U.S. Marines in autumn 1944, Palau had been the site of a Japanese advance naval base and assembly point for fuel, ammunition, and supplies moving between Japan and the

2   Chapter 1

southwest Pacific. Peleliu and Angaur, at the southernmost tip of the group, were the only islands developed for use by U.S. forces.[4]

Map 1-1

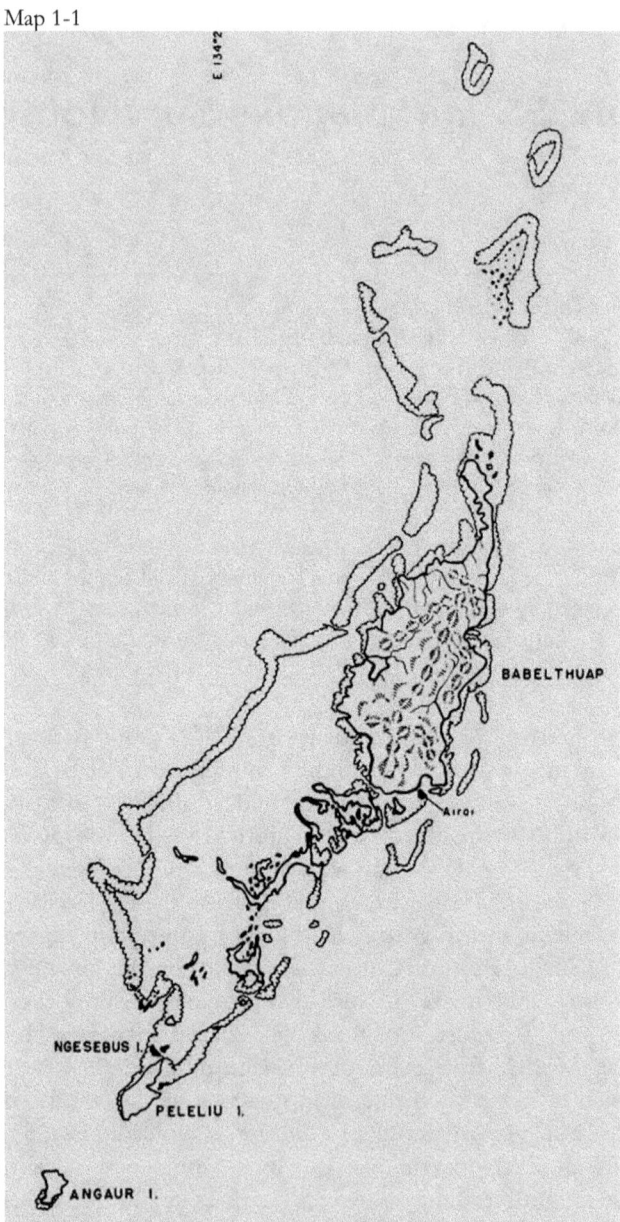

Palau Islands in the Western Caroline Islands

Squadron VPB-152 operated twin-engine Lockheed Ventura PV-1 patrol bombers, while VPB-23 flew the PBY-5A Catalina—an amphibious aircraft capable of landing either on the water or, using retractable beaching gear, ashore.[5]

Photo 1-1

PV-1 Ventura patrol bombers en route to Brunei, Borneo, circa 1945.
National Archives photograph #80-G-331264

Photo 1-2

A PBY-5A waits off Morotai (an island located 300 miles northwest of Sansapor, New Guinea) during amphibious landings by U.S. Army forces, 15 September 1944.
National Archives photograph #80-G-257979

VPB-152 had been carrying out routine anti-shipping searches and patrols from Peleliu. On 12 July, the squadron's mission changed to special weather flights and rescue missions. For this new task they were assisted by three PV-1 Ventura aircraft from VPB-133. The patrol bombers could either carry 3,000 pounds of bombs, six depth charges, or one torpedo, and were also armed with machine guns for self-protection. They had a cruising speed of 170 mph, and range of 1,360 miles if carrying one torpedo, or 1,660 miles with six 325-lb depth charges.[6]

While the Venturas could drop life rafts or other gear, or direct other aircraft or ships to survivors found at sea, they could not land to provide direct assistance. Thus for the rescue operation being described, the guidance, "PV will stay on station until Dumbo arrives" (in the report on survivors sighted in the water, north-northwest of Peleliu), referred to the PV-1 Ventura overhead the men in the water, remaining on the scene, awaiting a PBY-5A Catalina to take over the air-sea rescue operation. "Dumbo patrols" or "Dumbo aircraft" denoted open ocean rescue missions.

PBY CATALINA AIR-SEA RESCUE OPERATIONS

The venerable Catalina served in every theater of war, and carried out several types of missions. These included long-range reconnaissance, nighttime anti-shipping operations, and air-sea rescue. Later models of the PBY were equipped with radar and Magnetic Anomaly Detection equipment for use in the detection of enemy submarines.[7]

Organized "Dumbo" operations in the Pacific Theater was born during the Solomon Islands Campaign. Dumbo planes in the Solomons were seaplanes stripped of all non-essential heavy gear. Guns were loaded and manned, but no bombs were carried. The seaplanes rescued pilots whose planes were shot down, or failed for any reason. PBY-5s sometimes alone, or at other times escorted by fighters, were dispatched to pick up survivors. In many cases, the planes landed within easy range of enemy shore-based artillery in order to rescue downed pilots. In such circumstances, fighter aircraft employing diversionary tactics endeavored to keep the enemy as busy as possible away from the rescue operations.[8]

DISCOVERY OF SURVIVORS

LTJG Wilbur G. Gwinn, piloting a Ventura at 3,000 feet on a regular sector search, had spotted the survivors by accident. Noticing a large oil slick on the sea below, he descended to 900 feet, and followed the slick (with a radius of about 30 miles) to its origin, hoping to find a damaged Japanese submarine. At 1118, aircrewmen sighted splashing swimmers in the water, and Gwinn immediately send a message that thirty survivors had been sighted about 280 miles north of Peleliu, and requested assistance. As a result, Navy ships were ordered to the scene, but the closest one would not arrive for some time. Meanwhile, LCDR George C. Atteberry (commanding officer, VPB-152) took off in another PV-1 and sped toward Gwinn's position, with LT Robert A. Marks (from VPB-23) following in a PBY-5A.[9]

Following discovery of the thirty survivors, Gwinn searched the entire area. After finding up to 150 personnel in the water, he dropping emergency rations and equipment to the men who seemed to be without rafts. When later asked about his impression upon finding the first survivors, Gwinn responded:

> I don't know—it was a funny feeling. The oil slick was large, seeming to indicate a large vessel having been sunk, but I didn't know of any large craft being lost or going down, and didn't know just what kind of vessel it was.[10]

After dropping the rescue gear, Gwinn circled around until Atteberry arrived mid-afternoon, then left immediately. Marks reached the scene well after the other plane. He had launched a minute earlier than Atteberry but PBYs, with a cruising speed of 105-125 mph, were much slower than PVs.[11]

FINAL VOYAGE OF THE USS *INDIANAPOLIS*

Unbeknownst to the aircrews overhead, the men in the Philippine Sea below were all that remained of the USS *Indianapolis* (CA-35). The loss of the heavy cruiser to torpedoes fired by the Japanese submarine *I-58*, followed the completion of a secret Allied mission that directly contributed to the end of World War II.[12]

Following receipt of orders to make a top-secret delivery at high speed to Tinian, in the Mariana Islands, *Indianapolis* had left Mare Island, California, on the morning of 16 July. The heavy cruiser's commanding officer, CAPT Charles Butler McVay III, did not know the contents of a large wooden crate under guard in the ship's empty port hangar, nor that of a metal canister in the empty flag staff's quarters, only that every hour saved in transport would reportedly shorten the war by a corresponding amount.[13]

Averaging 28-29 knots, *Indianapolis* delivered her cargo to Tinian on 26 July, setting a speed record for the Pearl Harbor leg in the process. After the delivery of cargo and passengers, she made an overnight transit to Guam for routing instructions to her next assignment. *Indianapolis* was en route from Guam to Leyte on 30 July, when she was struck by two torpedoes from the *I-58* (Comdr. Mochitsura Hashimoto) and sent to the bottom of the Philippine Sea. This occurred in darkness in the first fifteen minutes of the new day.[14]

Photo 1-3

USS *Indianapolis* (CA-35) preparing to leave Tinian, after delivering atomic bomb components to the island in the Central Pacific, circa 26 July 1945.
U.S. Naval History and Heritage Command photograph #NH 73655

No distress signal left the *Indianapolis*, and 200-300 sailors and Marines went down with the ship. Up to 800 of her remaining crew abandoned, most with either pneumatic life belts or kapok lifejackets. Few life rafts were deployed. The fortunate survivors in the rafts had canned meat, malted milk balls, and some water. Most of the crew was scattered, having left the moving ship in piecemeal fashion. Seven different groups of varying sizes became spread out over approximately twenty-five miles. It is thought around 100 men in the water died within the first few hours due to wounds sustained in the torpedo explosions. The others faced death from dehydration, overexposure, exhaustion, shark attack, or simply giving up and sinking down into the sea.[15]

HEROIC ACTIONS BY THE PBY CATALINA'S CREW

As he drew nearer the scene, Marks began picking up signals from Atteberry's PV-1 at 1503, and made visual contact and established voice radio communications forty-seven minutes later. Atteberry advised

Marks that there was a large number of survivors scattered around, and asked him not to drop any equipment until he had observed the whole area. Otherwise, he might drop the whole lot to the first group, not realizing there were others in more desperate need. Marks later described his ensuing actions after VPB-152's commander led him on a 30-minute tour to look over the situation:

> Scattered small groups were everywhere, without any help except life jackets, and I thought that they needed the assistance more than the ones on the rafts. I knew that no ship would be on the scene until midnight, so after looking the area over, at 16:05 I commenced dropping survival equipment with the end in view of getting assistance to the small groups who had nothing but life jackets.[16]

Marks dropped all carried survival equipment, as well as his plane's own equipment, except one life raft, necessary for emergency if his own crew had to ditch. At 1625, he sent a message to the base advising of the number of survivors, and asking for additional survival equipment. He then decided a landing would be necessary to gather in survivors, alone in the water. This decision was based on their numbers and the immediate threat posed by sharks, after he observed bodies being eaten by the fearsome predators.[17]

Marks notified Atteberry at 1630 that he was going to attempt an open sea landing, and made all preparations to do so. He chose an area that would facilitate maximizing picking up survivors, and asked Atteberry to fly above him to observe and direct in assistance with the operation. The conditions were extremely unfavorable, and damage to the aircraft resulted from the landing as Marks explained:

> At 17:05 a power stall was made into the wind. The wind was due North, swells about 12 feet high. The plane landed in three bounces, the first bounce being about 15 feet high. Immediately after landing a survey of damage done to the plane exposed rivets pulled loose and some seams ripped open. My plane captain and navigator effected emergency repairs, plugging rivet holes with pencils and stuffing the seams with cotton. The radio compartment took on water slowly and would have to be bailed out during the night - 10 to 12 buckets of water per hour. The hull of the ship survived very well.[18]

While the navigator was inspecting the damage, the co-pilot went aft to organize the rescue party. The PBY then proceeded to locate the single survivors, aided by Atteberry advising Marks via voice radio exactly where to go.

> It was very difficult to see good because of the high swells and without a doubt we would have missed many if it hadn't of been for Commander Atteberry directing our actions. We tried to bring the survivors close to the port side and throw a life raft to them. Considerable difficulty was had because of the speed of the plane taxiing and the survivors were dragged through the water. We had to cut the plane's motors quite a few times and considerable time was lost in starting and stopping. We got better at picking the people up as time went by. We had the ladder out and I had a man on the ladder to grab any men who drifted by. The survivors could not help themselves very much, as most of them were weakened terribly and could not grab the ladder and climb up by themselves. Further difficulties were caused by the fact that the men were burned and every time we grabbed them it caused extreme pain.[19]

Some of those plucked from the sea had broken arms or legs, requiring extreme care in handling them. Between the PBY's landing and darkness, her crew picked up over thirty single survivors; most were in critical condition and would probably not have survived the night. Men brought aboard were given water and limited first aid treatment. Just before total darkness, Marks headed for a group of men on life rafts, which had been dropped to them. He later learned from the doctor of the *Indianapolis*, that they were the ones in the worst shape, and that he had put them on the rafts because of their serious condition.[20]

Marks brought the plane alongside of the rafts and aircrewmen took the survivors aboard. About twenty-five men were placed on the wing, issued water, and covered by parachutes. During the process of bringing the wounded up on the wing, the fabric covering the trailing edge of its aluminum skin was broken in many places. Several shouts for help were heard nearby and the radioman and another crewman volunteered to go out in a rubber boat to try to find them in darkness. Before long they returned with two additional men. It had been difficult for them to relocate the plane, because the auxiliary unit had gone out and with it, aircraft lights. There were some carbide lights that provided some illumination.[21]

DESTROYER ESCORT *CECIL J. DOYLE* ARRIVES

Photo 1-4

Undated wartime photograph of destroyer escort USS *Cecil J. Doyle* (DE-368). Courtesy of NavSource

Around 2315 that night, 2 August, the plane crew sighted the searchlight of the destroyer escort USS *Cecil J. Doyle* (DE-368). At the same time, an aircraft circled the PBY and dropped parachute flares nearby so that the *Doyle* could locate her. The carbide lights were thrown overboard to provide additional illumination. Upon arrival on the scene at about 0015, the destroyer escort dispatched a motor whaleboat with a doctor and first aid party and commenced transferring survivors. Due to heavy swells, the process was difficult and lengthy, lasting from 0045 to 0330. There were a number of stretcher cases and most of the survivors were only able to stand with assistance.[22]

In order to execute the transfer of the fifty-six survivors, it was necessary for the *Doyle*'s motor whaleboat to lay alongside the PBY, and as a result, the boat did considerable damage to the plane. After inspecting his aircraft, Marks decided that a take-off would be extremely hazardous and should not be attempted. Accordingly, he ordered that all salvageable gear be removed and the aircraft be destroyed. At 0600, Marks and his aircrew boarded the boat for transfer to the *Doyle*. At 0720, after taking aboard all survivors, crew, and salvageable gear, *Doyle*'s 40mm guns sank the Catalina.[23]

The destroyer escort, commanded by LCDR Graham Claytor Jr. (a future Secretary of the Navy), departed for Peleliu midday with ninety-three survivors on board. Three additional ships had arrived in the early morning hours of 3 August. It was not until they and *Doyle* started bringing survivors aboard that they learned they were rescuing the crew of *Indianapolis*.[24]

AFTERMATH

Of the 1,195 men who had been aboard the *Indianapolis*, 316 survived. The other 879 lives were lost. The A-Bomb, for which *Indianapolis* delivered components to Tinian, was dropped on Hiroshima on 6 August as survivors convalesced in Naval Base Hospital No. 20 Peleliu and Fleet Hospital No. 114 Samar. Japan surrendered on 15 August 1945, bringing World War II to a close. Lieutenant Marks had the Air Medal pinned on him by ADM Chester W. Nimitz, the commander in chief of the Pacific Fleet. He died on 7 March 1998, in his hometown of Frankfort, Indianapolis, at the age eighty-one.[25]

Photo 1-5

Marks and his crew pose in front of a VPB-23 PBY. Front row, L-R: AMM3c Richard W. Bayer, ENS Morgan F. Hensley, ENS Irving D. Lefkowitz, LT Robert A. Marks, unidentified VPB-23 officer and (back row, left to right) AMM2c Donald M. Hall, ARM Robert G. France, S1c Warren A. Kirchoff, and AOM Earl Duxbury
Courtesy of VPNavy.com; source unknown

2

Coos Bay Leaves Builder's Yard and Prepares for War

Photo 2-1

USS *Coos Bay* (AVP-25) off Houghton, Washington, on 15 May 1943.
U.S. National Archives photograph #19-N-47253

Twenty-eight *Barnegat*-class small seaplane tenders served in WWII. Prior to the war, in light of its Pacific Ocean strategy, the Navy recognized that it would require a large expansion in numbers of seaplane tenders. In the event of hostilities, war plans called for the availability of widely-deployed seaplane detachments to scout for enemy naval forces—and its existing tender force offered only modest capabilities.[1]

Converted World War I vintage *Lapwing*-class minesweepers could carry small seaplanes, such as Curtiss SOC Seagull scout observation biplanes and Vought OS2U Kingfisher observation floatplanes, but could not host newer, much larger Consolidated PBY Catalina flying boats. Fourteen other ships had been drawn from different classes of World War I destroyers for use as tenders. The 314-foot "four pipers" pressed into these duties could tend more aircraft, carry greater quantities of aviation fuel, and provide more services for planes and their crews than could the small "minesweeper" seaplane tenders. They were also much faster, and more heavily armed, but were considered marginally adequate pending the availability of replacement purpose-built ships.[2]

As large seaplane tenders (AVs) capable of supporting two squadrons of flying boats each, would be expensive to build, and had a deep draft precluding their use in shallow harbors, naval architects drew up plans for a class of small seaplane tenders (AVPs) capable of supporting only a single squadron. They would be less expensive to construct and be able to operate in shallow waters.[3]

The *Barnegat*s (named after bodies of water, mostly bays and inlets, but some islands as well) were the first purpose-built AVPs. The other ships carrying the AVP designation were the ex-*Lapwing*-class minesweepers previously mentioned. Construction of seven of the newly-designed small seaplane tenders was authorized in 1938. The lead ship incorporated desired capabilities identified during the service of the earlier tenders. These included:

- aircraft and weapons repair shops
- aircraft crew and support crew facilities
- weapons and fuel storage capacities
- operating range and ability to conduct operations independent of other fleet vessels, air defense and fire support capabilities
- ability to navigate and maneuver in shallow or restricted waterways[4]

Drawing only thirteen feet of water, the 311-foot ships could navigate or anchor in shallow waters, and possessing twin screws (propellers), could maneuver easily in restricted waterways as well.[5]

SERVICE OF THE *BARNEGATS*

Thirty-one *Barnegat*-class small seaplane tenders entered service between 1941 and 1946, all but three in time for war duty. Two were built at the Boston Navy Yard, the others at Pacific Northwest shipyards:

- Lake Washington Shipyard, Houghton, Washington (21 ships)
- Puget Sound Navy Yard, Bremerton, Washington (4 ships)
- Associated Ship Building Inc., Seattle, Washington (4 ships)
- Boston Navy Yard, Boston, Massachusetts (2 ships)

Summary information about the twenty-eight tenders that served in the war (the identities of their first commanding officers, theaters in which they served, and number of battle stars earned) may be found in Appendix C. *Greenwich Bay*, *Timbalier*, and *Valcour* (which were also built at Lake Washington yard, but not completed in time for war service), are not listed. Nor are four ships laid down as seaplane tenders at the yard, but reclassified as motor torpedo boat tenders while under construction. These ships were the *Mobjack*, *Oyster Bay*, *Wachapreague*, and *Willoughby*.

BATTLE STARS GARNERED

Asiatic-Pacific Campaign Medal

Europe-Africa-Middle East Campaign Medal

American Campaign Medal

Twenty-two of the small seaplane tenders served all or a portion of their service in the Asiatic-Pacific Theater. The officers and men of eighteen of these ships earned one or more battle stars to affix to the Asiatic-Pacific Campaign medal or ribbon on their uniform blouses. A handful of AVPs also served in the American Theater and/or the Europe-Africa-Middle East Theater.

Barnegat-class Small Seaplane Tender Battle Stars in World War II

Ship	Battle Stars	Ship	Battle Stars
Barataria (AVP-33)	—	*Half Moon* (AVP-26)	★ ★
Bering Strait (AVP-34)	★ ★ ★	*Mackinac* (AVP-13)	★ ★ ★ ★ ★ ★
Casco (AVP-12)	★ ★ ★	*Onslow* (AVP-48)	★ ★ ★ ★
Chincoteague (AVP-24)	★ ★ ★ ★ ★ ★	*Orca* (AVP-49)	★ ★ ★ ★
Cook Inlet (AVP-36)	★	*San Carlos* (AVP-51)	★ ★ ★
Coos Bay (AVP-25)	★ ★	*San Pablo* (AVP-30)	★ ★ ★ ★
Duxbury Bay (AVP-38)	★ ★	*Shelikof* (AVP-52)	★ ★ ★
Floyds Bay (AVP-40)	★	*Suisun* (AVP-53)	★ ★
Gardiners Bay (AVP-39)	★ ★	*Yakutat* (AVP-32)	★ ★ ★ ★

To qualify for a battle star, a ship, aircraft unit, or shore-based force had to participate in actual combat with the enemy. In instances in which the duty performed did not result in actual combat, but was considered equally hazardous, the chief of Naval Operations could award a star to the units concerned. The catch was, that no matter how many instances of combat, not more than one star could be earned for a single operation or engagement.[6]

NAVAL OFFICER ACQUISITION PROGRAMS

With American shipyards churning out all manner of military ships to support expanding fronts in several theaters, the time from ship commissioning to entry into a war zone was typically short, putting a strain on procurement and training of necessary personnel. The wardroom of a bright and shiny new seaplane tender was usually comprised of a few regular officers (who had been serving at war's commencement, or were procured from the Annapolis Naval Academy or NROTC training programs) and many newly "minted" reserve officers. The latter were sometimes disdainfully referred to as "90-day wonders."

By 1940, in anticipation of possible war with Germany and/or Japan, Navy ROTC programs had been established at twenty-seven universities. Their combined enrollment of 3,096 midshipmen at the universities greatly supplemented the Academy's annual production of approximately 425 officers. However, planners had no idea that the Navy's requirement for officers would eventually exceed 300,000, necessary to lead a greatly expanded Navy of three-and-a-quarter-million personnel.[7]

Fortunately, even without fully appreciating the enormity of the coming shortfall, planners had foreseen the looming problem and began to take action in the spring of 1940. Northwestern University established a Reserve Midshipmen's School (on its Chicago campus). Other schools, including the Naval Academy, Columbia University, Cornell University, Notre Dame, and Smith College—which trained junior officers of the Women's Reserve of the U.S. Naval Reserve (WAVES)—established similar programs. A U.S. Naval Reserve Midshipman School was opened on board the converted battleship Illinois (BB-7), moored in the Hudson River to serve as a training ship. The V-7 Program provided ninety days of training.[8]

On 1 July 1940, the U.S. Navy had 13,162 officers and 744,824 enlisted men. On 31 August 1945, its manning had more than quadrupled to 316,675 and 2,935,695, respectively. Enlisted manning requirements, like those of the officer community, was met by taking in large numbers of men new to the sea. They came from urban cities and rural towns and communities all across the nation.[9]

MANNING OF *BARNEGATS*

The ship's complement aboard *Coos Bay*, a representative AVP, was 14 officers and 201 enlisted. Authorized manning of an embarked patrol squadron was 59 officers, and 93 enlisted men. The higher officer-to-enlisted ratio of a squadron reflected the numbers of flight officers needed to operate and maintain its PBY Catalinas.[10]

NEW SHIPS AND CREWS HURRIED INTO WAR DUTY

Thrust into war, following the Japanese attack on Pearl Harbor on 7 December 1941, with too few seaplane tenders, the authorized number of *Barnegat*-class vessels, like that of almost every other type of military vessel, was increased dramatically. Ships leaving builders' yards were transferred to operational commanders after their crews were put through accelerated training and inspection judged them satisfactory.

Coos Bay, needed to serve as an afloat airfield in the South Pacific, was commissioned on 15 May 1943. Her crew was assembled at 0930. A half hour later, CAPT William J. Malone, supervisor of Shipbuilding, accepted the ship for the Navy Department, placed the ship in full commission with appropriate ceremony, and delivered her to CDR William Miller, USN. Miller read his orders, and assumed command of her. Eight minutes later, the first watch was set aboard ship. Her officers and crew had moved aboard the vessel the previous evening.[11]

Miller had reported for duty at the Lake Washington Shipyard five months earlier. Under the supervisor of Shipbuilding (a Navy captain representing the service's interests at area shipyards), his responsibilities were in connection with the fitting out of the seaplane tender.[12]

Commander Miller had extensive experience in aviation, and in particular with scout aircraft. After graduating from the Naval Academy in 1926, he had received orders to the light cruiser *Concord* (CL-10). Later, while serving aboard the flagship of commander, Destroyer Squadrons, Scouting Fleet, he had applied for training as a Naval Aviator and after a year at the Naval Flight School, at Pensacola, Florida, earned his Navy wings.[13]

A series of aviation assignments followed: two years in Scouting Squadrons Three and One, a year of shore duty, then assignment to the battleship *Arizona* (BB-39) for flying duty. In 1935, Miller returned to Scouting Squadron One, followed by a two-year tour at Naval Air Station, Pearl Harbor. He was then assigned to Patrol Squadron 42, and after that to the Thirteenth Naval District (headquartered at Seattle, Washington), until detached to command *Coos Bay*.[14]

Photo 2-2

Battleship USS *Arizona* (BB-39), circa 1939-1940, with SOC aircraft on deck. Naval History and Heritage Command photograph #NH 77072

Newly commissioned ships are rarely idle. Between 15 May and 14 June, *Coos Bay*'s officers and crew were immersed in trials and evolutions to help prepare the ship for more advanced fleet training. The first order of business, in late morning on the same day of commissioning, was to cast off all lines, get under way, and proceed to Naval Air Station, Seattle, per direction from commander, Thirteenth Naval District. Late that afternoon, RADM Frank D. Wagner and staff came aboard for a visit. The following day, the ship moved to the fueling pier at Orchard Point, Manchester, for bunkering and then on 17 March, to Navy Yard, Puget Sound, at Bremerton, for outfitting.[15]

This began the process of preparing a seaplane tender built in the Seattle area for war—one that sister ships, commissioned earlier and those that followed, would endure. The next step on 1 June was movement to Illahee, Washington, for deperming; a one-time process in which cables are temporarily wrapped around metal-hulled ships, and electricity passed through them to decrease remnant magnetic fields and make the ships less susceptible to damage by a magnetic mine. Later that same day, *Coos Bay* berthed at the Naval Pacific Coast Torpedo Station at Keyport, to load ammunition. The following day, 2 June, she shifted to the Indian Island Bomb Depot, to load bombs, fuses, and depth charges.[16]

After leaving Indian Island on 4 June with ammunition on board, guns could be tested. Following calibration of her radar, test firing of the ship's four 5"/38 mounts commenced, followed by gunnery drills. Later that day, after arriving off the Jefferson Point Degaussing Control Tower, *Coos Bay* made several runs on the degaussing range to test/calibrate her installed degaussing equipment which, together with deperming, helped protect against mines. At completion, she berthed at Naval Supply Depot, Seattle.[17]

Coos Bay got under way the following morning for more evolutions and training. Upon securing the special sea detail, Condition III was set to exercise the crew in firefighting drills. Securing from these drills, the ship steered a variety of courses at various speeds to calibrate radio direction-finding equipment and the gyro compass. Later, as aircraft made mock attack runs on the ship, pointer and trainer drills were held for all gun crews. This involved practice in manually slewing (moving a gun left or right to a particular bearing) and changing elevation (by movement up or down) to engage a target. Additional runs on the degaussing range preceded the return of the ship that evening to the Supply Depot.[18]

Next morning, 6 June, the tender left her berth at 0800 for more evolutions in the Puget Sound. The first was running a measured mile at top speed off Vashon Island to validate her propulsion system. Subsequent maneuvering on various courses at various speeds, to compensate the magnetic compass (counteract the effects of ship's magnetism), took a little over two hours to complete. The final activity was calibration of her rangefinder. At completion, *Coos Bay* once again berthed at the Supply Depot. She remained there until early afternoon on 10 June, when she stood out, bound for San Diego. One final requirement remained before transiting the Strait of Juan de Fuca to gain the open Pacific. Bore sighting of her guns in Admiralty Bay, involved aligning their sights to the axis of barrel bores, so that the mounts would shoot where they were sighted.[19]

While *Coos Bay* proceeded independently down the coast, bridge watch team practiced zigzagging, and lookouts challenged friendly planes sighted. Once in the war zone, maneuvering in accordance with zigzag plans would be employed in areas where enemy submarines might be operating, and every aircraft sighted would be challenged to determine whether it presented a threat to the ship. At 0746 on 14 June, *Coos Bay* passed Point Loma abeam to port, passed through the outer anti-submarine net protecting the harbor, and entered San Diego Harbor. At 0905, after clearing the inner anti-submarine net, she berthed at the destroyer base, secured boilers, and began receiving steam, water, and electricity from the pier.[20]

FLEET TRAINING

Fleet training began the following day. Leaving her berth and after passing through the outer anti-submarine net, *Coos Bay* continued steaming westward to an operating area off San Diego. At 1030, she commenced long-range battle practice, firing at a target towed by the fleet tug *Mataco* (AT-86). Also taking part were the destroyers *Baldwin*

(DD-624) and *LeHardy* (DE-20). In early afternoon, *Coos Bay* simulated repelling high-altitude bombing attacks, then took up position for anti-aircraft battle practice. She then maneuvered on various courses and at various speeds in tactical exercises with *Baldwin* and *LeHardy*. At 2100, the long day was still not over, as *Coos Bay* proceeded to her assigned area for night battle practice.[21]

Three minutes past midnight on 16 June, *Coos Bay* began firing at a target towed by the minesweeper *Pursuit* (AM-108). The seaplane tender was the last ship in a column formation, with *Baldwin* as the guide, *LeHardy* the second ship, and she the third. *Coos Bay* ceased firing at 0036, but continued night steaming off of San Diego (remained under way). As an aside, it is noted that minesweepers and tugs, "maids of all duties," were often tasked with towing target sleds, which their crews did not particularly enjoy. Especially, when working with ships whose gunnery accuracy needed much improvement.[22]

At 1125 on 17 June, *Coos Bay* began anti-aircraft practice while lying to. Under way again, twenty minutes later, she proceeded to take up position for an anti-aircraft battle exercise.[23]

Photo 2-3

Fleet tug USS *Kewaydin* (AT-24) at anchor, with a target hove in astern, circa 1943. While supporting gunnery exercises, she would have towed it much farther behind her. National Archives photograph #80-G-411692

At this point, readers will recognize that Navy pre-deployment fleet training was then, and remains today, quite extensive and intensive. Instruction and inspection by commander Fleet Operational Training Command, Pacific Fleet (COTCPac) staff was designed to stress a ship and its crew, to ensure combat readiness for duty in a theater of war.

(As necessary, additional instruction and reinspection was rendered, something not enjoyed by a ship's company.) Remaining training for *Coos Bay* included fueling PBYs of Squadron VP-91 from astern and bowser tank of her aircraft fueling boat, and West Coast Sound School operations involving sonar training, including making runs on a submarine target.[24]

The most excitement during training occurred in late afternoon on 5 July, when a plane was sighted adrift in the water. *Coos Bay* hove to, and launched her crash boat. A crash boat, bower (refueling) boat, and cranes to lift aircraft aboard was standard equipment on a seaplane tender. Once brought alongside, plane U18-J5F #00750 was hoisted aboard. The wayward aircraft was passed that evening to "Mary Ann," Naval Air Station, San Diego's salvage derrick barge.[25]

Coos Bay's last under way training day was 9 July. From 10-21 July, she remained moored at the destroyer base for a post-shakedown availability to correct material deficiencies identified during her training and to effect engine room repairs.[26]

Photo 2-4

Destroyer Base, San Diego, with a boat park along the waterfront, the marine railway in the upper right, and new buildings under construction, 9 April 1941. Naval History and Heritage Command photograph #NH 95609

Around first light on 23 July, *Coos Bay* twisted clear of her berth and stood out of the harbor, bound for Pearl Harbor, en route to Espiritu Santo, New Hebrides Islands, in the South Pacific.[27]

CREW A LITTLE MORE SEASONED, BUT UNTESTED

Less than ten weeks after their ship had been commissioned, the officers and men of *Coos Bay* were headed westward, en route to combat duty in the Pacific. In addition to himself, Commander Miller's wardroom included Lieutenant Commander Michael (the executive officer) holding regular commissions (USN), and the remaining officers with reserve (USNR) commissions. *Coos Bay*'s complement was fourteen officers. The thirteen listed below were identified by their signed deck log watch entries as officer of the deck, or of logs as navigator or commanding officer, prior to the ship's departure from San Diego.

USN
CDR William Miller
LCDR Fred D. Michael
USNR
LT William F. James
LT Burton R. Manser
LT Richard H. Randall
LTJG Robert N. Hargis

USNR
LTJG Alexander Iungerich
LTJG William K. Paynter
LTJG Robert W. Read
LTJG Robert West
ENS William Black
ENS Fritiof L. Ekholm
ENS Norman S. Hemphill

One of the most junior officers then carrying out very significant duties was Fritiof L. Ekholm. He had reported to the *Coos Bay* as an ensign, but on 1 July, took over the duties of ship's navigator, previously those performed by LCDR Fred D. Michael, the executive officer. Ekholm had majored in Forestry in college, but had done a lot of work for the U.S. Coast and Geodetic Survey in Alaska. As many aspects of surveying and navigation are similar, he was able to perform the duties and earned the captain's trust for the safe navigation of *Coos Bay*, which, in the South Pacific, would include operations in poorly charted, reef-laden waters.[28]

Photo 2-5

Ekholm on the bridge wing of *Coos Bay* (AVP-25).
Stephen Ekholm collection

While making passage from San Diego to the South Pacific, and loaded out with fuel, food, munitions, supplies, and equipment, *Coos Bay* had one item on board not part of her authorized inventory, an Army jeep. Two months earlier, while loading stores at Pier 40 in Seattle, an Army major had parked his jeep in front of the ship's brow. Asked politely to move it, he refused, then departed on foot to carry out whatever errand brought him to the area.[29]

As soon as the Army officer was out of sight, members of *Coos Bay*'s crew diverted the employment of the ship's crane then bringing stores aboard, to remove it for him. They carefully hoisted the vehicle aboard, and deposited it in a location where it would not be noticed. When the major returned, he was understandably concerned about the disappearance of his vehicle. When asked what happened, the sailors told him that they thought they had seen someone drive off in it. Later, these enterprising individuals painted it grey, and made a canopy to cover it. The captain did not find out about the jeep until well away from Pier 40. He did not ask where the ship's new vehicle had come from, although, undoubtedly, he quickly surmised the source. The whole crew took great pride in the jeep, which they were free to use, if the officers were not utilizing it.[30]

Photo 2-6

ADM Chester W. Nimitz, USN, commander in chief, Pacific Fleet, being toured around Tarawa, Gilbert Islands, in a jeep on 27 November 1943, inspecting the damage done during capture of the Japanese-held atoll by the 2nd Marine Division only days earlier. Naval History and Heritage Command photograph #NH 58530

There is an old saying in the Navy, "If it moves, salute it, if not, bring it aboard and paint it haze grey." This refrain reflected the belief of many sailors that it was their job to help their ship and shipmates. Anything left unattended on the pier was fair game, and other ships could be victims of "commandeering" of particularly desired items. There are stories about paint floats departing on deployment aboard ships whose boatswain's mates wanted their sides to look exceptionally good. (A paint float is a large piece of harbor equipment, that sailors stand on to paint the sides of their ship.) Particularly difficult was the acquisition of an anchor at the dip (hanging down near the water's surface) from a neighboring ship for one needing a replacement. Anchors are quite heavy, so their stealth procurement would have required the use of a barge with a crane under the cover of darkness.

Much audacity was required to take a jeep from a public pier in broad daylight, but *Coos Bay* counted among her enlisted ranks many seasoned sailors who were used to such shenanigans. Some had long fleet experience; others had been present at Pearl Harbor during the attack, and gained combat experience. The crew was led by ship's bosun John Tees, torpedoman Charles Norman, carpenter Alexander Campbell, and other senior enlisted men. All were regular Navy.[31]

TRANSIT TO PEARL HARBOR

Photo 2-7

Recruiting poster depicted sailors on the beach at Waikiki, Honolulu, Hawaii. Famous Diamond Head (on Waikiki's eastern coastline) is in the background. Naval History and Heritage Command photograph #NH 86427

Coos Bay proceeded independently to Pearl Harbor, mostly at standard speed of 15 knots, with periods of zigzagging in accordance with Plan No. 6 (one of several classified zig-zag diagrams designed to disguise a ship's true course and confuse the enemy). Two days out from reaching Pearl Harbor, she maneuvered on various courses and at different speeds during gunnery practice against a target balloon. The following day, manned "ready guns" opened fire during a surprise practice.[32]

On the morning of 28 July, Molokai Island was sighted at 0823, twenty-three miles distant. *Coos Bay* passed Diamond Head abeam at 1111, while proceeding inbound toward the entrance to Pearl Harbor Channel. She then passed through the entrance buoys and anti-submarine net, and continued past Bishop's Point. The harbor pilot boarded at 1301. Twenty-two minutes later, she berthed at Naval Air Station, Pearl Harbor, on Ford Island, with six standard mooring lines.[33]

Photo 2-8

Aerial photograph of Ford Island taken 10 October 1941. The Naval Air Station occupies most of the island, with the seaplane base on the point at the near right. National Archives photograph #80-G-279375

VOYAGE TO ESPIRITU SANTO

Two days after her arrival, *Coos Bay* proceeded to a gunnery area off Hawaii, to conduct exercises in company with the destroyer *Lansdowne* (DD-486) and the large seaplane tender *Wright* (AV-1). Practice in surface gunnery and firing at a target sled towed by a tug, preceded automatic weapons fire, at a target sleeve towed by an aircraft. After this honing of their gun crews' skills, the three ships set course for Espiritu Santo.[34]

During the transit, *Coos Bay* maintained station, 3,000 yards on the port bow of *Wright*, the formation guide. Fitted with sonar, *Coos Bay* was positioned thus to detect any enemy submarines that might be in the vicinity of the formation. A submerged diesel submarine is not particularly fast so, in order to catch a ship on the open ocean (unless dead ahead), one had to approach a ship from off its port or starboard bow. *Lansdowne* was positioned in an anti-submarine screening station on *Wright's* starboard bow. The ships zigzagged periodically in accordance with the plan in effect. Ships operating at slow speeds were at greater risk of a submarine attack. Accordingly, on 6 August *Wright* and *Lansdowne* slowed to 6 knots, and while *Wright* refueled the destroyer, *Coos Bay* took position 2,000 yards ahead, patrolling back and forth between 30 degrees relative on either bow of the *Wright*.[35]

In late afternoon on 6 August, Tutuila (the largest and main island in American Samoa) came into view thirty-five miles distant as the group proceeded on its way. Three days later, *Coos Bay* test fired all her guns. On the morning of 11 August, twelve days after departing Hawaiian waters, landfall was made at 0634, when Pentacost and Abrim islands were sighted. An hour later, the ships began passage of Selwyn Strait in single file. Exiting it, *Lansdowne* left the others and proceeded on her assigned duties. *Coos Bay* once again took up her screening station, 3,000 yards ahead of *Wright*.[36]

In early afternoon, the two tenders passed Pallikulo Point abeam to port, entered Diamond Channel and anchored in Pallikulo Bay, Espiritu Santo Island. Upon arrival, Commander Miller reported for duty to commander, Fleet Air Wing One (CAPT Harry S. Kendall, USN). That evening, *Coos Bay* went to General Quarters for an air raid alert. Her crew secured from battle stations upon an "All Clear" signal. Three days later, on the 14th, *Coos Bay* relocated to Segond Channel, the site of the seaplane base at "Santo." Sailors and Marines often use fleet shorthands to refer to places or locations. Two such were "Santo" (Espiritu Santo) and "the Canal." The latter was a favorite of Marines who had fought on Guadalcanal.[37]

3

Duties of Tenders at Espiritu Santo

Before Guadalcanal the enemy advanced at his pleasure—after Guadalcanal he retreated at ours.

—ADM William F. Halsey Jr., USN, commander,
South Pacific Force and South Pacific Area.[1]

Photo 3-1

Allied shipping in the Segond Channel, 3 April 1944.
National Archives and Record Administration photograph #80-G-K-1987

Following the Battle of Midway, the Pacific Fleet's primary focus in the second half of 1942 was supporting operations in the Solomon Islands, and specifically the island of Guadalcanal. The Guadalcanal Campaign was spurred by the Japanese occupation of Tulagi Island (a small island nestled in a bay at Florida Island opposite Guadalcanal) on 3 May 1942, which occurred just before the Battle of the Coral Sea. The enemy's advances in the region had provided the impetus for the Battle of the Coral Sea, which preceded the Battle of Midway. Fought during 4-8 May between the Imperial Japanese Navy and naval and air forces from the United States and Australia; it would be the first great naval action between aircraft carriers and the first naval battle in which no ship on either side sighted the other.[2]

Map 3-1

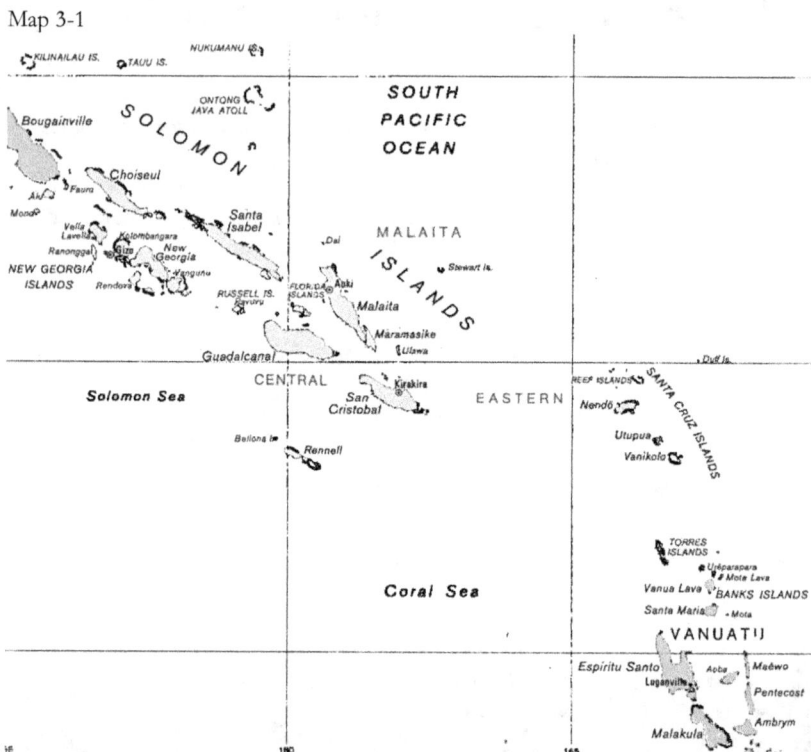

Central and Eastern Solomon Islands, and Santa Cruz and New Hebrides Islands (now Vanuatu)

Japan wanted an air field in the Solomons from which its land-based bombers could provide air cover for the advance of Imperial land forces to Port Moresby, the capital city of Papua and the site of an Allied base (the territories of Papua and New Guinea were combined after World War II into a single territory that today is known as Papua-New Guinea). The thousands of troops based at Port Moresby were the Allies' last line of defense before Australia. On 5 July, having found Tulagi fit only for a seaplane base, Japanese forces landed on Guadalcanal, twenty miles across the New Georgia Sound (which Allied servicemen referred to as "The Slot") from Tulagi, and began the rapid construction of Lunga Point Airfield (later, Henderson Field, following its capture by U.S. Marines) from which the empire's planes could menace the shipping lanes to Australia.[3]

In an effort to prevent the Japanese from using Guadalcanal to launch air attacks on Allied shipping in the South Pacific, 11,000 members of the 1st Marine Division landed at Guadalcanal on 7 August and captured the airstrip at Lunga Point, as well on the following day, the Japanese encampment at Kukum on the west side of Lunga Point. The afternoon of 8 August, after fierce fighting, Marines discharged at Tulagi took the Japanese-held island, as well as the smaller islands of Gavutu and Tanambogo. The airstrip on Guadalcanal was renamed Henderson Field, and its occupation and use by Allied forces temporarily halted Japanese expansion in the South Pacific.[4]

Throughout the prolonged, bloody Battle of Guadalcanal from 7 August 1942 to 9 February 1943, Espiritu Santo served as a Naval Air Base, staging area and supply point. From Bomber One (a significant airfield at Santo capable of hosting large land-based aircraft), Army, Navy and Marine heavy and medium bombers made around-the-clock runs on enemy troops, installations, and fleets in the greater Solomons area. Units of the patrol squadrons in the area reconnoitered the enemy's movements. From her anchorage in Segond Channel, the large seaplane tender *Curtiss* (AV-4) tended planes of VP-11 and VP-23 squadrons engaged in conducting surveillance and anti-submarine patrols during the landings. (*Curtiss* was the flagship of RADM John S. McCain, commander, Aircraft, South Pacific, and his successor, RADM Aubrey W. Fitch. More about her follows in the next chapter.)[5]

Patrol Squadrons VP-24, -71, and -91 also flew missions from the Segond Channel seaplane base in support of the American forces on Guadalcanal and later, on islands farther up the Solomons chain. Beginning in February 1943, Dumbo work was added to VP-24's activities. In June of 1944, these missions became the squadron's primary function. During the New Georgia Campaign, the entire squadron functioned in this role, and is believed to be the first squadron to assume Dumbo work as their primary task. Four hundred and sixty-six men were either rescued or evacuated by this patrol squadron during this campaign, sometimes in the face of enemy opposition.[6]

DANGER TO TENDER CREWS AND AIR CREWS

Patrol Squadron 71, present when *Coos Bay* arrived at Santo, had been operating PBY-5 Catalinas from Segond Channel since 27 June 1943. Commanded by LCDR Cecil Kelly Harper, USN, its pilots and crews flew 800-mile day searches to the north. For extended range, an advance base was maintained at Vanikoro (230 miles north of Espiritu Santo), with the *Mackinac* (AVP-13) tending planes.⁷

Mackinac had established the base at Vanikoro on 12 November 1942. The Santa Cruz Islands included the big islands of Santa Cruz, Utupua, and Vanikoro; the Swallow or Reef island group; and the Duff group. The Swallow Islands were all quite small, the largest being not more than four miles in length. In 1766 an English vessel, the HMS *Swallow* commanded by Capt. Philip Carteret visited the islands, which he named after his ship. Resembling reefs, they were also called the Reef Islands.⁸

Positioning a tender at Vanikoro extended the ranges of PBYs based aboard the *Curtiss* (AV-4) anchored in Segond Channel at Espiritu Santo. Using the advance base, patrol planes from Espiritu Santo were able to fly searches to the north scouting for enemy forces with stops at Vanikoro for fuel and crew rest. On 9 July 1943, the large seaplane tender *Chandeleur* (AV-10) relieved the *Curtiss* of her multiple duties as seaplane tender, flagship for commander Aircraft South Pacific, and repair and supply ship for destroyers and small craft.⁹

Chincoteague (AVP-24) relieved *Mackinac* of her duties at Saboe Bay, Vanikoro on 6 July 1943. (Both were sister ships of *Coos Bay*.) After she dropped anchor in Saboe Bay, operations with PBY-5 Catalinas of VP-23 and -71 squadrons proceeded without incident for a week or so. The 2,800-ton *Barnegat*-class small seaplane tender was a mere two months old. Another product of Lake Washington Shipyard in Houghton, Washington, she had been commissioned on 12 April 1943 and following shakedown training, had sailed from San Diego in June for Saboe Bay.¹⁰

Chincoteague's first encounter with enemy aircraft came around noon on 14 July, when an enemy plane believed to be on a photographic mission was sighted at about 15,000 feet overhead. The ship got under way while firing at the aircraft, passed through the reef at the entrance to the bay and stood out to sea. She remained outside for about an hour before returning upon the departure of the aircraft. The results of her 5"/38 battery firing at the high flying "snooper" were undetermined.¹¹

Operations returned to normal through the remainder of that day, and all of the next until early evening when, at 1847, a Japanese twin-engine bomber was picked up on radar. The plane was believed to be a Sally (Mitsubishi Ki-21 Type 97 heavy bomber). It circled for some minutes and then dropped two brilliant white flares about two thousand yards off the ship's port quarter, near the entrance to the bay. Fortunately, these did not illuminate the tender's position. Taking full advantage of the darkness, the *Chincoteague* withheld anti-aircraft fire in order to not disclose her position and the enemy did not attack.[12]

Enemy aircraft attacks against the *Chincoteague* did begin in earnest the morning of 16 July, and continued through the afternoon of the 17th. The personnel casualties and material damage inflicted on her over this period would result in the Navy abandoning the Vanikoro advance base. In a post action report, CDR Ira E. Hobbs, *Chincoteague*'s commanding officer, described the opening action. The other quoted material in the next few pages is from the same report.

> On the morning of the sixteenth at 0717, five Jap twin engined bombers were picked up by radar circling the bay and then sighted overhead. They released their bombs from an altitude of approximately eight thousand feet. The sticks fell about fifteen hundred yards from the ship, bursting in the jungle east of Saboe Bay Head. Only the one run was made, the planes continuing on out of contact. A few hours later, at 1110, another formation was sighted overhead making a dummy run on the ship. Again these were thought to be the Sally type and this time there were nine of them flying in a close V of V's. On this run they passed on the starboard side heading out to sea. After circling about they came in from the sea, passing this time on the portside.[13]

During the preceding and all subsequent actions, the *Chincoteague*'s guns were fired by local control. She had no radar-controlled gun director and without one, anti-aircraft fire against distant targets was not of great value. The next morning, 17 July, a flight of six PBY-5s of VP-71 returned to Vanikoro from their mission. These planes, led by Harper, had struck Nauru Island, 675 miles north of Vanikoro, dropping six tons of bombs and scoring hits on the airstrip, dispersal areas, barracks, and the phosphate plant. The Japanese had invaded and occupied Nauru, in part because they intended to take over the island's fertilizer industry.[14]

The small island hosted a Japanese airfield, and was home to the 67 Naval Guard Force, the Nauru Expeditionary Force of the 3rd Special Naval Base Force, and the Nauru Special Construction Unit of the Fourth Fleet. The PBYs had encountered intense but inaccurate anti-aircraft fire, and suffered no resulting damage. The last plane had just landed in Saboe Bay, when at 0725 the first stick of bombs from five twin-engine bombers bracketed the fo'c'sle of the *Chincoteague*. The enemy had come in at about eight thousand feet from the northwest, apparently having followed the patrol aircraft in.[15]

Chincoteague got under way, and attacks continued. Finally, after contact with the enemy was lost, and due to VP-71 requiring fueling prior to taking flight for Espiritu Santo, she returned to the bay. Upon receiving orders for *Chincoteague* to depart as well, Hobbs dispatched all patrol planes with the exception of three still needing fueling. The seaplane tender weighed anchor at 1020 on 17 July, leaving bowser boats to refuel the remaining planes. By 1120 the ship had cleared the reefs and was standing out to sea. However, open water did not mean safety. An aircraft attack twenty-five minutes later killed ship's personnel and caused significant damage to the tender. Hobbs later described the sequence of events:

> Five Sallys [Mitsubishi Ki-21 heavy bombers] overhead released sticks which bracketed the ship. We were maneuvering to the extreme degree again which helped to prevent any direct hits although some splinter damage was sustained. A few minutes later, having circled about into the sun they came in again. The stick of bombs again bracketed the ship. This time we received a direct hit from what is believed to be a 100 k.g. delayed action bomb which pierced three decks, the super, and main and second, before exploding in the after engine room. All after engineering personnel including the Chief Engineer were killed, as well as a fireman in the forward messing compartment. Number four gun went out of commission at the same time due to water damage.[16]

Following herculean efforts by her crew to save their ship, *Chincoteague*'s departure under tow by the destroyer seaplane tender *Thornton* (AVD-11), signaled the abandonment of Vanikoro as an advance base. Returning to Espiritu Santo, VP-71 flew 650-mile day searches with occasional Dumbo, escort, and night anti-submarine missions until 13 October. On that date, the squadron was ordered to Halavo seaplane base at Florida Island.[17]

Coos Bay spent two months at Santo (11 August-11 October), while waiting assignment. During this period, much time was devoted to ship's force work, drills, and training. A highlight was a three-day trip to Vanikoro to retrieve boats and other gear belonging to *Chincoteague*. On the night of 18 August, signal lights were observed across the reef bordering the bay. No ships were known to be in the area. *Coos Bay* exited the bay the following morning. Upon doing so, Commander Miller received a report from the soundman, "sonar contact." Having no room to maneuver to attack, because of reefs, he ordered flank speed and left the area to return to Santo.[18]

DECLINE IN THE IMPORTANCE OF ESPIRITU SANTO

U.S. Army forces had first occupied Espiritu Santo on 28 May 1942, when Force "A" landed there, following a landing at Efate ten days earlier. Located approximately 140 miles to the south, Efate was also a part of the New Hebrides (which is today known as Vanuatu.) The geographical location of Espiritu Santo was of great strategic and tactical value in the early phases of the war. The first evidence of this came in July 1942, with the commencement of bombardment of Japanese positions on Guadalcanal, less than 650 miles from bomber and fighter airfields on Santo. (Bomber One was at Palikulo Bay, and the fighter strip, Turtle Bay.) The construction of Bomber Two was begun in November 1942, and officially opened on 6 January 1943.[19]

Photo 3-2

Major Gregory Boyington, commanding officer, Marine Corps Fighter Squadron 214 (VMF-214, the "Black Sheep"), gives pre-flight instructions to his pilots, at Turtle Bay Fighter Strip, Espiritu Santo, New Hebrides, 11 September 1943.
National Archives photograph #80-G-54303

From that first strike until movement of the front well up the Solomons, aircraft from Espiritu Santo played an important role in supporting Allied forces. Facilities were expanded so that the base might serve as a major supply and shipping point. By December 1943, Espiritu Santo had grown into the largest Naval, Naval Supply, and

Naval Air Base west of Pearl Harbor. Broad expansion took place until 1944, the peak year. As the principal phases of the war moved north and west of the New Hebrides, it became apparent that Espiritu and Efate were too far in the rear and off the line of attack to be of any further service, other than in a minor way. Units based there began moving forward to Manus, Admiralty Islands; Emirau, Papua New Guinea; Green, Solomon Islands; Guam and Saipan, Mariana Islands; Peleliu, Palau islands; and the Philippines.[20]

Meanwhile, tenders and patrol squadrons, including *Coos Bay*, Patrol Squadron VP-71, and Royal New Zealand Air Force No. 6 (Flying Boat) Squadron, would relocate a more modest distance from Espiritu Santo to the Halavo Seaplane Base at Florida Island.

4

RNZAF No. 6 Squadron

Photo 4-1

Royal New Zealand Air Force No. 6 (Flying Boat) Squadron PBY-5 Catalina and crew. Courtesy of Jenny Scott

Royal New Zealand Air Force No. 6 Squadron was initially formed at Milson Aerodrome in New Zealand, in February 1942, as an Army co-operation squadron. (The Milson facility was later renamed RNZAF Station Palmerston North, in January 1943, reflecting its location near that city on the country's North Island). Operating Hawker Hinds biplane light bombers, the squadron's main role was ground support exercises with units of the New Zealand Army, as well as those of the U.S. Marine Corps stationed at Paekakariki, a coastal town in the southwestern area of North Island.[1]

No. 6 Squadron was disbanded in New Zealand, on October 1942, and later officially reformed, on 25 May 1943, at Laucala Bay, Fiji, as a flying boat unit. One of its earliest operations, carried out before the squadron was officially formed, was a search for survivors of SS *William K. Vanderbilt*. The Liberty ship was torpedoed southwest of Suva, Fiji Islands, on 16 May 1943, by the Japanese submarine *I-19*. A Catalina, captained by Ronald B. L. MacGregor, located eight survivors bobbing about on a raft in very rough seas. After standing by for three hours, without a rescue ship appearing, he made a daring mid-ocean landing and recovered them all. The rough seas did some damage to the aircraft, but he was able to take-off safely and returned to Laucala Bay.[2]

When the squadron was recommissioned, Wing Commander G. G. Stead, DFC, RAF, was appointed commanding officer and squadron leaders McGregor and Alan V. Jury, flight commanders. Before the recommissioning, McGregor and several aircrews had been attached to the seaplane tender USS *Curtiss* (AV-4) in Segond Channel at Espiritu Santo, to learn American naval methods and the operation of Catalinas. This training proved extremely valuable in the formation and training of RNZAF flying-boat squadrons. As soon as sufficient aircraft arrived from the United States, No. 6 Squadron undertook an intensive crew-training program with the objective of becoming fully operational by August 1943.[3]

Photo 4-2

Oil painting of USS *Curtiss* (AV-4) by RADM John W. Schmidt, USN (Ret).
Naval History and Heritage Command photograph #NH 55536

The *Curtiss*, commissioned on 15 November 1940, was the Navy's first purpose-built seaplane tender. Her primary function was tending planes and providing living accomodations for the officers and men of embarked squadrons. After serving from June 1942 at Noumea, New Caledonia, with additional duty as flagship and headquarters for RADM

John S. McCain, (commander, Aircraft, South Pacific Force), she relocated, in June 1943, to Espiritu Santo, New Hebrides, where RADM Aubrey W. Fitch relieved McCain of his duties. *Curtiss* served as seaplane tender and headquarters during the Solomon Islands campaign, and also functioned as a repair/supply ship for destroyers and small craft engaged in the battle for Guadalcanal.[4]

Curtiss left Espiritu Santo in July 1943, bound for San Francisco for overhaul and repairs. No. 6 Squadron subsequently operated from the seaplane tenders *Wright* (AV-1) and *Coos Bay* (AVP-25).[5]

Photo 4-3

Rear Admirals John S. McCain (left) and Aubrey W. Fitch shake hands on the occasion of Fitch relieving McCain as commander, Aircraft, South Pacific Force. National Archives photograph #80-G-43065

DETACHMENT SENT TO TONGA

In mid-August 1943, a detachment from No. 6 Squadron at Fiji was sent to Tonga. Six aircraft and crews were stationed at the Nukualofa Harbor Seaplane Base, and became responsible for the protection of shipping in Tongan waters. Tonga, a small archipelago in the South Pacific, lay directly south of Samoa and about two-thirds of the way from Hawaii to New Zealand. The capital city of Nukualofa was on Tongatapu, the largest island.⁶

Photo 4-4

Vought OS2U Kingfisher half-way out on the seaplane ramp at Nukualofa, Tongatapu, Tonga Islands, on 8 June 1942.
Naval History and Heritage Command photograph #NH 91171

RELOCATION OF SQUADRON TO ESPIRITU SANTO

No. 6 Squadron left Fiji at the beginning of October when the whole unit, except the detached Tongan flight, was posted to Espiritu Santo. There, the unit came under the operational control of USS *Wright*. It carried out its first operational patrol from *Wright* in Segond Channel, three days after its arrival. The detached flight from Tonga rejoined the squadron in November. While based at Santo, the squadron carried out anti-submarine and anti-shipping patrols, and took part in the rescue of survivors of the SS *Cape San Juan* (a passenger-cargo vessel under U.S. Army charter for use as a troop transport), torpedoed, on 11 November 1943, by the Japanese submarine *I-21*.⁷

SINKING OF THE TROOP SHIP SS *CAPE SAN JUAN*

Unclipping the port browning machine gun I swung the barrel over the side, pulled the breech block back and let it fly forward taking the first half-inch armour piercing bullet into the breech, then I braced my legs as we went into a vertical bank. Suddenly the sharks were plum in the middle of the reflector sight, no lead was necessary as we were now doing a tight turn around them. I pressed the firing mechanism, putting three bursts of 25 rounds into the pack. One or two of the wounded sharks leapt right out of the water and when they fell back in the other sharks just tore them to pieces, the water coloured and turned pink, the carnage I had caused below was completed. We moved on quickly shooting up some ten packs of sharks. Some packs were swimming too deep, these we passed by.

—Actions taken by Sgt. Walter Leadley, RNZAF, following the order, "Air Gunner, stand by with the 05s will you, we'll give the sharks something else to occupy them." The RNZAF Catalina flying boat, of which Leadley was an aircrewman, was taking part in rescue operations for survivors of the SS *Cape San Juan*, which had been torpedoed by a Japanese submarine.[8]

Photo 4-5

SS *Cape San Juan* listing to starboard, and down by the bow, on 12 November 1943: she had been torpedoed by Japanese submarine *I-21*. She sank on the following day. Naval History and Heritage Command photograph #NH 89892

At 0530 on 11 November 1943, the US flagged SS *Cape San Juan* (Master Walter Mervyn Strong) was attacked 300 miles southeast of Fiji by the Japanese submarine *I-21* (Comdr. Hiroshi Inada). The sea was slightly choppy, with a few white caps, and weather clear. The sun had not yet risen, but visibility was good. The troopship was sailing

38   Chapter 4

unescorted from San Francisco, California, to Townsville, Australia, with a total of 1,464 personnel on board.[9]

Comprising this group were: 57 Merchant Crew; 42 Navy Armed Guardsmen; 3 radio operators; 1,340 military passengers; 21 permanent Army personnel (3 officers and 18 enlisted men) commanded by Maj. Robert A. Barth, who were responsible for the military personnel while being transported; and 1 civilian. The passengers were primarily from units of the U.S. Army Air Corps—the 855th "All Negro" Engineers (Aviation) Battalion, the 1st Fighter Control Squadron, and the 253rd Ordnance (Aviation) Company.[10]

The ship had been steering 270° true, speed 14.7 knots, and was changing course to starboard (in accordance with the zigzag plan being used) when a torpedo exploded in her No. 2 hold. A second one fired passed harmlessly astern. Ships commonly zigzagged when transiting areas where enemy submarines might be operating, to reduce the accuracy of torpedo fire. This tactic involved making periodic course changes to the left or right of track, in order that a submarine could not predict exactly where a ship would be by the time a torpedo reached it. Low-speed ships, such as the *Cape San Juan*, used very little rudder and short legs (relatively modest course changes and remained on a new course for only a short time) before zagging in the opposite direction.[11]

Photo 4-6

Heave Ho! My Lads! Heave Ho!

Give us the oil, give us the gas
Give us the shells, give us the guns.
We'll be the ones to see them thru.
Give us the tanks, give us the planes.
Give us the parts, give us a ship.
Give us a hip hoo-ray!
And we'll be on our way.

Heave Ho! My Lads, Heave Ho!
It's a long, long way to go.
It's a long, long pull with our hatches full,
Braving the wind, braving the sea,
Fighting the treacherous foe;
Heave Ho! My lads, Heave Ho!
Let the sea roll high or low,
We can cross any ocean, sail any river.
Give us the goods and we'll deliver,
Damn the submarine!
We're the men of the Merchant Marine!

U.S. Navy World War II recruiting poster, and a verse and chorus from *Heave Ho! My Lads! Heave Ho!*, the official song of the U.S. Maritime Service.

The officer of the deck, Navy Armed Guard officer, and bridge lookout had reported to the master sightings of a whale or periscope. The visual contact was approximately 3,000 yards distant, on the ship's starboard side, fifteen degrees abaft her beam. The Armed Guard (a U.S. Navy gun crew assigned to merchant ships in convoy or sailing independently) was at General Quarters stations with all gun stations manned.[12]

Two torpedo wakes were soon sighted. One of the torpedoes struck, and exploded. *Cape San Juan* transmitted a distress message, "torpedoed, ship sinking fast." Communication codes were thrown over the side in a weighted box to prevent possible capture, and radio equipment was smashed by the Army operator without orders to do so. This action precluded further radio communications. The troops and Merchant crew abandoned, leaving the ship by boat, life rafts, diving over the side, and descending lines.[13]

The ship was still afloat but there were heavy seas, and it was raining when, at 1100, the Liberty ship SS *Edwin T. Meredith* (Master Murdock Daniel MacRae) arrived at *Cape San Juan*'s location. She immediately put motor lifeboats in the water, and began taking off the casualties and remaining survivors (mainly Army transport personnel) still aboard her. The *Cape San Juan*'s master, the majority of the crew, and Navy Armed Guard stayed on board until later that day. The water around the ship was littered with men on rafts, others clinging to wreckage, and some swimming. Sharks were present. *Meredith*'s chief mate, Alex Appelbaum, witnessed some men being pulled off partly submerged rafts by the savagers of the sea.[14]

After taking casualties and survivors off the *Cape San Juan*, the *Meredith* circled around, picking up other survivors. She departed that evening about an hour before darkness, and delivered 438 survivors to Noumea, New Caledonia, on 16 November.[15]

Midafternoon on 11 November, a Naval Air Transport Service PBM-3R Mariner (flight V2163) sent out from Fiji, arrived on scene. The flying boat had stopped there to refuel, while en route to Noumea from Pearl Harbor, when word of the sinking was announced. Approaching the *Cape San Juan*, about ten miles dead ahead, the PBM's aircrew first sighted life boats and rafts, then noticed small clusters of black dots, which they soon realized were men wearing life vests.[16]

Under less than ideal conditions, William W. Moss Jr., put the seaplane down on the water's surface. PBMs were designed to take off and land in sheltered bays, not the open ocean with confused swells (differing height and wave lengths) and 10-12 knot winds. Moss had difficulty finding an area smooth enough to attempt a landing. The large amount of oil in the water dampened the waves somewhat in the immediate vicinity, which Moss tried to use to his advantage. He came in crosswind at 70 knots, to avoid the pounding that would result from landing into the seas, or downseas.[17]

Photo 4-7

A Martin PBM-3R Mariner transport aircraft taking off, circa 1942-1943. National Archives photograph #80-G-K-13496

Nevertheless, the first wave crest encountered slammed the plane 30 to 50 feet high into the air. Moss throttled back the engines and held the elevators full back, as the nose hit successive sides of waves in an extremely violent manner. So forceful were the shocks experienced that Moss did not believe the plane would remain in a floatable condition, let alone flyable. However, a quick inspection seemed to indicate that everything was intact, and efforts began to rescue men in the sea.[18]

Moss decided to concentrate on saving the most vulnerable men, those not in rafts. It was too risky to try and maneuver through the survivors in the water, and they were too exhausted to swim for the plane. Therefore, it was decided they would troll for survivors by towing a string of small rafts behind the plane as they zig-zagged upwind along the edge of the rescue area. (The wind set the aircraft down toward rescuees.) Moss stayed at the controls, blipping the throttle and steering while the rest of the crew were aft retrieving men in the water, one at a time, and lifting them into the plane. Forty-eight survivors were recovered. Oily, water-soaked clothes were stripped off, and thrown back into the sea, and the men were given blankets, food, hot coffee, water, and/or a shot of whiskey.[19]

Photo 4-8

Cockpit of a PBM-3R Mariner, August 1943.
National Archives photograph #80-G-377237

The liquor had been brought aboard the plane by a Navy captain (a passenger bound for Noumea, left ashore at Fiji)—five bottles of Kentucky Straight Bourbon Whiskey destined for William F. "Bull" Halsey Jr. Admiral Halsey was then commander, South Pacific Forces and South Pacific Area. In June 1944, he would assume command of the Third Fleet, and operate against the Japanese in the Palaies, Philippines, Formosa, Okinawa, and South China Sea. Following the Okinawa campaign in July 1945, his carrier aircraft struck Tokyo and the Japanese mainland before war's end.[20]

Photo 4-9

Admiral William F. Halsey, USN, circa 1942-1945.
National Archives photograph #80-G-205279

After two hours of grueling recovery work, with his crew exhausted, the plane at maximum capacity, and a heavy rain squall closing in, Moss decided it was time to depart. A circling RNZAF Hudson bomber had also fired two red flares, which he assumed meant danger. After dropping both auxiliary tanks to lighten the load, Moss estimated his gross take-off weight was 45,258 pounds. He later described the take-off which, lasted nearly a minute, as the longest fifty seconds of his life, explaining:

> At 55 knots the plane bounced off the top of a wave to a height of 30 to 50 feet, setting up a series of five or six bounces until the plane finally became airborne at approximately 70-75 knots. On the second bounce the left wing dropped approximately twenty degrees, but full aileron control brought it up before the plane touched the water again.[21]

RNZAF PBY-5 CATALINA ARRIVES ON SCENE

Photo 4-10

Royal New Zealand Air Force PBY-5 Catalina (No. 4017 XXT), circa 1945. Courtesy of Jenny Scott

The PBM Mariner's take off was witnessed by a Royal New Zealand Air Force PBY-5 (No. 4017 XXT), which arrived in the area in late afternoon, amid a tropical rain storm. Coming in on a low level run at seventy feet because of the poor visibility, there was suddenly a clear patch in the weather. The Catalina's aircrew then saw below, hundreds of men in the sea, and many more crammed onto bits of timber, life rafts, Carley floats, duckboards (walkways of wooden slats joined together), and pitifully few life boats. Off to the left, on the edge of the rain, the *Edwin T. Meredith* with cargo nets over the side, was steaming very slowly through the survivors, to enable those able to do so, to climb up to safety. Her master did not dare stop, for fear of being torpedoed. After clearing the rain squall, he proceeded at full speed for Noumea.[22]

A few seconds later, the New Zealanders sighted the sea-gripped PBM Mariner struggling to get free. Chief Air Gunner Sergeant Walter Leadley explained the difficulties they observed:

> We estimated a 16 foot swell was running. It had picked up some forty-plus men and was not happy about taking off as he had damaged his starboard float.... The Mariner turned into the wind, looking like a massive bird with her gull wing and twin tail, and then I became alarmed that she was riding too low in the water. I racked my brains on aircraft recce [recognition] - yes, she had a carrying capacity of only about three tons, with 40-plus men on, at least an extra 1 1/2 tons!!
>
> She was now riding the swell and gaining speed into the wind, leaving a white trail of foam behind her. Suddenly she altered course to port, thus giving a little more lift to the damaged float on the starboard wing. At this point she started to go through the tops of the swells, the tips of her propellers striking the sea, sending up great clouds of spray. The revs of the motor would drop rapidly, then as she went into the trough she would build up to full revs again, repeating the performance again and again, until sufficient speed and wind built up under the wings to give her lift, then she started hitting the tops of the swells with a mighty thump, leaving a trail of evenly-spaced white patches of foam behind her. Suddenly they ceased and, thank God, she was airborne.[23]

At this point the New Zealand plane received orders to conduct a radar search and keep the Japanese submarine down or sink her, but lacking depth charges it was first necessary for her to return to Tonga to procure them. She therefore climbed to 2,000 feet and left the area, bound for the flying boat base at Tonga. There, she refueled and received four depth charges, two in mounts under each wing. Arming the plane took some time, as a rising sea and 6-foot chop kept tossed the barge about, upon which the 250lb depth carges were cradled for hoising up onto the aircraft. Once finally readied, it was dark, requiring another long wait while a take off flare path was laid out. This required use of a number of miniture boats, each with a battery-powered electric light affixed atop a pole. These boats were anchored about a chain apart (66 foot spacing) in a straight line directly into the wind.[24]

The RNZAF Catalina was finally airborne at 2210. Within half an hour, the *Cape San Juan* came into view, 20 miles away and burning brightly. The hulk served as a navigation becon throughout the night, as the plane's twin Pratt and Whitney engines droned on during a protracted radar search for the submarine. Two hours before dawn, Sgt. Harry Farmiloe decoded a message, informing them to expect a destroyer and two sub-chasers in the area by daylight.[25]

At first light on 12 November, blotches could be made out on the sea surface which, as the aircraft drew nearer, turned out to be men standing on rafts, some waist deep in water. Detecting a blip on its radar screen, the Catalina climbed upward and set course for the contact some fifty miles away. Upon sighting the source—a destroyer and two smaller ships—twenty minutes later, Flight Sgt. Larry Heath plugged in the aldis lamp, and signalled the course and distance to the survivors to the destroyer USS *McCalla* (DD-488), yard minesweeper USS *YMS-241*, and submarine chaser USS *SC-654*.[26]

Returning to the vicinity of *Cape San Juan*, the aircraft discovered that during the night, the survivors had spread out over an area of approximately five square miles. The PBY flew around the perimeter, then turned into the center where there was a great number of life rafts, and several packs of sharks. The first group of survivors appeared below and to the right of the aircraft, where there were some 25 to 30 sharks moving inwards just below the surface. As the PBY descended and drew near the first rafts, a second pack of sharks loomed into view, right on the surface, with the center of the pack thrashing the water. At 100 feet in altitude, the plane passed over the first of the rafts. Leadley described the desperate state of men, and their amazing show of appreciation for the presence of the seaplane overhead:

> Twenty to twenty-five men were standing in a tight bunch shoulder to shoulder up to their waists in the sea and the outline of rafts could be seen below them, the sheer weight having submerged it. Floating around the rafts were from four to nine men, some face down, then an astounding thing happened. After being in that exhausting position for twenty-one hours they each raised an arm very carefully not to upset the next fellow on the raft and waved to us. In the next few minutes we had passed some twenty rafts in a similar situation and they all waved. I believe it was their way of saying "thank you" for staying with us all night.[27]

Upon order, as previously described, Leadley began putting bursts of machine gun fire into packs of sharks, whereupon, those not killed or wounded, began tearing apart the others in a frenzy. At last, the

McCalla, under the command of LCDR Halford A. Knoertzer, arrived. She had reached the area the previous night, but was not able to home in on the survivors until daylight when joined by the *YMS-241* (ENS Emery L. Burgess) and the *SC-654* (ENS John C. Boutall).[28]

The Catalina circled the destroyer while she made the first pick-up. This was accomplished by putting landing nets over her leeward side, with three-tiers of three sailors each on the nets, starting at the water line to assist exhausted men up to the deck. The destroyer stopped, and rolled in beam seas, with the first two tiers of sailors disappearing under water. Rolling the opposite way, up came the sailors, each one hanging onto a man. She rolled again and another 15-20 men were on the nets, eager hands helping them over the lifelines to safety. Most were covered in oil or diesel fuel; after their clothes were stripped off, they were taken below for a shower. The retrieval of survivors was made difficult by the high swell running. As *McCalla* headed toward a raft visible on a wave crest, individual survivors bobbing in the trough, directly in her path, could not be seen. Under these conditions, these men had little chance of grabbing a net on the moving ship and were at risk of being run over by churning screws.[29]

The PBY's Sergeant Farmiloe's quick thinking saved the day. He suggested the dropping of navigation smoke flares to mark clear passage, but also to indicate those who desperately needed to be picked up. Flight Sergeant Abb Ormesby manned the aldis lamp, as the PBY lay a smoke flare every ten minutes, and used it when necessary to communicate with the destroyer. After an hour of providing this assistance, the aircraft broke away for another sweep around the rafts and bits of wood—shooting up the odd packs of sharks, which were still around in large numbers.[30]

At last, there were few survivors in the water. The sub-chaser and minesweeper, loaded with survivors, had left. The destroyer followed when it appeared there was no one else to rescue. However, as the PBY made a final low run around the wreckage, the airmen were startled to see below, three men in a rubber raft pulling a fourth one in. The raft had been dropped by the RNZAF Hudson the previous day. *McCalla* had already departed and was by then twenty miles away, but quickly returned and plucked the men from the sea.[31]

After the *McCalla* had commenced the recovery, the destroyer escort *Dempsey* (DE-26) had arrived on scene from anchorage at Viti Levu, Fijian islands, and joined in the rescue efforts, picking up five Army personnel from a life raft at 0925. Upon orders from *McCalla*, she hoisted in her motor whaleboat and secured from rescue work at 1032, then departed with the destroyer.[32]

The evening of 14 November, *McCalla*, *Dempsey*, and *YMS-241* arrived at Suva, Fiji, at 1715 and landed in excess of eight hundred forty survivors. The submarine chaser *SC-654*, not having enough fuel to make it to Suva, had manuevered along the starboard side of *McCalla*, transferred her 152 survivors to the destroyer, and departed the group to proceed back to Tongatapu independently.[33]

The PBY Catalina touched down at Laucala Bay, Fiji. Squadron Leader Richard Makgill and Crew 5 had flown to Ile Nous Naval Air Station, Noumea, New Caledonia, to pick up flying boat, No. 4017 XXT. Following completion of the rescue operation, and some crew rest at Fiji, they returned to Espiritu Santo with the plane.

Crew 5 of No. 6 Flying Boat Squadron

Position	Individual	Position	Individual
Captain	Flying Officer John Macgrane	Chief Air Gunner	Sgt. Walter Leadley
2nd Pilot	Sgt. Harry Farmiloe	Chief Flight Engineer	Sgt. Ralph Rigger
1st Wireless Operator	Flight Sgt. Abb Ormesby	2nd Flight Engineer	Sgt. Noel Melvill
2nd Wireless Operator	Flight Sgt. Larry Heath	3rd Flight Engineer	Sgt. Jack Wakeford
Navigator	Pilot Officer Ross Laurenson[34]		

5

Operations from Halavo Seaplane Base

Photo 5-1

Japanese Destroyer *Kikuzuki* under salvage in Halavo Bay, Florida Island, August 1943. She had been sunk in the bay by USS *Yorktown* (CV-5) aircraft on 4 May 1942. Naval History and Heritage Command photograph #USMC 59615

Coos Bay joined the list of ships participating in the Solomons Campaign when she left anchorage in Segond Channel, on 11 October 1943, bound for Florida Island. Proceeding into Halavo Bay, the smoke and flames of the last Japanese air raid of the war on Guadalcanal could be seen twenty miles away across the Slot. As soon as the ship was anchored, the Air Officer and his men aboard her began establishing the *Coos Bay*'s first seadrome (A floating landing field). Patrol Squadron

VP-71 personnel then moved aboard and began carrying out search patrols.[1]

Map 5-1

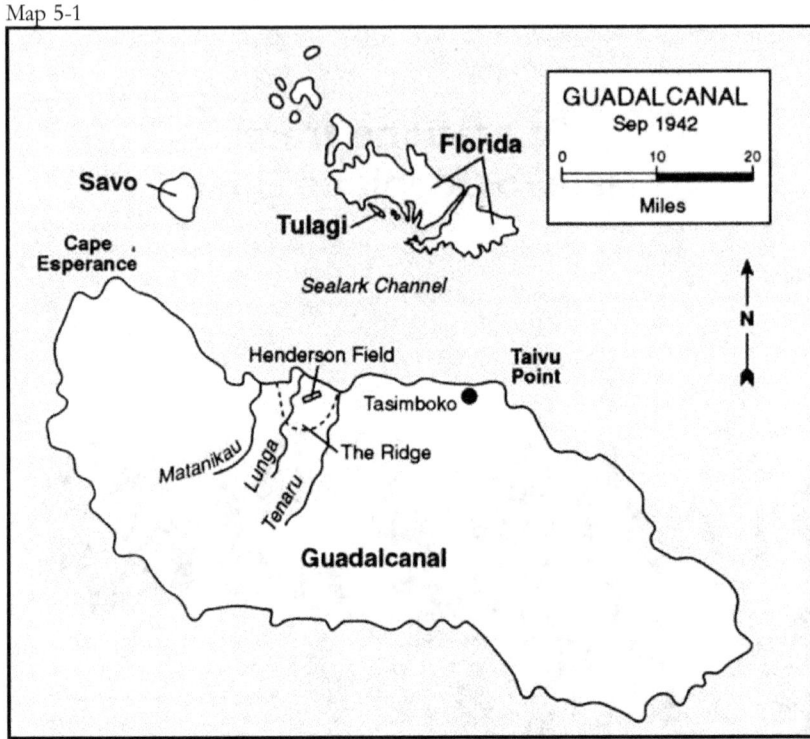

Guadalcanal, Savo, and Florida islands
(Source: www.nps.gov/history/history/online_books/npswapa/extContent/usmc/pcn-190-003130-00/sec6.htm)

Florida Island (the largest island of the Florida Group) was located across the New Georgia Sound (of which the Sealark Channel was a part) from Guadalcanal. As previously mentioned, (see pages 26-27) on 3 May 1942, Japanese forces occupied Tulagi, a small island nestled in a bay at Florida Island. They utilized Tulagi Harbor as an anchorage and a former Royal Australian Air Force seaplane base on adjacent Gavutu and Tanambogo islands (see map on next page).[2]

OPERATIONS FROM HALAVO SEAPLANE BASE

Tulagi Harbor was bounded by Florida Island and Tulagi Island. Eastward, and adjacent to it, was Gavutu Harbor bordering Florida Island, and smaller Gavutu, Tanambogo, and Palm islands. Farther to the east was Halavo Bay, bordering the southern coast of Florida Island.

The U.S. Navy Halavo Seaplane Base was in Halavo Bay, with facilities on Florida Island. PBY Catalina take-offs and landings took place in Gavutu Harbor. (The Halavo seaplane base was also known as "Tulagi Seaplane Base" for nearby Tulagi, and not to be confused with the previous RAAF/later Japanese Tulagi seaplane base at Gavutu.)[3]

Map 5-2

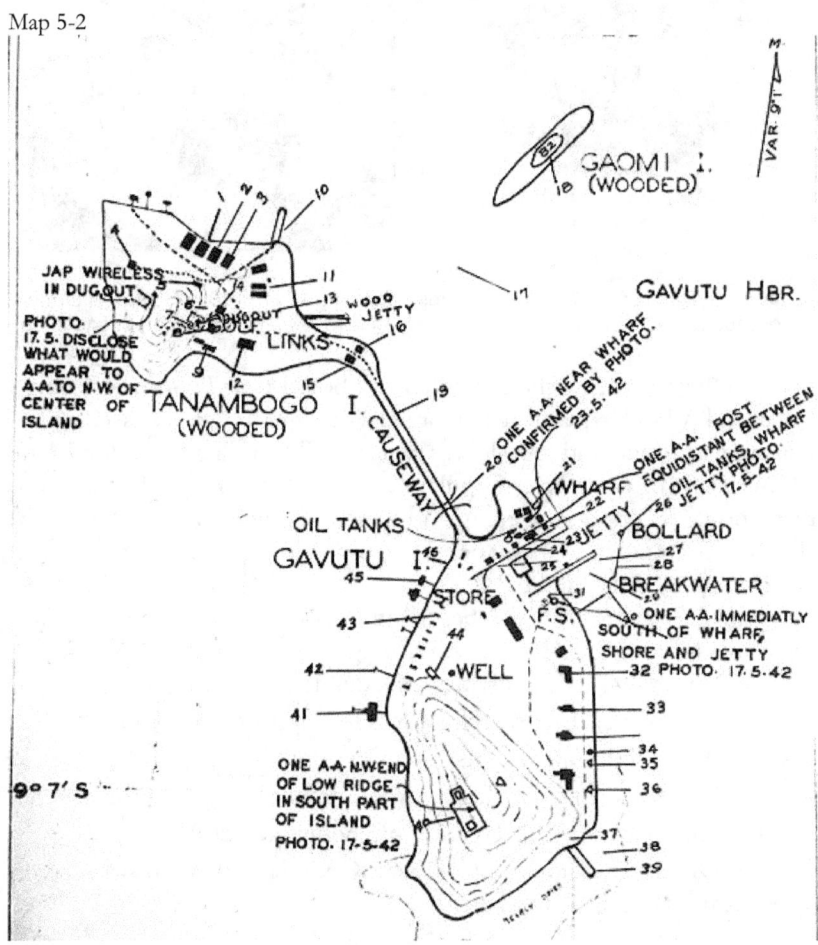

Intelligence map of Tanambogo and Gavutu islands, near Tulagi, prepared in July 1942 for use during the invasion of the Guadalcanal and Tulagi area by the 1st Marines. U.S. Naval History and Heritage Command photograph #NH 97748

PATROL SQUADRON VP-71 (LATER VPB-71)

Squadron insignia of VP-71 (Thunderbirds), and later VPB-71 (Black Cats) after the patrol squadron was recommissioned on 25 April 1944 as a patrol bomber squadron

The living and working conditions aboard *Coos Bay*—where ship's force and VP-71 squadron personnel were quartered—was crowded. Leisure was found and morale was maintained by utilizing Gaomi Island (shown on prior map), popularly known as "Palm Island," for recreation. Its good swimming and diving conditions were enjoyed by all, and an abundance of marine life produced some very good spearfishermen and fish dinners.[4]

Patrol Squadron VP-71 began operations from Halavo Seaplane Base on 14 October 1943. In the ensuing months (until relieved by VP-81 on 6 March 1944), its activities included the following missions:

- 650-mile surface search patrols to the north and northeast
- Anti-submarine patrols
- Night spotting for shore bombardment by surface forces
- Snooper missions (location of enemy facilities)
- Dumbo missions out of New Georgia and Rendova Islands, and later Treasury Island[5]

Coos Bay was the advance base for VP-71 Dumbo missions from Rendova and Treasury islands. The seaplane tender moved up the Central Solomons to Rendova Island, New Georgia Islands, and took over the seadrome in Rendova Harbor on 23 November 1943. There, she tended six to nine PBY-5s belonging to Patrol Squadrons VP-14, -23, and -71 (discussed in a subsequent chapter.)[6]

TREASURY-BOUGAINVILLE ISLAND LANDINGS

Enthusiasm for the plan was far from unanimous, even in the South Pacific, but, the decision having been made, all hands were told to 'get going.'

Worse than anything ever encountered before in the South Pacific.

—Comments by ADM William F. Halsey Jr. regarding scheduled amphibious landings on Bougainville Island, and the beach conditions and terrain encountered. Most areas of the landing beaches were so narrow that two bulldozers could not pass abreast between jungle and sea.[7]

By July 1943, the South Pacific Command had decided that, as part of the final phase of the Solomon Islands Campaign, an assault should be made on Bougainville Island. In enemy hands, the island constituted the next major obstacle to Halsey's forces driving northwest through the New Georgia Islands group. In the Allies hands, it would provide a base for future operations against Japan's powerful base at Rabaul to the northwest. Japanese-held Bougainville hosted numerous airfields: two at the north end, two on the southern, a field and a seaplane base on the east coast, and a field on Shortland Island to the south. Enemy ground forces were concentrated in and around southern Bougainville, with garrisons on the east coast and northern part of the island.[8]

South Pacific allied forces had virtually completed occupation of the New Georgias with the capture of the Japanese airfield at Munda Point on New Georgia (the largest island in the group), on 5 August 1943, and landings on Vella Lavella ten days later. Occupation of Vella Lavella, also one of the New Georgia Islands, was completed less than two months later. In October, the famed Black Sheep Squadron, VMF-214, led by fighter ace Major Gregory "Pappy" Boyington, moved forward from the Russell Islands to Munda. From there, F4U Corsair fighter aircraft could strike the next big objective, Japanese bases on Bougainville. (Boyington's VMF-214 was not the same squadron that had provided air cover for the *Chincoteague* in July. The Swashbucklers had been broken up in September and replaced by the Black Sheep, an entirely different unit that inherited its number.)[9]

Map 5-3

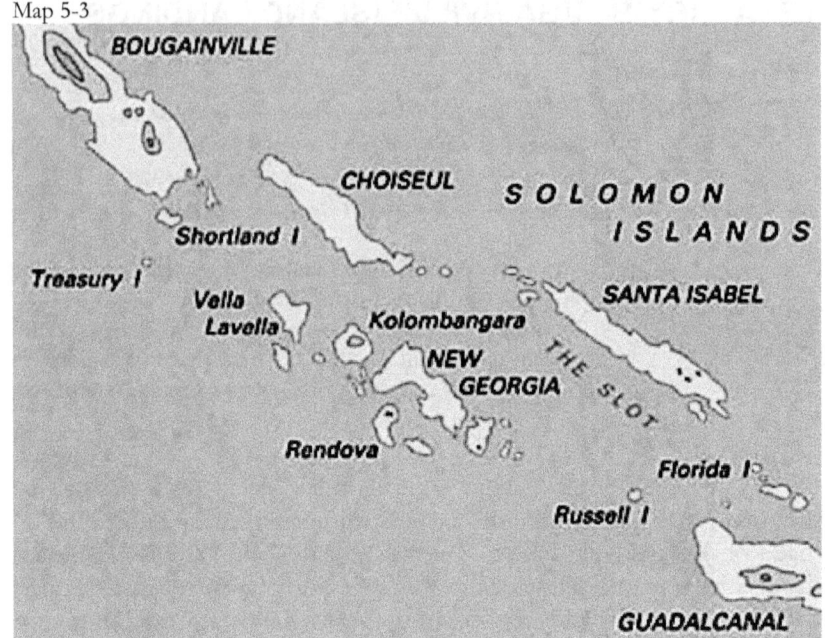

Central Solomon Islands

TREASURY ISLANDS (OPERATION GOODTIME)

Earlier in the autumn of 1943, as a prelude to the much larger Bougainville expedition, the Allies had taken the Treasury Islands. Over 5,000 members of 8th Brigade Group of the 3rd New Zealand Division landed, on 27 October 1943, with 1,900 U.S. Marine support troops making up the balance of the amphibious force. It was a one-sided operation, with nearly 7,000 troops in total facing fewer than 250 Japanese. The Japanese chose to fight to the death and only eight were taken prisoner. (This landing was the basis of the famous scene in the John Wayne movie *The Fighting Seabees*, in which a tank landing ship was under heavy fire on the beach when a Seabee raised the blade on a bulldozer as a shield, drove off the LST's ramp, and bulldozed and buried the Japanese fighting positions along with their occupants.) Allied casualties were forty New Zealanders and twelve Marines killed, with additional wounded.[10]

Map 5-4

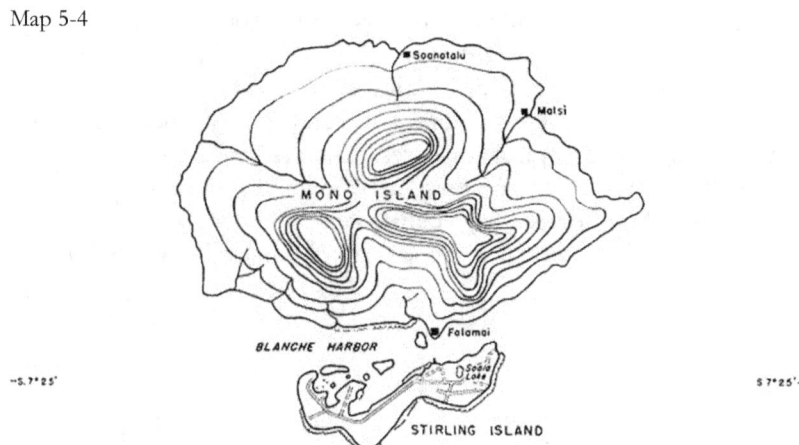

Treasury Islands, Solomon Islands

Photo 5-2

Camp of the 7th Australian Infantry Battalion on Stirling Island, 21 March 1945. Some members are playing volley ball in the foreground.
Australian War Memorial photograph 079956

BOUGAINVILLE (OPERATION CHERRY BLOSSOM)

In the much larger Bougainville assault, the 3rd Marine Division came ashore at Cape Torokina, on Empress Augusta Bay, on 1 November 1943. The purpose of the invasion of the largest and northernmost of the Solomon Islands was to gain and hold a strategic area around the beachhead only, since Lt. Gen. Harukichi Hyakutake, the island commander, had some 40,000 soldiers and 20,000 sailors available under him for the defense of the entire area. Construction of airstrips in this specific area of the long southwest coast of Bougainville would enable U.S. planes to strike Rabaul, and defend the beachhead area base. Many miles of dense tropical rain forest separated the Empress Augusta Bay area from Japanese garrisons on Bougainville. The plan was for U.S. troops to build the airstrips before large numbers of enemy forces on the island could mount a sustained counterattack from land.[11]

Photo 5-3

Landing craft circle while awaiting the invasion of Bougainville on 1 November 1943. U.S. Marine Corps photograph taken from aboard the USS *American Legion* (APA-17).

This rapid operation (code named CHERRY BLOSSOM) was a success and by February 1944, Japanese air strength at Rabaul had been eliminated. However, fighting on Bougainville would be protracted and last until the war's end. In November 1944, twelve Australian Army brigades replaced six U.S. Army divisions that were performing defensive roles in Bougainville, New Britain and the Aitape–Wewak area in New Guinea. This freed up the American units for operations in the Philippines. While the American units had largely conducted a static defense of their positions, Australian commanders mounted offensive operations designed to destroy the remaining Japanese forces in these areas. The value of these campaigns was controversial at the time and remains so today. Critics believe that they were unnecessary and wasteful of the lives of Australian soldiers involved, as the Japanese forces were already isolated and ineffective.[12]

PATROL AIRCRAFT AND THEIR TENDERS

On 1 November 1943, a Catalina of Patrol Squadron VP-71 provided anti-submarine coverage for the landing of Task Force 31-5, which established a beachhead at Cape Torokina, Empress Augusta Bay. Other PBYs later made numerous landings in the open water off Cape Torokina beachhead, bringing in medical supplies, transporting personnel, and evacuating casualties. Several open-sea rescues were made by the squadron in this vicinity under fire from enemy shore batteries and under attack by enemy aircraft.[13]

Aircraft of VP-71 were staged through advance bases at Rendova Harbor and Hathorn Sound, adjacent to Ondonga Airfield in the New Georgia group. The patrol squadrons and tenders that earned battle stars for supporting the landings at Treasury and Bougainville, and subsequent actions, are identified in the following table. The letters FE afer a VP signify that a flight echelon (subunit of it) took part. Two commanding officers are listed for some ships, because a change of command took place during the qualifying period.[14]

Treasury-Bougainville Operation: Supporting Air Actions

Chandeleur (AV-10)	27 Oct 43-1 May 44	CAPT Albert K. Morehouse
		CAPT Walter V. R. Vieweg
Pocomoke (AV-9)	27 Oct 43-1 May 44	CAPT Curtis S. Smiley
		CAPT Edward L. B. Weimer
Wright (AV-1)	27 Oct 43-1 May 44	CAPT Frank C. Sutton
		CAPT James E. Baker
Chincoteague (AVP-24)	27 Oct 43-1 May 44	CDR Robert A. Rosasco
Coos Bay (AVP-25)	27 Oct 43-1 May 44	CDR William Miller
		CDR Delbert L. Conley
VP-23 FE	27 Oct-15 Dec 43	LCDR Frank A. Brandley
VP-54 FE	27 Oct-15 Dec 43	LCDR Carl W. Schoenweiss
VP-71 FE	27 Oct-15 Dec 43	LCDR Cecil Kelly Harper
VP-81 *P-7*	10 Dec 43-1 May 44	LCDR E. P. Rankin
VP-91 FE	9 Nov-15 Dec 43	CDR E. L. Farrington

FE: Flight Echelon *P-7*: Single plane of VP-81

A PARTICULARLY NOTEWORTHY AIR-SEA RESCUE

On 24 November, a Catalina of Patrol Squadron VP-23 made a daring rescue, requiring extreme skill and courage. The plane commander, LT Milton R. Cheverton, USNR, and his crew were based at Ondonga, as a Dumbo standby (plane and crew) maintained there. A B-25 medium bomber, hit by anti-aircraft fire and set aflame, while on a strike over Kahili, had ditched in the harbor at Shortland Island. Shortland (the largest island of the Shortland group) lay directly south of Bougainville Island. The downed aircraft had been one of twenty U.S. Army Air Force B-25s that hit the Japanese airfield at Kahili, located near Buin on the southern coast of Bougainville.[15]

Commander, Aircraft Solomons (comprised of aviation units of the U.S. Army, Navy, and Marine Corps as well as two New Zealand fighter squadrons) ordered Cheverton to the flight crew's rescue with a fighter cover of twelve RNZAF P-40s. After getting airborne, Cheverton set a course for Shortland, maintaining an altitude of 50 feet with the fighters "stepped up" in pairs to 2,000 feet overhead.[16]

East of Ballale Island (one of the Shortlands), heavy anti-aircraft fire was encountered. A flare from the life raft containing the six members of the B-25 crew revealed their position to be one mile south of Maifu, a small island between Bougainville and Shortland Island. The Dumbo landed under intense fire from gun emplacements on Shortland and adjacent Faisi Island. After recovering the grateful Army Airforce members, Cheverton returned to Ondonga. He was later awarded the Distinguished Flying Cross, and his first and second pilots and every crew member were commended for their share in the rescue by the Secretary of the Navy.[18] A copy of Cheverton's medal citation follows:

The President of the United States of America takes pleasure in presenting the Distinguished Flying Cross to Lieutenant Milton R. Cheverton, United States Navy, for extraordinary achievement while participating in aerial flight as Commander of a Patrol Plane near Bougainville on 24 November 1943. Lieutenant Cheverton courageously landed his plane in hostile waters to rescue survivors from a plane crash. Under intense fire, he boldly picked up the crew and effected a take-off.[19]

6

JFK Rescues Marine Paratroopers on Choiseul Island

When all seemed lost, you can well appreciate my relief to see the landing craft returning, escorted by PT boats, one commanded by you. As I recall, we both had our hands full and there was little time for amenities. Please accept again my heartfelt thanks.

—Letter to President John F. Kennedy from retired Marine Corps Colonel, Warner T. Bigger.[1]

Photo 6-1

Salvation from the Sea depicting USS *PT-59*, under the command of LT John F. Kennedy, taking U.S. Marines pinned down by Japanese forces at the mouth of the Warrior River off Choiseul Island to safely. Cover boat *PT-236* is in the background. Painting by Richard DeRosset

Although not related to air-sea rescue, a daring PT (motor torpedo) boat rescue of U.S. Marines is associated with the Bougainville assault, previously described. The skipper of *PT-59* was LT John F. Kennedy. Before describing the action, background regarding Kennedy's war service and the circumstances leading to his specific involvement in the Choiseul Island rescue, is provided.

The account of John F. Kennedy's command of *PT-109* and her loss in the early morning darkness of 2 August 1943 is well known. While on patrol in Blackett Strait off the Kolombangara coast in the Solomon Islands, she was sliced in two by the Japanese destroyer *Amagiri*. He made great effort to get crew members safely to a small island and to secure their rescue, for which he earned the Navy and Marine Corps Medal and was also awarded a Purple Heart.[2]

Less familiar is Kennedy's subsequent tenure as the skipper of the *PT-59*. He accepted command at Tulagi on 1 September of the motor torpedo boat, which was to be converted to a gunboat to serve as a more formidable weapon against armed enemy barges. Following its modification to PTGB-1 (PT Gunboat No. 1), he joined Motor Torpedo Squadron 19 at Vella Lavella. (Since the Navy did not ultimately adopt this designation officially, it was an informal term. However, a Squadron 19 war diary entry does refer to "Gunboat 1" commanded by LT J. Kennedy, USNR.)[3]

Changes made to the boat involved removal of her torpedo tubes to facilitate the installation of Bofors 40mm cannons fore and aft. She was also fitted with six .50-caliber machine guns (four twin mounts and two single barrel guns) to augment the two twin mounts she already had. This gave her an impressive fourteen mounted .50-caliber machine guns, and for more punch, the two Bofors guns on the bow and stern. The installation of more and heavier armament resulted in an associated increase in manning. Typically, there were one to three officers and twelve enlisted men of various ratings aboard a PT boat; the crew size of the three Solomons PTs converted to gunboats swelled to eighteen. Since there were only twelve bunks for the crew, the men had to "hot rack," meaning a man relieved on watch would occupy a bunk vacated by someone taking over the duty.[4]

Kennedy was cognizant of the greater combat capabilities his *PT-59* offered. An officer assigned to the same squadron later recalled that during briefing sessions, before the boats went out on patrol, he would frequently ask [the squadron commander]: "Skipper, why don't you let me go in first with my boat? This is going to be a strafing operation [presumably referring to barge hunting missions at night] and I can fire many more rounds than an ordinary PT-boat."[5]

U.S. MARINES REQUIRE IMMEDIATE ASSISTANCE

On 2 November 1943, the PT base at Lambu Lambu Cove on Vella Lavella Island, received a radio call from Lt. Col. Victor H. Krulak, commander, 2nd Marine Parachute Battalion, 1st Marine Parachute Regiment, for PT boat assistance to take off a group of Marines. The battalion had landed on the southeast coast of the island a week earlier in LCP(R) landing craft from the high-speed transports *Kilty*, *Ward*, *Crosby*, and *McKean*, to carry out Operation BLISSFUL. (Designed by George Higgins, these type of wooden landing craft were commonly called "Higgins boats.") This "Choiseul diversion" was a feint designed to make top Japanese brass believe that it was the objective of an impending amphibious assault, versus at Torokina, on Bougainville Island to the northwest, where the 3rd Marine Division would land on 1 November.[6]

The intent was to conceal the true objective of landing at Torokina for as long as possible and to also decoy enemy troops to Choiseul that might otherwise fight on the Bougainville beachhead. To conceal from the enemy that the Marines were considerably less in number than an invasion force, the 650-man battalion had gone ashore the night of 27–28 October on unguarded beaches near Voza, an abandoned village which lay between concentrations of Japanese forces at Sangigai and Choiseul Bay. Intelligence indicated there were some 4,000 to 5,000 enemy on the island, dispersed in small camps along the coast.[7]

In succeeding days after their landing, the Marines carried out raids to destroy installations, barges, and supplies. These activities, along with skirmishes with the enemy in the jungle, created the intended impression that a larger force was at work. On the afternoon of 2 November, a group of Marines executed Operation BLISSFUL. As part of the operation, Marines under the direction of the battalion executive officer, Maj. Warner T. Bigger, used a M2, 60mm mortar, to shell Guppy Island, which was the site of an enemy barge replenishing-center and fuel base in the middle of Choiseul Bay.[8]

At completion, the group made its way southward on foot to the west bank of the mouth of the Warrior River, which lay to the east of the entrance to the bay. The Japanese located the raiding party, and enemy forces attacked the Marines from both upstream and from across the river. Outnumbered, outgunned, pinned down, and cut off from the remainder of the battalion, Bigger desperately required extraction from the hostile beach.[9]

When the radio call came in to the Lambu Lambu PT base, there were only two PT boats available, the *PT-59*, commanded by LT John F. Kennedy, and the *PT-236* of ENS William F. Crawford. Kennedy was refueling and had only one-third of a tank of fuel in the *PT-59*, enough to make it to the Warrior River, about 65 miles distant, but not enough to get back. The urgency of the situation dictated that a decision be made that the *PT-59* and *236* would leave immediately, and when Kennedy's boat ran out of fuel, the other boat would tow it.[10]

Bigger's patrol, consisting of Marines mainly from Company G plus several from Headquarters Company, was, by then, involved in a firefight. The battle with the Japanese at the mouth of the Warrior River resulted in the wounding of corporals Schnell and May. During the fight, the Japanese spotted Higgins boats approaching the river mouth from seaward and opened fire on them. As the boats began taking fire from the east bank, Marines on the west bank attracted the attention of personnel in the two boats. Return fire from the boats, combined with that of the Marines ashore, caused the enemy to temporarily cease fire, allowing the boats to make the west bank. As the Marines waded out to board the craft, the enemy began to lob mortar rounds at them. Luckily, a storm was gathering—with a heavy sea rising and winds from the southwest creating rough surf—and a heavy rain began to fall, screening the Marines from view.[11]

The first landing craft, having taken aboard more than thirty Marines, reversed its engines and began backing to prepare for departure. Deeper laden by this load, the boat scraped against the shallow coral reef beneath its keel, bending the rudder and making steering difficult. Bigger and the remaining thirty plus men boarded the other wooden-hulled boat. As the boat retracted, it too worked against the coral, opening a seam, and the boat began to fill with water. The engine compartment flooded and the engine stopped and could not be restarted. It began to drift dangerously toward the hostile shore before finally grounding.[12]

The PTs appeared at dusk finding one Higgins boat hard aground and the other damaged. Although a sliver of a crescent moon and a few stars visible through patches in the overcast provided little illumination, the boats were clearly distinguishable as landing craft versus the Japanese barges. One Marine aboard the disabled craft, upon hearing gunners aboard the *PT-59* rack their machine guns, worried that they might be shot by their own guys.[13]

When the two PT boats reached the Choiseul coast, they began looking for a landing craft from the invading force to help guide them to Bigger's force. They were successful around 1800, when one was located. Kennedy found that, much to his delight, LTJG Richard E. Keresey Jr., a PT boat skipper who had accompanied the 2nd Marines to investigate possible sites for a PT boat base on the island, was aboard. Keresey immediately transferred from the landing craft to the *PT-59*, and they set off, headed full-out for the river mouth with the boat's fuel gauge reading almost empty and little daylight remaining.[14]

With Keresey guiding, Kennedy arrived on the scene, "coming in on mufflers" (this expression refers to PT boats, when wanting to run silent, feeding engine exhaust out under the water at low speeds). Positioning his boat between the shore and the immobile craft to shield the Marines, Kennedy drew fire from the enemy. Despite the rain squall, enemy mortar rounds and machine-gun fire had continued to seek the immobile Marines in the waterway.[15]

As the *PT-236* lay starboard side to the beach, she drove the Japanese back up into the jungle with 20mm and .50-caliber machine-gun fire. The crew of the *PT-59* began pulling Marines up on deck. Kennedy took aboard all of the Marines from the grounded boat as well as its Navy crew, and also the Marines from the other boat. The damaged, but operable boat had stood off, well clear of enemy fire until the arrival of the PT boats. Once all the Marines were aboard—crammed onto the fantail and packed into the crew's quarters below, leaving scarcely any room for sailors to move about—the grounded Higgins boat was scuttled. Approximately twenty-five Marines were subsequently transferred to the *PT-236*.[16]

Following their departure, the two PTs boats, with the remaining Higgins boat struggling to keep up, transited around the southwest coast to Voza on the southern coast of Choiseul (to return the Marines to their parent battalion) without incident. Upon arrival there at 2130, Kennedy transferred the Marines to waiting craft that took them to shore. Their work done, the PT boats set a course south back across the New Georgia Sound to Vella Lavella. Halfway home, *PT-59* finally ran out of fuel and was taken under tow by the *PT-236* until the arrival around dawn of the two boats at Lambu Lambu Cove. A few hours earlier, Corporal Edward J. Schnell had died quietly in Kennedy's bunk at 0100. As a result of his critical injuries, he had been kept aboard, instead of being transferred ashore at Voza, in the hopes of getting him better medical treatment at the field hospital on Vella Lavella.[17]

On Choiseul Island, the entire 2nd Marine battalion was withdrawn by three infantry landing craft at 0130 on 4 November, leaving behind only minefields and booby traps set by Marines to delay an anticipated enemy attack. During embarkation, the sounds of exploding mines (triggered by enemy craft) signaled a timely departure. The battalion returned later that morning to their encampment on Vella Lavella.[18]

As events turned out, Kennedy had only a little over two weeks to continue his command. On the night of 5 November, he led three PT boats to the Moli Point/Choiseul Bay area, where they attacked Japanese barges. During the next week and a half, his *PT-59* prowled off Choiseul Bay looking for barges, concluding with his final action, on the night of 16 November, with an uneventful patrol. On 18 November, a doctor directed Kennedy, who was physically and mentally exhausted and had lost twenty-five pounds, to go to the hospital at Tulagi. He gave up his command of *PT-59* that same day to ENS John N. Mitchell (who would later be United States District Attorney under President Richard M. Nixon) and left the Solomons on 21 December 1943.[19]

Back in the States, Kennedy appeared to have lost the edge that drove him on *PT-59*. He jumped back into the nightlife scene, which he had pursued before his naval service, and assorted romantic dalliances. Assigned in March to Submarine Chaser Training Center, Miami, Florida, Kennedy joked about the easy duty, "Once you get your feet upon the desk in the morning, the heavy work of the day is done." He was discharged from active Navy duty due to a physical disability in March 1945.[20]

7

Duty at Rendova Harbor

Photo 7-1

USS *PT-164* with demolished bow, following a Japanese air attack on Rendova Harbor, 1 August 1943.
Naval History and Heritage Command photograph #NH 44494

Arriving at Rendova Island, on 23 November 1943, to take over its seadrome, *Coos Bay* functioned there until 28 December as an advance base for Dumbo operations originating from Rendova and Treasury Islands. In this role, she berthed personnel and tended six to nine aircraft of VP-23 and VP-71 squadrons, while operating as part of Task Group 33.1 (Air Solomon Islands). The harbor, located on the northwest coast of Rendova, which lay about halfway up the Solomons chain, was the best natural anchorage in the New Georgia group. The much larger New Georgia Island was located northward, across a narrow waterway, Blanche Channel.¹

Map 7-1

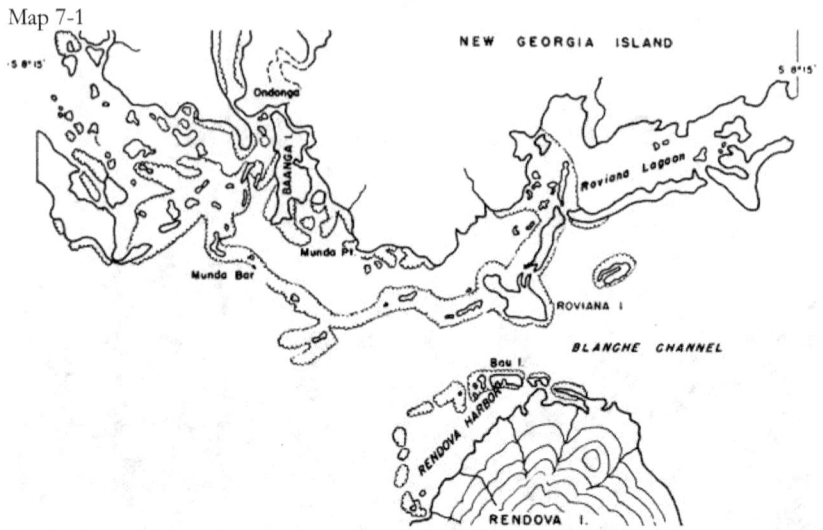

Southern coast of New Georgia Island, and the northern coast of Rendova Island
Building the Navy's Bases in World War II Volume II (Part III)

The PBY-5s under *Coos Bay*'s care were employed for rescue missions, night snooper missions, anti-submarine missions, and general utility flights. Tending operations included feeding and housing plane crews; seadrome maintenance; rearming and refueling seaplanes; minor repairs to aircraft hulls, engines, and electronics gear; and serving as a communications center for both ship and aircraft.²

The first rescue attempts from *Coos Bay* occurred the afternoon of 29 November, when planes No. *6* and No. *3* of VP-71 left separately for Dumbo missions to Empress Augusta Bay, on the western side of Bougainville Island, and VP-23 plane No. *34* for Kieta Harbor on its eastern coast. The three Catalinas returned without any recoveries, apparently having failed to locate any survivors.³

Commander Delbert L. Conley relieved CDR William Miller as commanding officer of the *Coos Bay*, on 1 December 1943, and was soon in the thick of rescue efforts. On the morning of 5 December, plane No. *35* of Patrol Squadron VP-23 effected the rescue of Maj E. P. Paris Jr., USMC, and TSgt Allen E. Christiansen, USMC, from Tomass Island, about ten miles northeast from Ballale in the Shortland Group south of Bougainville (see map on following page).⁴

While the change of command was taking place on 1 December, sixteen SBD Dauntless aircraft of Marine Scout Bombing Squadron VMSB-236 were embroiled in a dive bombing and strafing attack on Japanese AA guns and installations in the vicinity of their Kara and Ballale Airfields. These fields were sited near Buin in the southern part of Bougainville Island, and on nearby Ballale Island.[5]

Photo 7-2

Douglas SBD Dauntless scout-bomber aircraft in flight, circa 1943-1944. Naval History and Heritage Command photograph #NH 105333

Map 7-2

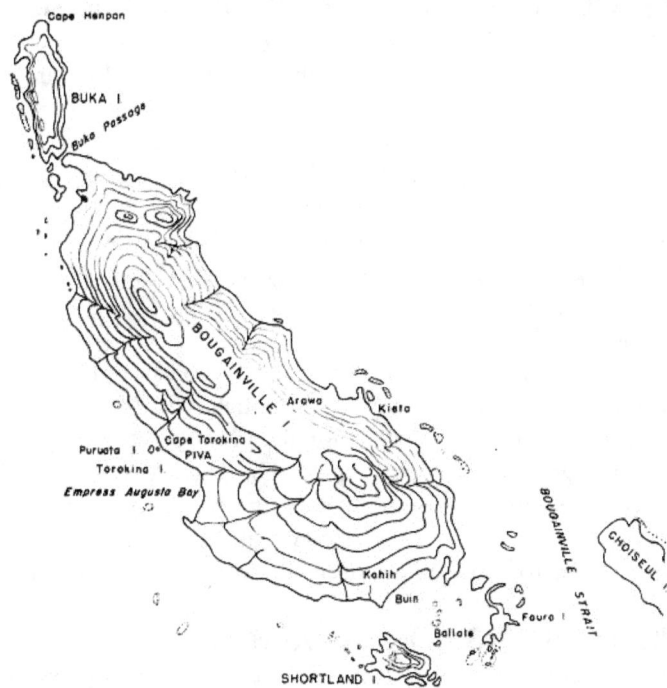

Bougainville Island and the Shortland Islands to the south

Two of the Dauntless scout-bombers were shot down by anti-aircraft fire while diving on AA positions on Ballale Island. Major Paris and his gunner, Technical Sergeant Christiansen, and Lieutenant Whiteley and his gunner, TSgt Charles A. Gotchling, were observed by other aircraft baling out. Three personnel were sighted in the water, two to three miles southwest of Ballale. A fourth was observed landing on the runway at Ballale.⁶

After freeing themselves of parachutes, Paris and Christiansen swam to Tomass Island. They were picked up on 5 December, by a Dumbo sent out on the basis of a report by a TBF Avenger torpedo bomber gunner. He had spotted them in the water while returning to base after a strike on Tonolei Harbor near Buin. On recovery, both men were found to be in fair condition.⁷

Another Catalina from VP-23 made a rescue on 14 December. PBY-5 No. *30* recovered 1stLt G. A. Davis Jr., USMCR, from a life raft about four miles south of Alu (also called Shortland Island, the largest island of the same named group). He too had been shot down by enemy AA fire from Ballale Island.⁸

RESCUE FOLLOWING A MID-AIR COLLISION

PBY-5 No. *30* made another rescue operation on 16 December, retrieving 1st Lieutenants R. Oughton and W. A. Monbart from the sea. The Marine Corps Reserve Officers were the only survivors of a mid-air collision between their respective planes. Following a strike by sixteen Marine Corps SBD Dauntless aircraft on enemy installations on Sohana Island, and Bonis airstrip on Bougainville, they were returning to base. (Sohana is an islet in Buka Passage, separating Buka and Bougainville islands. The airstrip was located on a same-named (Sohana) narrow peninsula in northern Bougainville, south of Buka Passage.)⁹

The accident occurred at about 1,500 feet, approximately six miles off the Bougainville coast, due to the failure of Monbart's engine while flying above Oughton. His plane fell into Oughton's, striking the latter near its right auxiliary wing tank. The aircraft went into violent spins and crashed into the sea. The pilots bailed out, but their rear seat gunners, Sergeant H. M. Mateja and Staff Sergeant W. F. Rollins, were both unable to extricate themselves and went down with their aircraft.¹⁰

NEW ORDERS FOR *COOS BAY*

On 17 December, *Coos Bay* received a dispatch from commander, Fleet Air Wing, ordering her, when relieved by USS *Wright* at Rendova, to "proceed Treasury [Island], plant moorings for six planes, base, and operate Dumbo unit."[11]

Coos Bay did not depart Rendova Harbor until 28 December. Meanwhile, PBY-5 No. *30* took flight on Christmas Day for another Dumbo mission, the third one for plane and crew since 14 December. They were headed to Cape St. George, the southernmost point on the island of New Ireland. The headland was named for the Battle of Cape St. George, fought only five weeks earlier on 25 November 1943. On that date, five American destroyers of famed Destroyer Squadron 23, under the command of CAPT Arleigh Burke interdicted and almost completely destroyed three Japanese IJN destroyer-transports escorted by two destroyers. The sea battle in waters between Cape St. George, New Ireland, and Buka Island (off the north end of Bougainville Island) marked the end of Japanese resistance in the Solomon Islands.[12]

Map 7-3

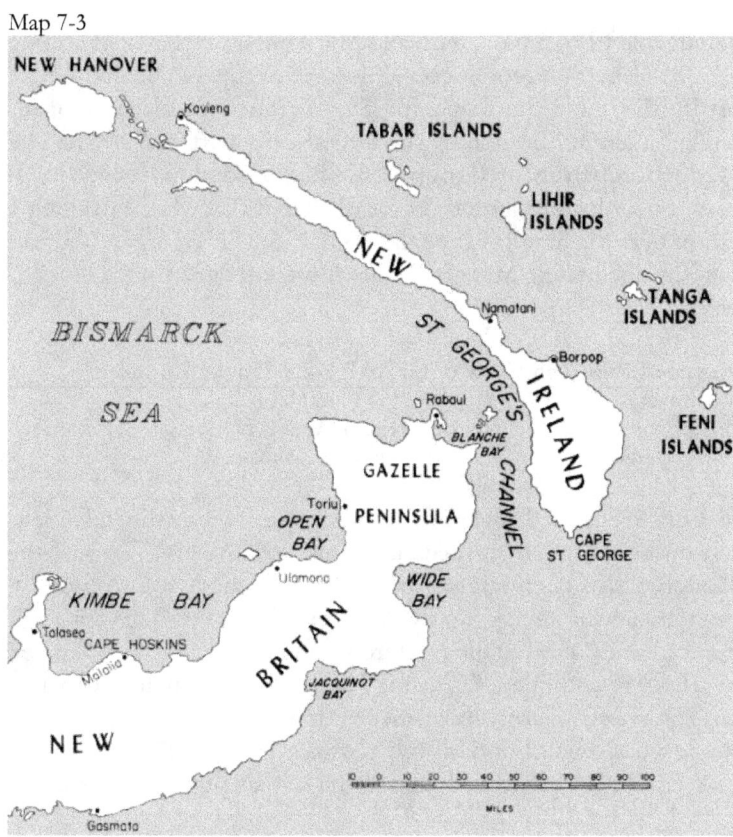

New Ireland and New Britain islands

RESCUE OF A B-25 MITCHELL BOMBER CREW

Photo 7-3

USAAF twin-engine B-25 Mitchell medium bomber, which crash landed on the sand bar of a river near Siokne, Fukien, China, 22 March 1945.

The indefatigable PBY-5 No. *30* on a Dumbo mission located a raft one mile south of Cape St. George containing six occupants, including 1st Lt Albert B. Marx, and the remaining five members of his crew of his B-25 Mitchell medium bomber. The aircraft was part of the 13th Air Force, 42nd Bombardment Group, 69th Bombardment Squadron. It had been shot up by anti-aircraft fire earlier that day while attacking a radar station near the lighthouse at the tip of Cape St. George. With his plane critically debilitated, Marx had been forced to ditch in St. George's Channel.[13]

Rescued Pilot and Crew of USAAF B-25 Mitchell Bomber	
1st Lt Albert B. Marx	Sgt P. A. Martin
2nd Lt Joseph R. Durkin	Sgt G. A. Paulson
2nd Lt W. A. Harden	Sgt P. A. Dumas

The Catalina landed and was able to rescue the Mitchell crew, despite heavy shore fire from machine guns and artillery. Two of the Mitchell aircrew had been injured, but all six had made it into the raft, from which they were picked up by Dumbo *I Love You*. (This apparent nickname for the PBY-5 Catalina, referenced in a magazine article dated February 21, 1944 and titled, "Catalina Kids Robbing Rabaul Reaper of American Flying Men," likely indicates that it was this same Dumbo that inspired a rescued aviator aboard the *Coos Bay* to pen FIFTY BAKER TWENTY EIGHT, a poem included in the book's preface.) If so, that Dumbo was good old PBY-5 No. *30*.[14]

8

Coos Bay Steps Forward to Blanche Harbor, Treasury

Map 8-1

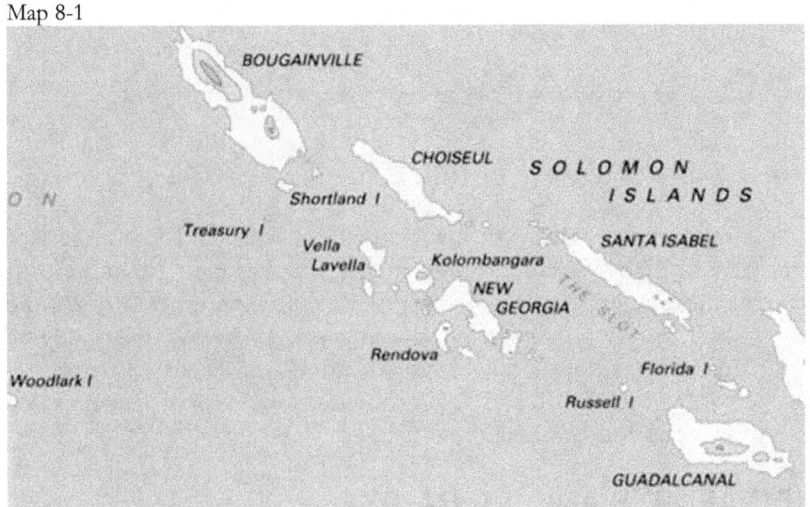

Central Solomon Islands

On 28 December, following relief by the *Wright*, *Coos Bay* moved forward to Blanche Harbor, Treasury Islands. This small group, comprised of Mono and Stirling islands, was commonly referred to as "Treasury." Allied occupation of the Treasury Islands two months earlier on 27 October, had been undertaken to facilitate construction of a radar station on Mono and use of Stirling as a staging area for the then pending assault on Bougainville. From Blanche Harbor, *Coos Bay* operated a Dumbo unit, comprising Patrol Squadrons VP-14, -71, and -91, and the Royal New Zealand Air Force No. 6 (Flying Boat) Squadron, during her four-and one-half-months at Treasury. Up to nine seaplanes were present in the seadrome at any given time. [1]

Map 8-2

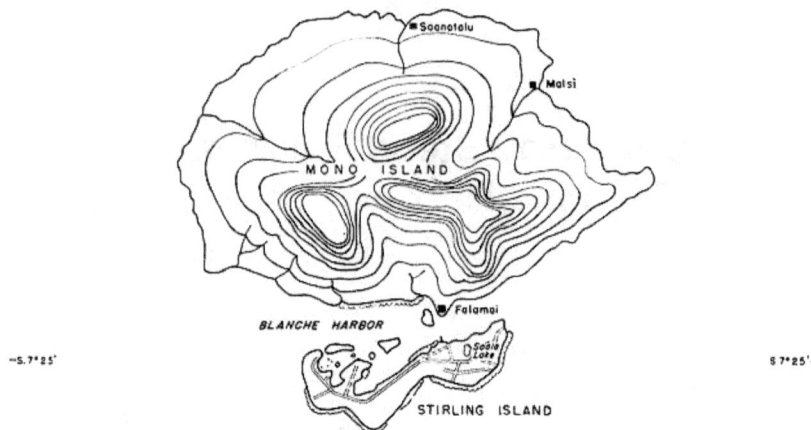

Treasury Islands in the Solomon Islands

Extensive air-sea rescue activity steadily increased the numbers of survivors brought aboard the ship during this tasking. In addition to the aircraft rescues, the *Coos Bay* made one rescue with her boats in the harbor at Treasury. A fighter plane plunged into the bay due to engine failure during take-off from Stirling Field. Plane crash and fire and rescue parties were called away (summoned), and the pilot was retrieved from the water and brought aboard.[2]

RED ALERTS AND AIR RAIDS

While at Treasury, *Coos Bay* was then the most forward-based surface ship in the Solomons Campaign and as such she was frequently subjected to enemy attacks. Red Alerts (warning of an impending air raid) and actual air raids were so common that .30- and .50-caliber aircraft machine guns were employed to supplement the ship's armament. During an enemy air raid, on 13 January 1944, she was straddled by bombs but not damaged. For her defense in this action, the island command credited her with one Japanese plane shot down. Three weeks later, on 3 February, two enemy aircraft carried out a surprise early morning attack, but the bombs again missed their target. This attack was followed by one on the night of the 5th. On this occasion, ship's radar proved its worth by detecting three planes which island shore-based radar had failed to identify.[3]

"Tokyo Rose" later asserted the *Coos Bay* had been sunk in Blanche Harbor. Interestingly, this pronouncement came just after the tender had been relieved by the *Chincoteague*, on 10 May 1944, and had departed the Treasury Islands. Tokyo Rose, whose real name was Iva Toguri, was an American-born Japanese woman who hosted a Japanese propaganda radio program titled "Zero Hour," whose purpose was to demoralize U.S. troops during World War II.[4]

RESCUES MADE BY *COOS BAY*-TENDED PBYS

The first rescue was made on 20 January 1944 by a VP-14 Catalina, captained by Lieutenant B. Smith, USNR. Patrol Squadron Fourteen was based aboard the *Wright*, with an advance echelon of four PBY-5s and five crews (rotating duty) operating from the *Coos Bay*. The *Wright* remained at Rendova until 17 January when, under orders from commander, Fleet Air Wing One, she moved to Hathorn Sound (north and west of Munda Airfield on New Georgia) to take advantage of the more protected seaplane landing area there. In January, VP-14 patrol planes from the *Wright* and *Coos Bay* flew thirty-four Dumbo missions, during which thirty survivors were picked up in open-sea rescues. Only the rescues made by VP-14 *Coos Bay*-tended aircraft are described herein.[5]

On 20 January, Smith had taken off from Treasury at 1300, with instructions to circle over Torokina, Bougainville Island, and await orders for a rescue mission. He arrived at Empress Augusta Bay on the western side of the island at 1420. A directive came an hour later to attempt the rescue of a B-25 bomber crew reported down in the water at 4°50'S, 152°40'E. Complying, he left the area with an escort of six P-39 Airacobra fighters after they joined him at this location. Fifteen minutes after arriving on station and after beginning a square search, the life raft was found thirteen miles southwest of the specified position. Despite a 12-knot wind, parallel to the 15-foot swells, Smith made a satisfactory landing in a trough between swells. The six survivors, two of whom were injured, were taken aboard and the Dumbo took off at 1700. The rescue plane left its fighter escort at Torokina and proceeded to Treasury, where the survivors were transferred to the *Coos Bay*.

USAAF B-25 Mitchell Bomber Crew
(70th Bombardment Squadron, 42nd Group, based at Russell Islands)

1st Lt John Edward Warner	Sgt George H. Purcell
2nd Lt Herbert Allen Hoffman	SSgt John C. McClure
2nd Lt Jack Bogart Schwartz	SSgt Henry King[6]

Two days later, VP-14 pilot LTJG O. H. Patterson, USNR, was orbiting over Torokina, when he received instructions from Fighter Command to attempt the rescue of pilots reported in the water at three different locations. At 1530, his Dumbo plane left Torokina with an escort of three P-39s to pick up a F4U Corsair pilot down fifteen miles southwest of Buku Passage. After an unsuccessful search, a report was received from a RNZAF plane of a man in the water, forty-seven miles west of Buku Passage. Finding a raft at that location, Patterson landed and picked up 2nd Lt E. M. Studley, USAAF, 339th Fighter Squadron, who had crashed the P-38 Lightning he was piloting.[7]

A report of this rescue was made to Fighter Command, which directed Patterson to search at the other positions. This was done, but only an empty life raft sighted. The Dumbo then returned to Treasury, and the rescued pilot was taken aboard *Coos Bay*.[8]

Following these initial Dumbo missions, others followed rapidly, as the USN Catalinas continued to pick up USAAF, Marine Corps, and Navy pilots and aircrewmen forced to parachute from or ditch aircraft on the water, owing to combat damage, failure of engines, or other critical components. Dye on the water's surface aided in the location of rafts and, on one occasion, survivors in a raft attracted the attention of a Dumbo plane by use of a smoke bomb. The twelve personnel rescued between 23 January and 9 February 1944 (from four different aviation squadrons) are identified in the table.

USAAF
339th Fighter
Squadron
Lockheed
P-38G Lightning
fighter aircraft

Marine Fighting
Squadron
VMF-215
Vought
F4U Corsair
fighter aircraft

Marine Torpedo
Bomber
Squadron
VMTB-143
Grumman
TBF Avenger
torpedo bomber

Navy Fighter
Squadron
VF-17
Vought
F4U-1 Corsair
fighter aircraft

Date	Squadron/ Pilot	Personnel Rescued	Reason Ditched, Location Rescued
23 Jan 44	VP-14 LT B. Smith	1stLt Eugene Victor Smith, USMCR	Engine trouble; life raft at 5°20'S, 154°25'E
	VP-14 LT B. Smith	1stLt Richard Marsh	Plane forced down; life raft at 4°33'S, 153°15'E
24 Jan 44	VP-14 LTJG O.H. Patterson	1st Lt Glen E. Hart, USAAF, 339th Fighter Squadron	Plane forced down; life raft 30 miles S. of Cape St. George
	VP-14 LTJG O.H. Patterson	Maj R.G. Owen, USMC, Commanding Officer VMF-215	Plane shot up; life raft 15 miles W. of Torokina

24 Jan 44	VP-14 LT B. Smith	1stLt Arthur H. Johnston, USMCR; Pvt Vernon H. Villalbos, USMC; Pvt Raymond J. Czarnecki, USMC, of VMTB-143	Plane forced down after hit by AA fire over Rabaul; life raft at 5°50'S, 153°30'E
27 Jan 44	VP-14 LTJG F. B. Thorn, USNR	LTJG Howard M. Burris, USNR, VF-17	F4U forced down off Lakunai due to enemy action; life raft at 5°30'S, 153°30'E
28 Jan 44	VP-14 LT B. Smith	2nd Lt Thomas A. Benim, USAAF; 2nd Lt William H. Thomas, USAAF	Both from 339th Fighter Squadron; life raft at 4°55'S, 153°00'E
5 Feb 44	VP-14 LTJG R. R. McQuaid, USNR	2nd Lt Kenneth I. McCloud, USAAF, 339th Fighter Squadron	Plane shot down in St. George Channel on 28 January; life raft at 5°30'S, 154°00'E
9 Feb 44	VP-14 LT R. D. Oakley, USNR	PFC C. W. Vorachek, USMC, VMTB-143	Plane in which he was a turret gunner shot down on 28 January; life raft at 5°40'S, 154°20'E[9]

The rescue by Lieutenant Oakley on 9 February, was the last by the VP-14 advanced echelon staging from *Coos Bay*. Following the inception on 6 February of a new search plan for the South Pacific area, designed to provide maximum coverage for the Green Island landing operations, VP-14 was assigned patrol plane duties in addition to Dumbo rescue and all-night coverage missions. To ensure adequate aircraft availability, the Catalinas and crews with *Coos Bay* were withdrawn to the *Wright*, operating at Hathorn Sound, New Georgia. The reunited squadron (61 officers, 113 enlisted men, and twelve PBY-5 planes) was initially assigned one 650-mile sector to search. On 10 February, this was increased to two sectors and two days later, to three sectors.[10]

The VP-14 Catalinas withdrawn from *Coos Bay*, were replaced by ones from VP-71 and RNZAF No. 6 Squadron.

PATROL SQUADRON 71 AND RNZAF NO. 6 (FLYING BOAT) SQUADRON COMMENCE DUMBO MISSIONS

Patrol Squadron VP-71 was resourced with fifteen PBY-5 aircraft and eighteen flying crews at Halavo Seaplane Base. On 9 February 1944, with the transfer of two Catalinas and three crews to Blanche Harbor, where *Coos Bay* was then located, it began staging Dumbo missions through this forward position, from the squadron. Two aircraft from

RNZAF No. 6 Squadron also left Halavo that day to form a detached flight group operating with *Coos Bay*.[11]

The chain of command for these squadrons, others in the area, and supporting seaplane tenders is identified below:

- Commander, Third Fleet/South Pacific Force:
  ADM William F. Halsey Jr., USN
- Commander, Aircraft, Third Fleet (Task Force 33):
  VADM Aubrey W. Fitch, USN
- Commander, Aircraft, Solomon Islands (Task Group 33.1):
  MajGen Ralph J. Mitchell, USMC
- Commander, Air Munda (Task Unit 33.1.1):
  RADM Harry S. Kendall, USN
- Commander, Air Operations Henderson Field (Guadalcanal):
  Wing Comdr. Ivan Evan Rawnsley, RNZAF[12]

A few days prior to the transfer, Aircraft of VP-71 and No. 6 Squadron had worked together. On 1 February, LCDR Cecil Kelly Harper, and LT J. F. Jones in USN Catalinas, along with one from the RNZAF, spotted gunfire for destroyers shelling enemy targets on Choiseul Island. After dropping smoke lights to mark targets in the jungle, and strafing them with .30- and .50-caliber machine gun fire, the aircraft took up their duties over Choiseul Bay. Their voice reports to the USS *Waller* (DD-466), *Wadsworth* (DD-516), and *Halford* (DD-480) on the fall of their gun rounds enabled the ships to "walk" them onto the targets.[13]

During a Dumbo mission from the *Coos Bay* on 10 February, LT W. J. Johnson, USNR piloting a PBY-5 from Patrol Squadron VP-71, made a landing in St. George's Channel, ten miles south of Cape Gazelle, to rescue a man in a raft believed to be a friendly fighter pilot. Closer inspection revealed the survivor to be Japanese. He would not permit himself to be taken prisoner, and a fighter flying cover strafed the raft killing him.[14]

Four days later, Patrol Squadron VP-71 transferred five planes and crews temporarily from Halavo Seaplane Base to the *Wright* at Hathorn Sound, to conduct rescue and supply missions and long-range strategic searches from the nearby airfield at Munda. The Catalinas and crews on loan returned to Halavo on 19 February.[15]

Photo 8-1

Munda airfield, New Georgia Island, Solomons, circa 1943-1944.
National Archives photograph 80-G-K-1455

RNZAF NO. 6 SQUADRON AIR-SEA RESCUES

On 12 February 1944, a Royal New Zealand Air Force PBY-5 piloted by Flying Officer J. A. Hendry, rescued the occupants of a life raft at 5°54'S, 154°29'E. The six men picked up were the crew of a USAAF B-25 Mitchell (70th Bomber Squadron, 42nd Bomber Group). The bomber had been forced to ditch after being hit by anti-aircraft fire while over Vunakanau, a former RAAF airfield on New Britain's Gazelle Peninsula, ten miles northeast of Rabaul.

USAAF B-25 Mitchell Bomber Crew	
1st Lt John H. Vandchaick	SSgt Earl D. Kirkpatrick
2nd Lt James G. McClure	SSgt John F. Howell
2nd Lt Ralph C. Little	PFC Louis C. Langelies[16]

The following day, Flight Lieutenant J. R. McGrane, RNZAF, rescued 2nd Lt Martin H. Hannum, USAAF, after he had made a forced water landing. The Catalina found the pilot's life raft at coordinates 5°8'S, 153°40'E. On 16 February, Flying Officer J. A. Hendry rescued one of his countrymen, Sergeant H. P. Crump of the RNZAF 14th Fighter Squadron, who had put his Curtiss Kittyhawk into the sea after fuel pump failure.[17]

On 18 February, two Catalinas of the Squadron both made rescues at sea. Flying Officer G. C. Hitchcock, RNZAF, recovered USAAF Sergeants Lester A. Steinberg and Charles E. Cowan from a dinghy. Their B-25 bomber had crashed, after being hit by anti-aircraft fire. Flight Lieutenant McGrane rescued 2ndLt Jack G. Morris, USMCR, from a life raft at 4°53'S, 152°45'E. He had spent considerable time afloat, after his F4U Corsair (Marine Fighting Squadron VMF-218) had been shot down over Gazelle Peninsula nine days earlier.[18]

VP-71/RNZAF NO. 6 SQUADRONS WORK TOGETHER

As tabulated below, many other flyers and aircrewmen were plucked from the sea by VP-71 and RNZAF No. 6 Squadron Catalinas during their remaining time operating from the *Coos Bay* at Treasury. Long-range searches by VP-71 at Halavo were discontinued on 1 March, and the squadron was transferred from Fleet Air Wing One in the Solomon Islands to Fleet Air Wing Two in the Hawaiian Islands, and then to Fleet Air West Coast at San Diego, California. The staging of Dumbo planes through the *Coos Bay* was discontinued on 3 April.[19]

VP-71 and RNZAF-6 Rescues (20 February-31 March 1944)

Date	Squadron/ Pilot	Personnel Rescued	Reason Ditched, Location Rescued
20 Feb 44	VP-71 LTJG W.R. Martin, USNR	1st Lt E.G. Keefer, 1st Lt R.C. Thompson, 1st Lt F.E. Fredenburg, SSgt E.E. D'Amico, SSgt C.P. Creech, Sgt Mandel Sandell	B-25 hit by AA fire over Lakunai; life raft at 4°50'S, 152°50'E; all USAAF, 75th Bomber Squadron, 42nd Bomber Group
	VP-71 LTJG W.R. Martin	Lt R. A. Schaeffer, USMCR, VMF-222	F4U shot down the day before; life raft at 4°42'S, 152°30'E
21 Feb 44	RNZAF-6 FO J.A. Hendry	Maj H. A. Pehl, USMCR, VMF-218	Compelled to make a water landing; life raft at 6°00'S, 154°30'E
	RNZAF-6 FO J.A. Hendry	1stLt W.J. O'Brien, USMC, VMF-216	Water landing earlier in the day; life raft 8 miles SSW of Cape St. George
22 Feb 44	RNZAF-6 FO G.C. Hitchcock	Capt C.E. Eaton, USMCR, and Cpl R.E. Van Pelten, USMCR VMSB-244	Forced water landing; life raft at 5°40'S, 154°28'E;
26 Feb 44	VP-71 LT A.L. Mills, USNR	Capt M.H. Porterfield, USMCR, and Sgt C.D. Faella, USMCR, VMSB-244	Hit by AA fire over Rapopo and forced to make a water landing; raft at 4°25'S, 152°28'E
5 Mar 44	RNZAF-6 Ft/Sgt. J.B. Monk	2nd Lt John H. Bauwens, TSgt Clarence P. Marcy, Sgt Martin Lund, USAAF, 31st Bomb Squadron	B-24J exploded and crashed in water after being hit by AA fire over Tobera; life jackets at 4°25'S, 152°25'E

8 Mar 44	RNZAF-6 FO D.S. Beauchamp	Capt. E.M. Shanks, 1st Lt J.V. Fairley, 2nd Lt A.F. Alagna, 2nd Lt C.L. Johnston, SSgt N.A. Gidley, SSgt R.G. Akers, SSgt D.D. Bergman	B-25 hit by AA fire while retiring from mission over Rabaul; life rafts at 4°56'S, 152°53'E; all USAAF, 106th Reconnaissance Squadron, 42nd Bombardment Group
16 Mar 44	RNZAF-6 FL H.T. Francis	1stLt Joseph George Maraist, USMCR	F4U shot down while strafing a Japanese barge 5 miles S. of Feni Island; raft at 4°17'S, 153°44'E
31 Mar 44	RNZAF-6 FL R.L. Scott	LT Russel P. Lecklider, USNR; ACRM Arthur L. DeMott; AMM1c Joseph D. Zollinhofer	TBF Avenger (VC-40) forced down after strike on Lakunai airstrip; raft in St. George Channel[20]

PATROL SQUADRON 91 PLANES/CREWS REPORT

In April 1944, five Catalinas from Patrol Squadron VP-91 based at Halavo Seaplane Base arrived at Treasury to operate from *Coos Bay*. Halavo was home to the squadron, however, with five planes dispersed at Treasury, three at Green Island, and two at Emirau Island, it served only as a maintenance and upkeep base. The Squadron aircraft at advance bases, primarily employed for rescue and general utility work, were under the local control of commander, Aircraft Solomons.[21]

During April, VP-91 aircraft made fourteen open sea landings on Dumbo missions. Thirty-eight personnel were rescued, including eleven non-US Allied military. All rescues were made in the Rabaul, New Britain, Kavieng, New Ireland, and Shortland Island areas.[22]

Dumbo rescue aircraft expended extraordinary efforts, sometimes even rising to heroic levels, to retrieve airmen from the sea. Pilots and aircrew of combat aircraft shot or forced down by damage sometimes had to prevail in life and death struggles to escape from their aircraft and survive in the sea until, hopefully, help would arrive. The following account describes one such incident.

PV-1 VENTURA SURVIVORS RESCUED UNDER FIRE

On 14 April, LT E. F. Frazier of Patrol Squadron VP-71 rescued pilot LT William Thomas Henderson, ENS Billy Robran, and ARM3c (Aviation Radioman Third) Thomas Phillip Humphrey from a life raft about four miles east of Ballale Island. They were surviving members of the five-man crew of a PV-1 Ventura from Bombing Squadron VB-148. The squadron was based at Munda, New Georgia, providing escort for C-47s carrying paratroops into New Guinea, and was under the local operational control of commander, Fleet Air Wing One.[23]

Lockheed PV-1 Ventura patrol bombers, circa early 1945. National Archives photograph #80-G-331264

The plane had taken off from Munda at 0825 that morning for a "target of opportunity search" of the coast of Bougainville Island. The pilot proceeded to the southern entrance channel between Shortland and Pauro Island, flew northward east of Ballale Island, and then turning toward Kahili, he intended to follow the Kara road northwest toward Mutupina Point on the western coast of Bougainville. South of Kahili, in Moisuru Harbor on the southern coast, he saw what appeared to be active shipping. Closer investigation revealed only hulks and wrecks, so he headed back toward Kahili.[24]

Just west of Kahili, while flying at about 200 knots at an altitude of approximately 400 feet, the plane came under enemy fire and was hit by anti-aircraft fire in the after section of the plane. The first burst entered the camera hatch and hit the turret, the second hit the starboard cabin fuel tank, setting in on fire. Two aircrew, AMM2c John Raymond Elkey and AOM3c Norman Edwin Wood, were severely wounded by shrapnel and by light machine gun fire that accompanied the anti-aircraft fire.[25]

The copilot, Ensign Robran, made his way to the rear of the plane to administer first aid, as Henderson turned seaward to retire from the area. VB-148's war diary describes the difficulties encountered before he was finally able to set the plane down on the sea, aflame, propelled by a single engine, and losing altitude at a high rate:

> [Henderson] started to climb, and at that time his port engine started to race and the oil pressure dropped. The engine started to miss, and when it was found that it would not function properly, Henderson feathered the propeller. The engine finally cut out altogether, and the pilot started single engine operation. At this point, the plane was at about 120 feet. He attempted to maintain

his altitude by dropping his bombs on Maifu Island, but kept losing altitude, probably because of the suspected failure of the bomb-bay doors to close. As he was dropping rapidly, Henderson decided to make a water landing immediately and notified the crew to that effect. At 0951, by Robran's watch, Henderson put his flaps full down and made a water landing, the plane falling off on the port wing slightly. They struck the water at a point approximately 4 miles east of Ballale Island. The plane burst into flames and sank in 7 to 10 seconds.[26]

Henderson and Humphrey exited the plane via the pilots escape hatch, and Robran through the cabin door. It is unknown how Elkey excited, but he managed somehow. Henderson and Humphrey swam upwind, clearing the flames, where they saw Robran on the downwind side and Elkey in the midst of the smoke and flames. Henderson went to Elkey's side and was able to get him clear. The four survivors, once together, worked to remain afloat with their Mae Wests, which provided little buoyancy. No life rafts had been deployed. The door raft had not released, and the flames in the plane had been so intense that Robran had been unable to retrieve any of the other rafts. All of the survivors removed their clothing to facilitate swimming, but it was difficult to keep Elkey afloat due to his injuries, because he did not have a life jacket. Humphrey put his Mae West on Elkey.[27]

At 1000, a New Zealand PV-1 Ventura bomber dropped two life rafts which the survivors inflated and were ultimately able to struggle aboard. Elkey was by this time delirious, and almost helpless. About an hour later, a USN Catalina piloted by Lieutenant Frazier (summoned by the PV-1) landed and picked up three of the survivors. It was not possible to get Elkey into the plane, owing to his inability to help himself, and intense fire originating from Japanese shore batteries on Ballale, Nusave, and Nusakon Islands. A short time later, another US PBY, returning from patrol, sighted Elkey in the raft and landed to pick him up. Finding the raft's occupant deceased, and considering the severity of shore battery fire, it was inadvisable to remain stationary while attempting to recover the body, and the plane took flight.[28]

The survivors were taken to the *Coos Bay*, moored off Stirling Island at Treasury, where they received medical attention. All were suffering from second degree burns. Henderson was allowed to return to Munda on 15 April. Robran and Humphrey were transferred to the base hospital on the Stirling airfield for further treatment.[29]

The chapter closes with a summary of rescues by VP-91 Catalina rescues while staging from the *Coos Bay*.

VP-91 Rescues (2-25 April 1944)

Date	Squadron/Pilot	Personnel Rescued	Reason Ditched, Location Rescued
2 Apr 44	VP-91 LTJG J. A. Hayes	Capt J.E. Congrove, USMCR, VMF-114	F4U forced down after hit by AA fire over Cape St. George; life raft 5 miles E. of Cape St. George
5 Apr 44	VP-91 LTJG R. C. Woolery	FO A.F. Tucker, RNZAF, Fighter Squadron 16	Aircraft hit by AA fire over Kieta Peninsula, Bougainville; life raft 15 miles E-NE from Nussi Point, Bougainville
12 Apr 44	VP-91 ENS W. C. Cook	1st Lt R.W. Read, 1st Lt P.H. Watts, 2nd Lt W.W. Carlisle, SSgt W.N. Price, SSgt A.C. Valentin, Cpl W.G. Johnson, Cpl C.H. Cook, all USAAF, 69th Bombardment Sqd.	B-25 forced down after hit by AA fire over Rabaul on day of rescue; life raft 4°17'S, 152°27'E
14 Apr 44	VP-91	Identified in preceding written account	Described in preceding written account
17 Apr 44	VP-91 LT D. G. McLaren	1stLt F.E. Lee, Sgt E.M. Perry, PFC R.A. Sowers, all USMC, VMTB-242	TBF hit by AA fire over Lakunai airdrome, Rabaul; life raft 5°27'S, 153°58'E
18 Apr 44	VP-91 ENS W. C. Cook	1stLt Donald W. Smith, Cpl Horace W. Camp, Cpl Roland D. Chiaro, all USMC, VMTB-134	TBF forced down after hit by AA fire over New Ireland earlier that day; life raft 4°50'S, 153°15'E
25 Apr 44	VP-91 LT J. E. McLaughlin	Capt Robert W. Seely, USMC; VMSB-235	SBD-5 forced down after hit by AA fire over Rabaul earlier that day[30]

COOS BAY DISPATCHED TO GREEN ISLAND

On 10 May 1944, after being relieved by the *Chincoteague* at Treasury, *Coos Bay* got under way for Green Island. She arrived at the lagoon there the following day, and assumed control of the seaplane base supporting Dumbo and anti-submarine missions.[31]

9

Capture of Green and Emirau Islands

Photo 9-1

Tank landing ship USS *LST-446* beached at Nissan Island, during the Invasion of the Green Islands, 15 February 1944. Local people are watching from the foreground. National Archives photograph #USMC 75194

Following establishment of the beachhead at Torokina, Bougainville, Admiral Halsey had planned additional actions necessary to complete the conquest of the Northern Solomons in preparation for the neutralization of Rabaul on New Britain, but the Joint Chiefs of Staff decided that South Pacific forces would next invade and neutralize Kavieng, a Japanese stronghold on New Ireland in the Bismarcks. However, other prospective operations in the Central Pacific (Marshall, Caroline, and Mariana islands) appeared to preclude the availability of necessary fleet support for the proposed landings which were then delayed until approximately 1 May 1944.[2]

In support of these landings, Halsey desired shore-based fighter cover over the Kavieng area, and shore-based air strikes preliminary to an amphibious assault. This was because carrier air support was not available. As Kavieng was too far for planes operating from the Allies forward-most airstrip at Torokina, it was necessary to seize intermediate land objectives. These areas would have to be suitable for landing-strips, and be within range of planes operating from Torokina in order that a continuous land-based fighter cover could be maintained over the beachhead as well as the supply lines.[3]

Map 9-1

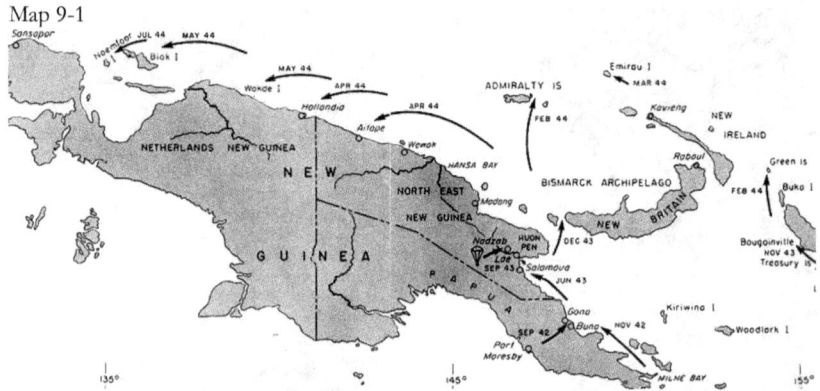

MacArthur and Halsey's Advancement. After defending Port Moresby, Papua, from a Japanese invasion, and establishing a base at Milne Bay, MacArthur's Australian and American forces advanced northwestward up the east coast of Papua, New Guinea along his road back to the Philippines. Halsey's forces moved northwestward up from Treasury and Bougainville to Green and Emirau Islands (upper right of map) (http://www.history.army.mil/books/AMH/Map23-43.jpg)

On 20 December 1943, at a conference between members of Admiral Halsey's staff and General MacArthur, it was suggested that the Green Islands be seized first, and, later on, islands of the St. Matthias Group. MacArthur made it clear that he did not intend to take Manus Island (a priority in his planned advance through the Admiralty Islands and return to the Philippines), until such time as South Pacific Forces were prepared to undertake operations against Kavieng or the St. Matthias Group. MacArthur suggested that Admiral Halsey could, in the interim, land in the Green Islands or engage in any offensive move that he saw fit that would meet the air cover objectives. Lying between Bougainville and New Ireland, crescent-shaped Green Island (Nissan) was the largest of the Green Islands group. Narrow passages between small islands guarding its western side opened up into a large, sheltered lagoon.[4]

Map 9-2

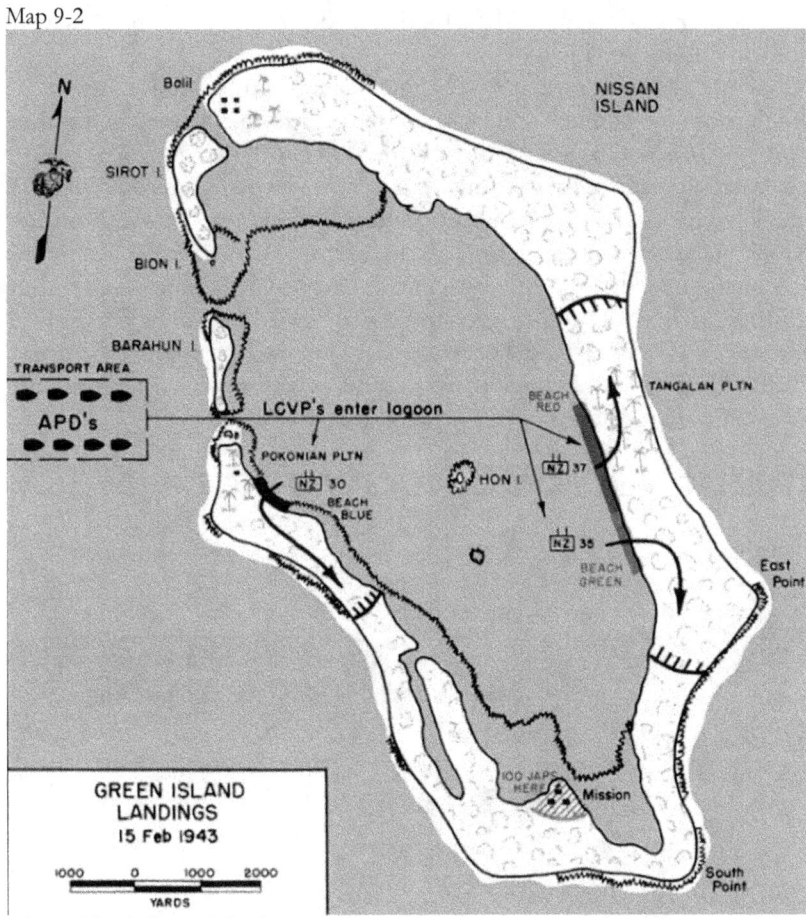

Nissan Island, Green Islands
www.ibiblio.org/hyperwar/USMC/USMC-M-NSols/maps/USMC-M-NSol-27.jpg

The seizure of Green Island fell to the 3rd New Zealand Division less the 8th Brigade under Maj. Gen. Harold E. Barrowclough with the USN providing transport and landing craft in the form of RADM Theodore S. Wilkinson's Task Force 31. At dawn on 15 February 1944, the troops landed on Green Island. As landing craft headed for the beach, F4U Corsairs of Marine Squadrons VMF-212 and VMF-216 patrolled the skies overhead. At 0645, about fifteen Japanese dive bombers were sighted setting up for an attack on the shipping lying offshore, but fighters pounced. First Lieutenant Phillip C. DeLong destroyed three Vals in quick succession, and Capt William C. Carlton, 1st Lt Theodore J. Horner, and 1st Lt Thaddeus J. Trojnar accounted for another one each. VMF-216 drove off the other attackers with bombs still in their racks.⁵

Third Division assault forces landed successfully on beaches inside the lagoon as planned and, within two hours, 5,800 men with their supplies and equipment were ashore. The following day, the New Zealanders, while moving southward from their positions, discovered a small Japanese garrison of about seventy troops on the south end of Nissan Island. On 19 February, operations there culminated in the elimination of the enemy resistance. Sixty-two dead Japanese were counted, while New Zealand forces sustained three killed and 11 wounded. Green Island became a forward base for the U.S. South Pacific Combat Air Transport Command, which supplied material and mail to combat soldiers and evacuated the wounded.⁶

4TH MARINES TAKE EMIRAU WITH NO OPPOSITION

Photo 9-2

Invasion forces off Emirau Island, as Marines land there unopposed on 20 March 1944. Landing craft mill about an attack transport, as a destroyer stands guard farther offshore. National Archives photograph #80-G-225254

Five weeks later on 20 March, the 4th Marine Division came ashore at Emirau, in the St. Matthias Group. The irregularly-shaped island lay north of Green Island, and only seventy-five miles northwest of the

New Ireland enemy fortress of Kavieng. Commanders in charge determined that a proposed invasion of Kavieng by the 3rd Marine Division, would be too costly in likely numbers of casualties suffered. It was determined that it was better to isolate Kavieng and let it "die on the vine" especially at this stage. Allied planes and ships operating from air and naval bases on Emirau could now deny Japanese forces, in the Solomons and Bismarck Archipelago, access to critical facilities and support previously available at Kavieng.[7]

Photo 9-3

ADM William F. Halsey Jr., commander Third Fleet, stands beside his plane at Emirau Island, 25 May 1944. The casual appearance and demeanor of the Marines, sailors and soldiers looking on suggest that this was an unannounced inspection tour of the airbase. U.S. Marine Corps photograph #USMC 81365

Continued Allied fighter and bomber attacks on Kavieng and Rabaul to ensure they "withered on the vine," came with a cost: aircraft and crews lost to AA shore battery fire. Dumbos saved as many as they could. In doing so, one PBY pilot earned the Medal of Honor.

10

VP-34 "Black Cat" Catalina Pilot Earns Medal of Honor

I just couldn't leave 'em back there.... I knew I had to go back and try.... The Japanese at that time were killing most of the flyers that they got out of the water.... I thought about that too. I have no doubt but what all these boys would have been killed if they hadn't been able to get out of there.... You have to do it. At least make an effort.

—Response by Nathan Green Gordon when asked by an interviewer, "Why did you feel it was so important to risk the lives of the men on board and also the people you already [saved] to go back for the last six?" This question refers to his actions in returning to Kavieng Harbor, New Ireland, and landing under heavy fire after learning about other downed airmen there.[1]

Photo 10-1

LT Nathan G. Gordon, USNR, shakes hands with VADM Thomas C. Kinkaid, USN (commander, Seventh Fleet) after Gordon was presented the Medal of Honor at Seventh Fleet Headquarters on 19 August 1944.
Naval History and Heritage Command photograph #NH 106206

"BLACK CATS" AIR-SEA RESCUE OPERATIONS

Patrol Squadron VP-34

- Established on 16 April 1942
- Redesignated as Patrol Bombing Squadron VPB-34 on 1 October 1944
- Disestablished on 7 April 1945

The PBY Catalina aircraft of some U.S. Navy patrol squadrons in World War II were painted flat black to help avoid detection by Japanese forces during nighttime armed reconnaissance anti-shipping patrols. These patrols were termed "Black Cat" operations, and their parent units "Black Cat" Squadrons. Some of these squadrons also performed air-sea rescue operations, one being Patrol Squadron VP-34.

VP-34 engaged mostly in action against enemy surface forces and shore installations. However, as noted in the squadron's war history:

> Air-sea rescue and evacuation missions and transportation of scouts, irregulars, and supplies, provided the most varied type of work and was frequently as interesting and satisfying as "Black Catting." These operations were conducted from Langemak Bay and Hollandia from February 12, 1944 to May 18, 1944 under various Type Group Commanders; and from San Pedro Bay, Leyte, Philippine Islands, from November 23, 1944 to December 22, 1944 under CTG 73.5.[2]

During its first period of rescue and related operations, VP-34 aircraft picked up a total of seventy-seven Army, Navy, and Australian flight personnel in open sea landings. All but one of these servicemen were retrieved near enemy bases or in waters regularly patrolled by enemy aircraft. Rescues were made in waters off New Guinea's north coast—at Hansa Bay, Wewak, and near Alexishaven—and in the harbor at Kavieng, New Ireland.[3]

Patrol Squadron Thirty-four picked up two hundred and four Army and Navy personnel during the second period of rescue and related operations. Like in the first period, the retrievals were near enemy bases or in waters regularly patrolled by Japanese aircraft. Rescues were made near Cebu Island in the Camotes Sea, near Ormoc town, off the Bondoc Peninsula on Luzon, and in other locations in the Philippine Archipelago.[4]

RESCUE OPERATIONS AT KAVIENG HARBOR

On 15 February 1944, in support of U.S. landings on Nissan (Green) Island, a large number of USAAF B-25 Mitchells took part in an attack on Kavieng, New Ireland. Aircraft of the 345th Bombardment Group struck the town, harbor, and shipping. Situated on the northwestern tip of New Ireland, Kavieng was "target rich" as an important logistical staging base for Japanese installations in New Guinea and the Bismarck Archipelago. It served as a major supply depot and boasted an adjacent harbor, an airfield, and an aircraft assembly facility.[5]

Map 10-1

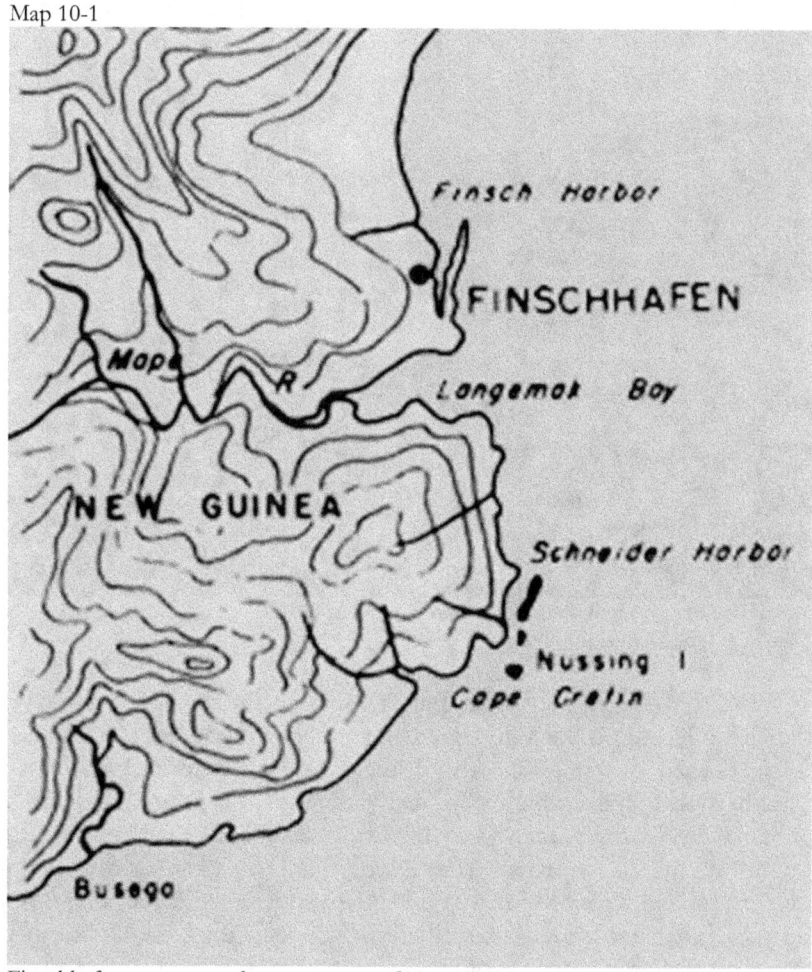

Finschhafen area on northeastern coast of New Guinea
https://www.ibiblio.org/hyperwar/USN/Building_Bases/maps/bases2-p293.jpg

That morning at Langemak Bay (located south of Finschhafen, on the northeast coast of New Guinea), Navy LT Nathan Gordon was busy making flight preparations for a Dumbo mission. Gordon, who hailed from Morrilton, Arkansas, had flown many "Black Cat" missions with Patrol Squadron VP-34. Today, he and the crew of PBY-5 Catalina No. 08434 (*Arkansas Traveler*) would standby the bombing mission. Their tasking was to orbit off Kavieng, to provide search and rescue cover for the strike on the heavily defended Japanese base.[6]

Photo 10-2

Aerial view of Langemak Bay, New Guinea, in April 1944.
Australian War Memorial photograph 071953

Taking off from Langemak Bay at 0823, the Catalina pilot first flew to Narega Island (in the Vitu Islands Group, north of New Britain). He rendezvoused ten miles north of the island with four Army P-47 Thunderbolt fighters which were to fly cover for his flying boat. (The pilots of the Army planes were 1st Lt. T. S. Wood, Lt. Maxwell, Lt. Frankfort, and Lt. Overby.) After a long flight to a position off the tip of New Ireland, Gordon idled at 2,000 feet and waited for a call for his services. Decades later, he described the scope of the USAAF bomber attack on the large Japanese base:

I have no idea how many planes were in the group, but there were three waves of 'em making low-level bombing runs, trying to destroy all the supplies there and everything else on that island.[7]

Gordon's call sign was "Gardenia 6." Almost immediately after arriving in position, a report came from one of the planes returning from the strike on Kavieng, that another one was down. Wood (the leader of the P-47 Flight flying cover, who heard the call on his radio) could not find the reported position on his grid map. Gordon also did not believe the coordinates provided were correct because, based on the flight path of the bombers, the plane would have been 45 degrees off course when it went down. Nevertheless, Gordon investigated the location, and found only empty sea. If it was the crash site, there should have been some tell-tale yellow-orange dye that downed aviators used to mark their positions, or oil on the water.[8]

Finishing this search, the Catalina received a report from another Army pilot of a plane down in Kavieng Harbor, with an offer to lead them to the crash site. The harbor was quite large and extended along the coastline for miles, not a protective place one might imagine.[9]

Photo 10-3

U.S. Navy carrier aircraft strike Japanese shipping at Kavieng, New Ireland, 25 December 1943.
Naval History and Heritage Command photograph #NH 91197

Led to the area where the plane had gone down, Gordon sighted a lot of oil on the water, and a lot of dye marker. There was also a large life raft capable of carrying six men and two life jackets below, but he could not determine whether or not there were any survivors, because of the adverse conditions existing at the time:

> We were in a real rough sea—the swells were some 15 to 18 feet. And it's difficult to see things in the ocean, particularly heads or something like that, even in a smooth sea; and it was a rough sea—why it's terrible hard. So anyway, I couldn't tell whether there was somebody there or not; they could have been hiding under the life jackets, which sometime they would do, if they thought they were enemy planes coming for them. Or they could have been hiding under that boat.[10]

It was essential that Gordon set his flying boat down with the bow high and the stern touching the surface first, to employ water resistance to slow the speed of the Cat and improve the accuracy and safety of the touchdown. Successfully pulling off this maneuver would be very difficult because of the necessity to land down slope of the big swells to avoid plowing into a wall of seawater.[11]

> I decided, I have to check it out, so I then started off. I dropped a smoke bomb where a plane was supposed to be, so I would have a place to orient myself by. Flew about a half mile and dropped another smoke bomb, and then flew another half mile and made my turn so I could come back in and make an approach for the landing, so I could get lined up on it 'cause, the swells as large as they were, I never would have found that plane again.

> I turned around and came back then, and I had my smoke bombs where I could orient myself.... In real rough seas like this was, it's a pretty delicate landing—you've got to keep your nose up; don't let it drop in, but be sure you don't let a wing drop.... Ordinarily you'd probably make a full-stall landing, but in a full-stall landing, you just cut your power off, and you just hold it right above the water until you drop out. And there's no way in the world to predict where you're going to land. And I knew I couldn't do that, 'cause I didn't want to get a long way from...the accident site. So, we did what's called a power-stall—you keep enough power so you don't drop out; and you can tell when you're about to lose the flight. So, you just keep it just a knot or two above that. And then, when you see the place you want to stop, you just pull the stick back - you should just drop like that, which I did.

But unfortunately, my vision wasn't very good because I had my nose up like this, and I was looking out the side of the window and couldn't see very much. And the waves and everything was so high, I overshot it some, so I dropped in a little bit high, which isn't unusual, I might say. And I busted a bunch of rivets in the bottom of the plane, which, again, isn't a serious problem, 'cause you can bucket the water out, you know, if you don't get too much. So, I landed and went by there, and we couldn't find anything. The dye markers and everything, even the jackets and the life boat were still there, but there wasn't anybody in it. So, all of them had already died, drowned, before we got there.[12]

RESCUE OF B-25 BOMBER CREW SURVIVORS

Gordon immediately took off. He had barely got airborne again when a plane (he believed the same one as before) reported, "I have another plane down and [can] lead you to it." The Catalina's position was then about a mile and a half offshore. At the site was a raft carrying three people.[13]

Gordon landed without any problems, and taxied toward the raft. The PBY crew threw a line to the water-soaked aviators, which they fortunately caught on the first attempt. However, he found that with his engine power as low as it could go, there was still movement of the Catalina through the water. The standard rescue procedure was for the plane to be stationary and to let the lighter object, in this case a raft, come to the plane or be hauled there.

> I said, "What's wrong, back there?" And they said, "Well, Nathan you're going to have to cut your engines." They said, "We can't pull the [raft] in." Of course, all of them [survivors] were injured, and they couldn't help you much, so anyway, he says, "You're gonna have to cut your engines." You don't like to do that, cause lots of times they don't start, for many reasons. Anyway, I called in my mechanic who was Wally. I said, "Wally, if we cut these engines, are they going to start for us again?" He said, "I believe so." That wasn't too much help, anyway. In order to do it, I had to cut my engines, which I did. And they got the men in, and there was three of them—pretty bunged up, but not real, real serious.[14]

The survivors of the downed B-25 bomber (nicknamed *Pissonit*) from the 71st Squadron were pilots 1st Lt. Eugene Benson and 2nd Lt. William J. Smith, and Hollie Rushing, the navigator. The radioman Claude Healan and tail gunner Albert Gross were casualties and their bodies were not recovered. Healan had apparently parachuted out of the aircraft at too low an altitude and perished. It was believed that

Gross had been thrown forward and down when the plane impacted the water, been swept or suctioned out through on an entry hatch on the bottom of the fuselage just behind the cockpit, forced open by the crash, and drowned.[15]

AIR-SEA RESCUE CLOSER TO SHORE BATTERIES

The Catalina took off and had hardly become airborne, when the same Army plane said, "I've got another plane down for you," and led the flying boat to the new site, which was considerably closer, less than a mile from the shore. On landing, the PBY encountered a lot of flak coming from shore guns, but it was not hit. Gordon and his crew retrieved six downed airmen from two rafts tied together. In spite of a challenging rough-sea take-off prefaced by a scary situation when the flooded starboard engine failed initially to start, the plane with the rescued airmen managed to take off. Gordon did not know before leaving the area that there were, or would soon be, other airmen down in the Kavieng Harbor.

> They didn't hit us. So, I went over there, and there were six men down in this raft. So, we made the same kind of approach and landed, picked them up without any real problems. Again, like I say, it's kind of difficult to get 'em out of lifeboats on to our plane in rough water, but they got 'em aboard. Then we started to take off again, and I started my port engine; it fired real good. And then I tried to start my [starboard] engine, and I couldn't get it started. I fooled around with it for 2 or 3 minutes; it still wouldn't start. So, I thought the only thing I could do is just to let it rest for a minute, 'cause it was over-choked with gas, I guess. Then I tried again; we got it started. We took off, and I thought that was all there were there, because I wasn't really able to look for things. Somebody had to show me 'cause I wouldn't have had time to scout it for anything. And these planes [reporting downed aircraft], I guess, were planes that had seen the other plane being shot down. So anyway, I didn't hear anything about another plane, and started for base.[16]

The six survivors were the five-man crew of a B-25 (*Gremlin's Holiday*) and an unauthorized passenger: Capt. Robert G. Huff, a ground officer and the squadron adjutant of the 498th Squadron, who had always wanted to witness combat. He had convinced the plane captain, 1st Lt. Edgar R. Cavin, to let him come along.

B-25 ("Gremlin's Holiday") Aircrew and 498th Squadron Adjutant

1st Lt. Edgar R. Cavin	SSgt David B. McCready
2nd Lt. Elmer "Jeb" Kirkland	TSgt Fred Arnett
SSgt Lawrence Herbst	Capt. Robert G. Huff[17]

The airmen were difficult to manhandle out of their rafts onto the PBY-5 Catalina, being helpless deadweight and hard to grip coated in oil from their aircraft. The crew had all survived the impact when their plane hit the water, but not without serious injuries. Huff had three broken lumbar bones and a deep gash to his leg. Arnett had suffered a broken shoulder and a deep slash down his face. McCready was in terrible pain with a compound fracture of the right ankle, a deep gash to his hipbone, and severely burned hands and arms.[18]

DECISION MADE TO TURN BACK

On his return flight to base with the survivors, Gordon believed that he had flown at least twenty miles, when he was contacted by the same observer plane, reporting another plane down. There were now seventeen men aboard the Catalina, nine rescued and eight crew members. But Gordon determined he could not leave other potential survivors behind, even knowing the grave risks to him and the others on board associated with another rescue attempt:

> I just couldn't leave them sitting up there. So, he turned me back, led me back to where this plane had been shot down about 600 yards off shore, and they were all out in their life raft, and I decided that really the only approach I could make where I'd have a really decent chance of making a good landing—not tearing my plane up—would be to fly over land, which I didn't want to do, of course. But, you know, I had to do it to make a landing. Anyway, I flew up over the land; didn't get shot at; came back down; landed, and then they started shooting at us pretty hard. But they never did hit us. You could see the regular fire all around us. I always thought that maybe because of the swells, that would cause 'em to miss us. I don't know, but anyway, I thought so. So anyway, there were 6 men in this life raft, and we got them all aboard pretty quick-like and took off again, and headed for base.[19]

SURVIVORS TAKEN TO LANGEMAK BAY

Overloaded with the additional personnel, Gordon managed to take off in heavy swells, under the enemy fire. In mid-afternoon, he reached his Langemak Bay base having rescued fifteen officers and men of the U.S. Army Air Corps 500th Bombardment Squadron (based at Dobodura, New Guinea). Second Lieutenant William J. Smith was brought aboard the USS *San Pablo* (AVP-30), tending the detachment of VP-34 Catalinas operating from the airdrome there, for first aid treatment. The other survivors were taken ashore and put in U.S. Army ambulances.[20]

GORDON AND PBY CREW LAUDED WITH AWARDS

> *Please express my admiration to that saga writing Cat crew. This rescue was truly one of the most remarkable feats of the war.*
>
> —Congratulations from ADM William F. Halsey Jr., USN[21]

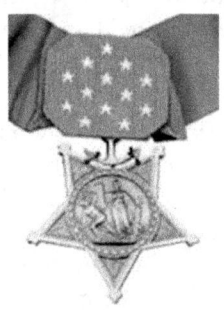

Lieutenant Nathan Gordon, U.S. Navy, was awarded America's highest military decoration, the Medal of Honor. He was the only PBY pilot thus honored in World War II and the first Navy man in the Southwest Pacific to receive the award. The seven members of his flight crew received the Silver Star Medal.

PBY-5 Catalina No. 08434 ("Arkansas Traveler")	
LTJG Nathan Green Gordon, pilot	AMM1c R.W. Routon
ENS Fulmer	AMM2c J. Paul Jeraeum
ENS Walter L. Patrick, co-pilot	AMM3c R.E. Brately
ENS Jack Kelley	ARM1c A.G. Alexander[22]

During Lieutenant Nathan Gordon's exemplary service in World War II, in addition to the Medal of Honor, he was awarded two Distinguished Flying Crosses, and six Air Medals. After the war, Gordon returned to his earlier profession of practicing law. He soon entered state politics and, in 1946, he ran for lieutenant governor. He won that election, and was reelected to that position an unprecedented nine further times, becoming Arkansas' longest-serving lieutenant governor, holding office until 1967. Gordon died on September 8, 2008, at age ninety-two.[23]

His Congressional Medal of Honor Citation follows:

CITATION:

The President of the United States of America, in the name of Congress, takes pleasure in presenting the Medal of Honor to Lieutenant [then Lieutenant, Junior Grade] Nathan Green Gordon, United States Navy, for extraordinary heroism above and beyond the call of duty as commander of a Catalina patrol plane serving with Patrol Squadron Thirty-Four (VPB-34), in rescuing personnel of the U.S. Army Fifth Air Force shot down in combat over Kavieng Harbor in the Bismarck Sea, 15 February 1944. On air alert in the vicinity of Vitu Islands, Lieutenant Gordon unhesitatingly responded to a report of the crash and flew boldly into the harbor, defying close-range fire from enemy shore guns to make three separate landings in full view of the Japanese and pick up nine men, several of them injured. With his cumbersome flying boat dangerously overloaded, he made a brilliant take-off despite heavy swells and almost total absence of wind and set a course for base, only to receive the report of another group stranded in a rubber life raft 600 yards from the enemy shore. Promptly turning back, he again risked his life to set his plane down under direct fire of the heaviest defenses of Kavieng and take aboard six more survivors, coolly making his fourth dexterous take-off with 15 rescued officers and men. By his exceptional daring, personal valor, and incomparable airmanship under most perilous conditions, Lieutenant Gordon prevented certain death or capture of our airmen by the Japanese.[24]

11

Duty at Green Island

Coos Bay arrived at Green Island lagoon on 11 May 1944 to service PBYs and provide quarters for their crews while facilities ashore were being constructed. Among the naval personnel at Green Island was future American president LT Richard M. Nixon, USNR.

Photo 11-1

LCDR Richard M. Nixon, USNR, date photograph taken unknown.

A graduate of Duke University, Richard Nixon had been employed as an attorney for the Office of Emergency Management in Washington, D.C. when he accepted an appointment as lieutenant junior grade in the United States Naval Reserve on 15 June 1942. Following completion of aviation indoctrination training at the Naval Training School, Naval Air Station in Quonset Point, Rhode Island, Nixon served as Aide to the executive officer at the Naval Reserve Aviation Base in Ottumwa, Iowa, until May 1943.[1]

Seeking more excitement, Nixon volunteered for sea duty and after reporting to commander Air Force, U.S. Pacific Fleet, he was assigned initially as officer-in-charge of the South Pacific Combat Air Transport Command at Guadalcanal and later at Green Island. His USN unit prepared manifests and flight plans for C-47 cargo plane operations and supervised the loading and unloading of the aircraft.[2]

Nixon was transported with hundreds of other servicemen to the South Pacific aboard the S.S. *Monroe*, one of the "President" liners pressed into service as a troopship. At age seventy, in an interview with Frank Gannon, he described the ocean crossing, on which he suffered chronic seasickness:

> It was a luxury liner fitted out for two hundred and fifty luxury passengers, and we had three thousand on it. We lived -- the officers were in bunks, three high on the walls and so forth and so on. [It] took seventeen days to get to Noumea, New Caledonia. And I remember the most unpleasant experience on that was not the fact that we had to wear life belts at times and so forth and so on, but was the fact that I was, of course, allergic to seasickness. And they used to bet -- we only had two meals a day. And my friends who were sitting at our table used to take bets among each other as to how long I'd stay at the table. I seldom got through a full meal.[3]

When asked by Gannon if he had seen any action "up the line," Nixon described being subjected to enemy air attacks shortly after he arrived at Bougainville Island in January 1944:

> I think the most lively place I was in was Bougainville. There were about thirteen or fourteen days when we had air raids every night. One night it was pretty close. The Japanese plane used to come over. The way you could tell it was a Japanese plane is the motors were not synchronized. They go, "Dee-dee-dee-dee." Even without the air raid, you knew it was a Japanese plane coming over. And they were really harassing us because our Air Force had knocked down most of their power. One night...we heard this plane. It had come in very low. And we heard the bombs dropping

as they came down the runway. "Rrrrrrrrr." They were dropping. And we dived out of our cabin into the foxhole. As soon as we got out, we saw that our whole tent had been sprayed with bullets. It was a close one.[4]

In the spring of 1944, Nixon (now a full lieutenant having been promoted on 1 October 1943) ran the base air cargo office on Nissan Island, commonly called Green Island. Green Island was the largest of a group of eight islands on the north end of the Solomon Island chain, just four degrees south of the Equator. After occupation, it was also the sites of U.S. Navy/Marine Corps, New Zealand, and Australian bases. Nixon had much respect for the Naval Construction Battalion personnel (Seabees) on the island, and it soon became apparent to him that they had the best chow. Seabees unloaded the infrequent reefer (refrigerated) ships that stopped at the island, they manned the cold storage warehouse, and they delivered the daily provisions to the many camp kitchens. Nixon related to Gannon:

> Well, the Seabees -- people wonder why I was so much for the hard hats. I talk about remarkable men, and they were remarkable. I remember one time on Green Island we were -- they were making an airstrip there, and it -- there was an air raid signal. But some of them were false. And these big Seabee guys, they'd be in the big bulldozers, they'd ignore the signals, and they'd keep working there, even in the middle of the night with their lights on in order to get the airstrip finished.
>
> Boy, they were something else. Most of them were from the east. This was the Twenty-Second Seabee Battalion. And I got to know them very well, and I ate with them because I was the head of a small detachment -- Army, Navy, Air Force were all members of it. I, being a naval officer, was the officer in charge, being a lieutenant and the ranking officer. And so I was able to select which mess we would use. Well, I turned down the Marine mess because the Marines can fight, but they couldn't cook. They were terrible cooks. I turned down the Army mess because they were almost as bad cooks...as the Marines. The only other mess was the Seabee mess, and it was the best.
>
> It wasn't because their cook was so good, but the Seabees, you know, they had access to a lot of things. They could put in a -- they could put in some flooring in your tent. They -- they could make various utensils and so forth. And so they would trade for meat and other vittles for their mess. And what they didn't trade, they stole. And they were very good.

After finishing his tour of duty at Green Island, Nixon returned to the United States. From August through December of 1944, he was assigned to Fleet Air Wing Eight. He then served through March 1945 at the Bureau of Aeronautics, Navy Department, Washington, D.C. In March, he transferred to the Bureau of Aeronautics General Representative, Eastern District, headquartered in New York City. Nixon was released from active duty on 10 March 1946. He had become a lieutenant commander on 3 October 1945, and would gain another stripe when promoted to commander in the Naval Reserve on 1 June 1953. Commander Richard M. Nixon, USNR, transferred to the Retired Reserve of the Naval Reserve on 1 June 1966, two years before becoming the thirty-seventh president of the United States.[6]

RESCUE OPERATIONS FROM GREEN ISLAND

The first three PBYs arrived at Green Island in April, dispatched by VP-91 to the forward area from Halavo Seaplane Base. The squadron had seaplanes operating from two other advance bases, plus five at Treasury and two at Emirau Island in the Bismarck Archipelago. Aircraft at the advance bases were primarily for rescue (Dumbo) and general utility work, under the control of commander, Aircraft Solomons.[7]

Coos Bay received orders from commander, Fleet Air Wing One, on 23 May, to perform the following actions at Green Island:

- Maintain two Dumbo aircraft each at Green and Emirau Islands
- Conduct Southwest Pacific Search Plan BAKER, consisting of five sectors 500 miles north of Green between 342° and 027° True
- Maintain on call anti-sub aircraft, sonobuoy equipped, as practicable for JASASA (joint air-surface anti-submarine action) use[8]

On 26 May, *Chincoteague* arrived at Green to assist in tending aircraft. During the latter part of May, the five VP-91 planes at Treasury were moved forward to Green Island to augment the three already there. On 28 May, tended aircraft began the mandated daily five-sector searches. VP-91 Catalinas flew five Dumbo missions in May, rescuing twelve airmen. On two occasions, the Catalinas came under enemy fire, but sustained no damage. On another different mission, LTJG R. J. Long's plane was hit as he attempted to go inside Blanche Bay, New Britain, to investigate a report that an American fighter pilot had been downed there. His second pilot, AP1c J. A. Parker, was wounded in the leg and required hospitalization for three days. Amazingly, after piercing the plane, the AA round exploded inside a steel helmet on the deck floor of the cockpit, dampening the shrapnel spray and sparing the crew greater injuries.[9]

Squadron VP-91 aircraft operating from the *Coos Bay* at Green Island conducted two Dumbo missions in May, along with one in June. All three involved the rescue of pilot and radio-gunners of downed SBD-5 Dauntless dive bombers. The aircraft were from Marine Scout Bombing Squadrons VSMB-241, VSMB-244, and VSMB-235. The two bombers downed in May were hit by anti-aircraft fire over Tobera Airfield, located on New Britain Island to the south of Rabaul.[10]

The Dauntless dive bomber of VMSB-241 was downed on 21 May. It was one of thirty planes that hit Tobera Airfield that day in an attempt to neutralize gun positions before TBF Avengers came in to bomb the runway. Five Dauntless were hit. All made it back from the mission, except for 1stLt William Noser's plane, which went down smoking. As highlighted in the squadron's war diary, "Although his [Noser's] vision was completely obliterated, he made an excellent water landing thirty miles from Cape St. George."[11]

On 28 May, twenty-eight SBDs loaded with bombs again struck Tobera. Gun positions were the targets. A high-speed approach was made from the southeast, and retirement after attack was seaward between Kabanga Bay and the Warengoi River mouth. Second Lieutenant Zehrung was hit by AA fire on the pullout and made a water landing two miles off the Warengoi. Rescue was made within thirty minutes of his landing by a VP-92 Dumbo Catalina.[12]

First Lieutenant Lindell, returning from a strike on gun positions near Rapopo Airfield on New Britain, was forced down west of Buku Island when the plane's gas selector valve froze, preventing shifting to a full tank and starving his SBD-5 of fuel. He was one of those rescued as enumerated in the following table:[23]

VP-91 Rescues from *Coos Bay* (21 May-13 June 1944)

Date	Pilot	Personnel Rescued	Reason Ditched, Location Rescued
21 May 44	LT T. E. Maurer	1stLt William Noser, USMCR; Sgt W.S. Enger, USMC	SBD-5 from VMSB-241 hit by AA fire over Tobera; life raft 4°57'S, 153°25'E
28 May 44	LT C. I. Hawkins	2ndLt R.D. Zehrung, USMCR; Cpl L.L. Selover, USMR	SBD-5 from VMSB-244 hit by AA fire over Tobera; from water 4 miles off mouth of Warengoi River, Gazelle Peninsula
13 Jun 44	LT T. E. Maurer	1stLt Carl W. Lindell, USMC; Cpl George J. Cortes, USMCR	SBD-5 from VMSB-235 made water landing earlier that day; life raft 5°15'S, 154°17'E[14]

On 16 June 1944, following her relief at Green Island by the seaplane tender *Half Moon* (AVP-26), *Coos Bay* left for Espiritu Santo for well-needed maintenance, and rest and relaxation. Before taking up with her once again, we will visit RNZAF No. 6 Squadron at Halavo Seaplane Base, Florida Island.[15]

12

RNZAF No. 6 Squadron at Halavo

> *In its air-sea rescue operations the squadron achieved considerable success. Nearly all the flying in the Solomons-Bismarcks campaign was done over water, and with hundreds of planes in the air every day it was inevitable that landings on the sea were fairly frequent, whether due to enemy action, engine failure or fuel shortage. Catalinas were stationed at various points throughout the Solomons, ready to take off immediately whenever they were called for. The efficient rescue service operated by the Allies contributed very largely to keeping up the morale of their aircrews. Pilots knew that if they 'ditched' or bailed out over the sea they had a good chance of being picked up within a few hours.*
>
> —Squadron Leader John Macaulay Sutherland Ross, in his seminal book, *Royal New Zealand Air Force*, published in 1955.[1]

Search patrols flown by No. 6 Squadron from the Halavo Seaplane Base at Florida Island were mostly uneventful. (These patrols began in late December 1943, when the squadron moved to Halavo from Espiritu Santo, and continued until it was disbanded on 9 September 1945.) As combat action moved with the Allies' advancement up the Solomons, such patrols involved eleven- or twelve-hour flights over hundreds of miles of empty sea, with no sightings of any enemy ships or submarines. Special searches for submarines, following reports by Allied shipping or aircraft, were equally unproductive.[2]

On the other hand, as indicated in the above quotation, rescue aircraft, of which No. 6 Squadron was an important part, were kept busy plucking grateful aviators from the sea. The crews for this flying boat squadron were trained at Hobsonville (Auckland region of the North Island of New Zealand), and at Laucala Bay Seaplane Base (on the southern coast of Viti Levu Island, Fiji). After a preliminary course on Walrus amphibians at Hobsonville, pilots were sent to Laucala Bay, where crews were formed and did their operational training on Catalinas. Until February 1944, crew training at Laucala Bay was done within squadrons stationed there, but in that month, No. 3 (Flying Boat)

Operational Training Unit was established and took over those responsibilities.³

Photo 12-1

Vickers Supermarine Seagull amphibious flying boat, later known as Walrus.
Australian War Memorial photograph 106071

FLYING OFFICER MACKLEY EARNS BAR TO DFC

The first Dumbo mission from Halavo Bay by No. 6 Squadron was launched on 26 January 1944, when an aircraft piloted by Flying Officer Winston Brook ("Bill") Mackley, DFC, was ordered to find and rescue ten men adrift on rafts some 220 miles to the north of the base. Approaching the reported site, southeast of Ontong Java Atoll in the Northern Solomon Islands, the crew first sighted a B-24 Liberator circling over two large patches of dye on the water, then as they closed in, they saw three rafts with downed flyers in them.⁴

The previous evening, nineteen USAAF B-24s of the Thirteenth Air Force (following three which dropped flares) had attacked Lakunai Airfield on New Britain Island. The airmen in the rafts were from one of the bombers, a USAAF Liberator (#420145) which, on the return flight, had run into a storm front. The Catalina landed and picked them up. With radar and flight equipment inoperative, and low on fuel, the plane captain had decided to ditch. Mackley managed to take-off in his Catalina and returned safely to Halavo despite rough weather, damage to his aircraft from wind and sea, and added weight associated with ten extra people on board.⁵

B-24 Survivors (72nd Squadron 5th Bombardment Group)

1st Lt. Nerva Malcolm Hayden (pilot)	SSgt John A. Plotczyk (2nd Engineer)
1st Lt. Richard B. Holdeman (co-pilot)	SSgt D. D. Duckworth (gunner)
1st Lt. John Francis Freet (navigator)	Temp Sgt Dallas Earle Steenis (gunner)
1st Lt. John W. Allsbury (bombardier)	TSgt Ray Edward Brubaker (1st Radio)
TSgt Leo Boyd Burton (1st Engineer)	Cpl Mattis G. Lofgren (2nd Radio)⁶

For his actions, Mackley earned a bar for the DFC which he already proudly wore with dress uniform; the bar signifying a second award of the Distinguished Flying Cross. Mackley's aviation career had begun in the Civil Air Reserve, which prepared its members for acceptance into the Royal New Zealand Air Force in the event of war. When World War II broke out, he was called up and began a training program at New Plymouth (on the west coast of the North Island of New Zealand) and Woodbourne (at the top of the South Island).⁷

Mackley's first duty prior to start of war in the Pacific was in Europe with RAF flying crews. He was sent to Britain in May 1940, and made his first RAF Whitley bomber raid over Rotterdam. In the next six months he took part in twenty-seven missions over Europe, and was awarded his Distinguished Flying Cross for his attacks on heavily defended targets such as Hamburg, Berlin, Cologne, and Bremen. The announcement of the award of the DFC to Pilot Officer Mackley for service with 58 Squadron, appeared in the *London Gazette* on 26 September 1940. (Such notification then, and still today is referred to as "being Gazetted.") Mackley returned to New Zealand in late 1942 for retraining on Catalinas as a member of No. 6 Squadron.⁸

FLYING OFFICER BEAUCHAMP EARNS DFC

Flight 4017 today spotted and picked up five survivors from a shot down liberator.

—Entry made by Flying Officer Donald Stanley Beauchamp, RNZAF, in his logbook regarding the rescue for which he would be awarded the Distinguished Flying Cross.⁹

Except under the best conditions, landings on the open sea were risky. Strong winds and/or heavy swells could damage flying boats, and even if undamaged, could impede their take-off runs so as to make it impossible for them to become airborne again. On 4 February 1944, a Catalina (NZ4017 XX-T) piloted by Flying Officer Donald Stanley Beauchamp, while on a routine patrol 100 miles south of Nauru, found five men in dinghies. (Nauru is an island in the west central Pacific, about 800 miles northeast of the Solomon Islands.) The airmen were survivors from a B-24 Liberator of the 26th Bombardment Squadron, 7th Air Force, based at Tarawa that was forced down while returning from a raid on Kwajalein.[10]

Photo 12-2

RNZAF No. 6 Flying Boat Squadron PBY-5 NZ4017 XX-T in flight.
Courtesy of Jenny Scott (provided her by RNZAF PBY Captain Dave Sheehan)

Beauchamp managed to land his flying boat in rough seas but its hull developed a leak due to the strain imposed, that required continuous operation of the bilge pump while the survivors were picked up. The take-off was even more difficult than the landing with chop impeding the aircraft and threatening to tear off the floats. After several attempts, the ocean-gripped plane sprang free into the air, and returned to base. For his fortitude and steadfast resolve in effecting the rescue, Beauchamp was honored with an immediate award of the Distinguished Flying Cross. The associated citation reads:[11]

By his skill and courage in landing a heavily-laden flying-boat in rough water in the open sea, this officer rescued five United States servicemen on 4th February, 1944. The men, who had been adrift for six days, were transferred to the aircraft, and in spite of rough water and a heavy swell, a take-off was made and the aircraft flown back to base. Flying Officer Beauchamp was captain of the New Zealand 'Catalina' flying-boat concerned in the rescue. On sighting the rafts adrift in the Pacific, he was unable to contact base for instructions, so he decided on his own initiative to attempt the rescue, well knowing that, should he fail, there would be little prospect of rescue for himself and his crew. Although the aircraft was slightly damaged in the open water landing and the subsequent take-off was cross-wind, Flying Officer Beauchamp displayed a high degree of skill and determination, and succeeded in a difficult and hazardous enterprise.[12]

FLIGHT SERGEANT JAMES MONK AWARDED DFM

The next member of No. 6 Squadron to earn an award for valor was Flight Sergeant James Benjamin Monk, who landed the Catalina he was piloting in dangerous Kabanga Bay. He remained there for over two hours exposed to enemy defenses, finding and rescuing dispersed survivors of a B-24 Liberator crew. United States military aircraft flying westward to strike Rabaul, routinely overflew the bay on the eastern coast of New Britain's Gazelle Peninsula while inbound, and outbound their targets.[13]

The challenges which Monk and his flight crew faced and overcame in successfully carrying out this mission are aptly described in the award justification for the Distinguished Flying Medal he received:

Within easy reach of powerful Japanese coastal defense batteries near Rabaul and dangerously close to enemy fighter strips, Flight Sergeant Monk, Captain of a Catalina, alighted and kept his aircraft on the water for 2 hours 20 minutes to rescue survivors of a United States heavy bomber that had been shot down on 5th March, 1944. Taxying among wreckage and exposed to the enemy, Flight Sergeant Monk completed his search and located and rescued three survivors known to have parachuted into the sea. Ordered to undertake the rescue, Flight Sergeant Monk landed close to the floating wreckage within two miles of the shore at Kabanga Bay. One survivor was located and picked up after taxying a mile but an extensive search was necessary to find two other men. In spite of the risk to which he and his crew were exposed, the pilot persisted in his endeavors and successfully concluded his mission, displaying throughout the courage and devotion to duty that have marked his operational career. For his exemplary brave conduct throughout a difficult and dangerous rescue, this N.C.O. is strongly recommended for the award of the Distinguished Flying Medal.[14]

Authorization for award of the DFM to "Flight Sergeant (now Flying Officer) James Benjamin MONK, 6 (R.N.Z.A.F.) Sqn." appeared in the *London Gazette* on 25 September 1945.

FLYING OFFICER REGAN RECEIVES DFC

During No. 6 Squadron's service at Halavo Bay, Florida Island, detached Flights operated at various times from Emirau Island; Los Negros, Admiralty Islands; Green Island, Solomons; and Jacquinot Bay, New Britain. Returning to Jacquinot Bay in August 1945, Flying Officer Christopher Regan (piloting Catalina NZ4039) was directed by his operations control tower to proceed to Ataliklikum Bay. The bay, which opened into the Bismarck Sea on New Britain's north coast, lay to the west of Rabaul.[15]

Map 12-1

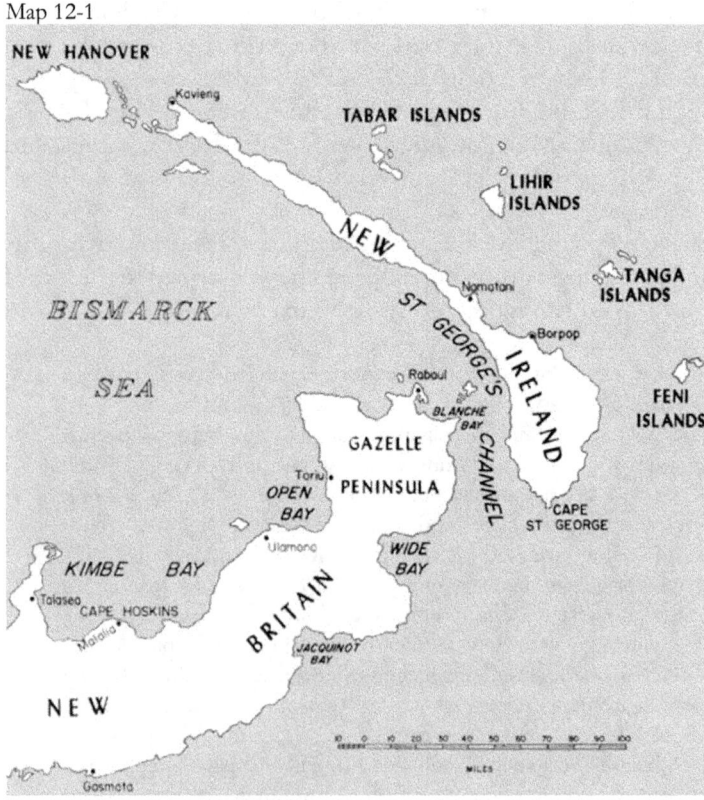

New Britain and New Ireland islands

Arriving there, Regan sighted two F4U Corsair fighter aircraft circling overhead a dinghy off the Kabanga River. Its occupant, Flight Sergeant C. E. Turner RNZAF, pilot of F4U-1D Corsair NZ5468 of No. 20 Squadron had ditched in the bay on 9 August after being hit by anti-aircraft fire.[16]

After circling, Regan landed in a heavy sea and Turner was brought aboard suffering from shock and exposure. The navigator administered first aid and the Catalina returned to the base at Jacquinot Bay.[17]

Turner's aircraft had been one of 424 Corsairs equipping thirteen RNZAF squadrons. The American-built fighters had replaced their SBD Dauntless scout dive bombers as well as the obsolete Curtiss P-40 fighter and ground attack aircraft. Most of the Corsairs arrived by ship at Espiritu Santo and were assembled there by RNZAF Base Depot Workshop Unit 60. One batch was assembled and flown at RNZAF Hobsonville (near Auckland on the North Island of New Zealand). Turner's Corsair had been assembled by Unit 60 on 2 September 1944. It arrived at No. 20 Squadron on 5 January 1945.[18]

Flying Officer Christopher David John Regan, RNZAF, was Gazetted on 11 December 1945, with award of his Distinguished Flying Cross, retroactive to 30 October 1945.

SQUADRON LEADER MACGREGOR HONORED FOR VALOR, LOST WITH OTHERS IN PLANE ACCIDENT

> *We couldn't find any survivors and we searched the place thoroughly. Later we found out that eight of them had been picked up off a life raft by a big PBY that landed in the water and almost broke itself up getting off.*
>
> —LTJG Paul R. Campagna in a 14 November 1943 interview about duty aboard the minesweeper USS *Dash* (AM-88).[19]

Before leaving this chapter, mention must be made of RNZAF Squadron Leader Ronald Bruce Leslie MacGregor's receipt of the Air Force Cross—for a rescue that occurred before No. 6 Squadron was officially formed at Fiji and of his subsequent sad loss. On 1 May 1943, a Catalina, piloted by MacGregor, located eight survivors from a torpedoed US merchant ship bobbing about on a raft in very rough seas of the South Pacific within range of his base. After standing by for three hours, without a rescue ship appearing, MacGregor made a daring mid-ocean landing—recovering four merchant seamen of the American

Liberty ship SS *Phoebe A. Hearst*, and four members of her Naval Armed Guard unit. Rough seas accounted for some damage to the aircraft during the operation, but he was able to take off safely and return to Laucala Bay. The uninjured men were landed at Suva, located west of the seaplane base on the southern coast of Viti Levu Island.[20]

As later discovered, the Japanese submarine responsible for the incident was *I-19* under Comdr. Kinashi Takakazu. She had pursued the 7,176-ton freighter for three hours, before torpedoing and sinking her at 20°07'S, 177°33'E. *Phoebe A. Hearst* was en route to Pago Pago from Noumea with a cargo of munitions and drummed aviation gas.[21]

MINESWEEPERS RECOVER OTHER SURVIVORS

> *I think it was about a week or two after that that a lifeboat was sighted out in that area and we were ordered to go out and pick up the people from the lifeboat. When we go there, we found that it was survivors that we had been hunting for two weeks earlier. They had been afloat in this lifeboat for two weeks and had sailed all over the islands and even made an island in the Tonga group but weren't able to go ashore because of a reef all the way around and anyway there was nothing on the island that they could see. It was a volcano so they decided not to go ashore there, but to try for another island. So, two weeks later, we finally found the people that we had been hunting for.*
>
> *There were two boats of survivors; the other boat had made one of the islands in the Tonga group but it was an island without much of anything on it. There were two pigs and a wild hog on the island. They ate the two pigs but they couldn't catch the hog. A friend of mine in the New Zealand Air Force flying over the island one day saw a red sail spread over the top of a tree. He was a little off his course on his way back from Tongatapu to Suvu and he thought that, as long as he was off his course anyway, he'd take a look at that island to see if there was anything on it and happened to sight this red sail. He flew low over it and saw the people on the island were waving and reported it. They sent out a patrol ship from Tongatapu to pick up these men. So, in that sinking, there weren't any casualties. Everyone was saved, that was a lucky break. The PHOEBE A. HEARST was the name of that ship.*
>
> —Continued account by LTJG Paul R. Campagna of the recovery of *Phoebe A. Hearst* survivors by a PBY Catalina and minesweepers.[22]

After the RNZAF Catalina found and landed eight survivors of *Phoebe A. Hearst*, there were no sightings of other crew members for nearly two weeks. This changed on 13 May. At a little past noon that day, pursuant to Port Director orders, the minesweeper USS *Dash* (AM-88) left King's Wharf, Suva, bound for the reported location (17°S, 179°11'W) of survivors of a torpedoed vessel.23

At dawn the following morning, a flare was sighted off the ship's port bow. *Dash* changed course to intercept, and soon a second flare broke the gloom. At 0605, she found twenty-five *Phoebe A. Hearst* survivors in a lifeboat at 17°S, 179°50'W, including the captain and chief mate. After embarking the men, who were all in fair condition, *Dash* cast the boat adrift, and set a return course for Suva. The minesweeper entered port and moored at her berth at 1555. Army ambulances and trucks were awaiting her arrival, and the transfer of survivors to them was completed by 1620.24

Following discovery by the aircraft of the remaining survivors of *Phoebe A. Hearst* on Tofua Island, the motor minesweeper USS *YMS-89* was sent out from Tonga. She collected the twenty-three men, and landed them on 16 May at Nukualofa (the capital of Tonga) on the north coast of the island of Tongatapu.25

Map 12-2

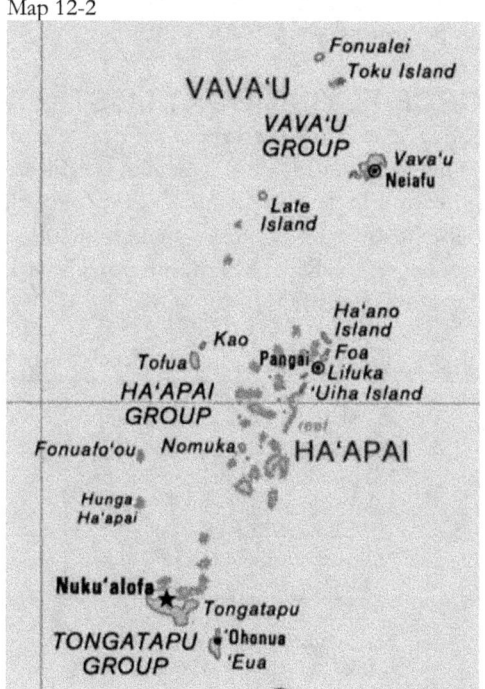

Tonga Islands (Tofua is a part of the Ha'apai group)

Photo 12-3

USS *YMS-89* in San Francisco Bay, circa 1945-1946.
Naval History and Heritage Command photograph #NH 79687

NEW ZEALAND AIR FORCE FLYING BOAT LOST

Squadron Leader Ronald MacGregor was awarded the Air Force Cross and Gazetted on 2 June 1943. Three days later, PBY-5 Catalina NZ4003 went missing on a sea flight from Fiji to Auckland, New Zealand. Extensive air searches failed to locate any of the aircrew, which included MacGregor, or the eight passengers (six RNZAF members) aboard. The missing, later presumed dead, are identified in the table:

PBY-5 Catalina NZ4003 Crew	Passengers
F/S Gordon Marsh Adie	FO Alan E. W. Bradmore
F/S Bruce Alexander Godley Bond	Sgt Henry Marton Kennedy
SL Ronald B. L. MacGregor	FL Maxwell William McCormick
LAC Hector A. G. McGregor	FL John Robert Mills Nicholson
FO Jack Eric Morison	LAC Eric Samuel Puddle
LAC Ivan James Waldie	LAC John Wilfred Russell
PO Douglass Edmund Wood	*Captain Norman John Paltridge
	*Mr. Marcellius James Scott[26]
SL: Squadron Leader	F/S: Flight Sergeant
FL: Flight Lieutenant	Sgt: Sergeant
FO: Flying Officer	LAC: Leading Aircraftman
PO: Pilot Officer	* Civilian passengers

13

Santo/Sydney/Noumea Interlude

Coos Bay arrived at Espiritu Santo on 21 June 1944, via stops at the Russell Islands and Halavo Beach, Florida Island, to embark Fleet Air Wing One personnel and materiel. Transport duty continued as she then proceeded to Munda, New Georgia, to bring the headquarters of Air Wing One and its personnel back to Espiritu Santo. This tasking completed, the seaplane tender went alongside the internal combustion engine repair ship USS *Mindanao* (ARG-3) for ten days for engine overhaul and antenna replacement. (*Coos Bay* was propelled by two General Motors 12-288 diesel engines.) She next entered Dry Dock No. 1 at Pallikulo Bay for the removal of a year and a half accumulation of marine growth on her hull, and minor repair work.[1]

Following this necessary work, under orders of commander, South Pacific VADM John H. Newton, (Admiral Halsey's successor), *Coos Bay* left Santo in the latter part of July bound for Sydney, Australia. She arrived in Woolloomooloo Bay on 29 July, after four days at sea. Within an hour after the ship was moored, two-section liberty was granted (half of the crew had shore leave each day), and a well-deserved ten-day rest and rehabilitation period had began.[2]

Photo 13-1

View of small Garden Island in Sydney Harbour, looking across Woolloomooloo Bay, showing the beginnings of a naval establishment. This image was incorrectly inscribed with the year 1889, by which time the tree-topped hillock on the right was excavated, and more buildings had been constructed.
Australian War Memorial photograph P02035.001

The time spent in Sydney were memorable days for her crew. One of the highlights was a ship's party staged in the expansive Paddington Town Hall. Built in the late 1800s, the impressive building sat at the highest point on the Oxford Street ridge in the inner eastern Sydney suburb of Paddington.[3]

Photo 13-2

Reunion of WWI Australian 2nd Battalion members at Paddington Town Hall, 1925. Australian War Memorial photograph P05182.116

Photo 13-3

Coos Bay leaving Sydney, 9 August 1945. She carried four 5-inch/38 dual-purpose guns, with the forward mounts in gun houses; two twin-mount 40mm guns, port and starboard side, amidships; and eight single 20mm guns located in the forward superstructure and abreast the mainmast. Mousetrap rockets on her stern (rocket-propelled depth charges) provided her a measure of anti-submarine capability. Australian War Memorial photograph 302575

STOP AT NOUMEA EN ROUTE TO ESPIRITU SANTO

On 9 August, *Coos Bay* took her leave of Australian hospitality, bound for Noumea, New Caledonia, to load aviation gear before returning to Espiritu Santo. Before continuing her story, an aside about New Caledonia is in order. It was a French Territory, located on the northeast approach to Australia. In the aftermath of the attack on Pearl Harbor, the U.S. government had decided in January 1942 to send troops to New Caledonia to fortify and defend it against a feared Japanese attack. From New Caledonia westward to Australia there was only the open Coral Sea separating it from Australia, making that country vulnerable to enemy forces wishing to sever the route by which ships were transporting men, materials, and munitions from America's west coast and the Panama Canal to reinforce her.[4]

Following occupation of Noumea on New Caledonia, on 12 March 1942, by U.S. Army Maj. Gen. Alexander M. Patch's Task Force 6814, development of it as the main Fleet Base in the South Pacific included a seaplane base. The *Tangier* (AV-8) actually arrived there from Hawaii prior to the task force that was to use the new base. Anchored in Great Road, she had functioned there since 4 March, 1942 as tender for a half squadron (six planes) of PBY-5s. She later also furnished services necessary to patrol aircraft passing through the area, and miscellaneous amenities to U.S. Army field forces in New Caledonia.[5]

Map 13-1

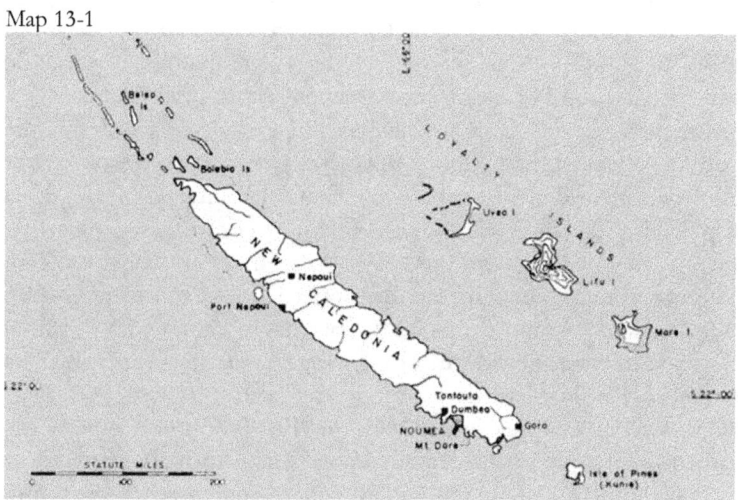

The French Colony of New Caledonia (comprising dozens of islands, north-northeast of Brisbane, Australia, across the Coral Sea)
Building the Navy's Bases in World War II History of the Bureau of Yards and Docks and the Civil Engineer Corps 1940-1946 Volume II

Photo 13-4

Pontoon Pier at the Seaplane Base, Noumea
Building the Navy's Bases in World War II History of the Bureau of Yards and Docks and the Civil Engineer Corps 1940-1946 Volume II

Coos Bay arrived at Noumea on 12 August, three days after leaving Sydney, brought aboard the aviation gear, and returned to Espiritu Santo. Her time at Santo was spent loading spare parts to support PBM Mariner seaplane operations in the Central Pacific. On 19 August, she proceeded to Lunga Point, Guadalcanal, to load more gear in preparation for tending Mariners. *Coos Bay* sailed from South Pacific waters on 27 August 1944, bound for Eniwetok, Marshall Islands.[6]

Commander Jesse B. Burks relieved CDR Delbert L. Conley as the commanding officer of the *Coos Bay*, upon her arrival at Eniwetok. After reporting to commander, Forward Area, Central Pacific (CTF 57), *Coos Bay* was placed under the operational control of commander, Fleet Air Wing Two (CTG 59.3) and ordered to Saipan, Marianas, for duty. She travelled there in company with the fleet oiler USS *Nantahala* (AO–60). Upon her arrival at Saipan on 8 September, eight crews of VP-18—a Mariner squadron engaged in long-range searches utilizing its complement of PBM-3D aircraft—moved aboard.[7]

The following chapter introduces the PBM Mariner, which the Navy had adopted as its standard patrol bomber. Although it was never as beloved as the PBY Catalina, the PBM with its increased range, bomb load, speed, and fire power, was considered better adapted than the former to meet the conditions in the forward areas of the Pacific.[8]

14

Martin PBM Mariner Patrol Bomber

The problem of maintaining the PBM-3D, particularly owing to the difficulties with the R-2600-22 engine, is the chief problem at advance bases. The performance and adaptability of the PBM for patrolling and anti-submarine work is unexcelled, but the maintenance has been consistently troublesome.

—Patrol Bombing Squadron VPB-18 War History.[1]

Photo 14-1

Martin PBM-3D Mariner patrol bomber awaiting launch into the water from a naval air station's seaplane ramp, circa 1942-43.
National Archives photograph #80-G-K-16065

Although less well known than the numerous PBY Catalinas, the PBM (Patrol Bomber, Martin) Mariner was one of the most successful patrol aircraft of World War II. While the Consolidated PBY Catalina proved to be highly versatile, by the end of the war, the larger, faster, and more capable PBM had supplanted it in many of its roles. Even though the Mariner entered service in 1940, there were not sufficient numbers of its type to begin replacement of the obsolescent Catalinas, until the latter part of the conflict.²

The Glenn L. Martin Aircraft Company delivered its first twenty-one production PBM-1s to the Navy in September 1940. The aircraft design incorporated several improvements over the PBY, including a deep hull mounting a gull wing, which kept the vulnerable engines above salt spray without the use of the Catalina's parasol wing and its external bracing. The incorporation of an internal bomb bay allowed for higher speeds when the aircraft was loaded with ordnance and the wing floats could retract into wing bays, further reducing drag. This streamlining, and the flying boat's two 1,600 hp Wright R-2600-6 engines, gave the much larger aircraft a speed advantage over the PBY. An upwardly-canted tail stabilizer resulting in an inward tilt of the twin vertical stabilizers, gave the aircraft a very distinctive appearance.³

In October 1940, the first PBM-1s went into service with Navy patrol squadrons VP-55 and VP-56. The Mariners were soon patrolling the Atlantic in search of the German U-boats that threatened to disrupt the trans-ocean supply of vital war material to Great Britain. The patrol aircraft could attack the enemy with its 4,000lb bomb load (or torpedoes in later versions). The aircraft, with its crew of nine, was also capable of defending itself against enemy aircraft with five .50-caliber Browning machine guns. One gun was mounted in a flexible position in the tail, with the gunner lying prone; two guns were fitted in a flexible-mounted position in the waist gimbals; one gun was fitted in a rear dorsal turret; and one gun was fitted in a nose turret.⁴

Photo 14-2

PBM Mariner starboard waist gunner manning his .50-caliber machine gun. National Archives photograph #80-G-K-15041

Several variants of the follow-on PBM-3 were designed for specialized missions. The PBM-3C boasted a powerful APS-15 radar and twin .50-caliber machine guns in its three turrets, in addition to the original single .50 caliber guns mounted in the waist windows on either side. The PBM-3S was a stripped-down version for long-endurance anti-submarine patrols. Armor and armament were removed from the PBM-3R transport version to increase its payload and range. The transport versions supplied distant submarine bases lacking conventional airstrips, and later flew medical evacuation missions, a job for which their spacious hulls were particularly well-suited.[5]

Changes incorporated into the PBM-3D version, the aircraft flown by Patrol Squadron VP-18, included more powerful 1,900hp Wright R-2600-22 engines, a Norden bombsight, and additional armor and self-sealing fuel tanks for regular patrol bomber missions. (VP-18, introduced in the previous chapter, was re-designated VPB-18 on 1 October 1944 to reflect its change from patrol squadron to patrol bombing squadron.)[6]

Diagram 14-1

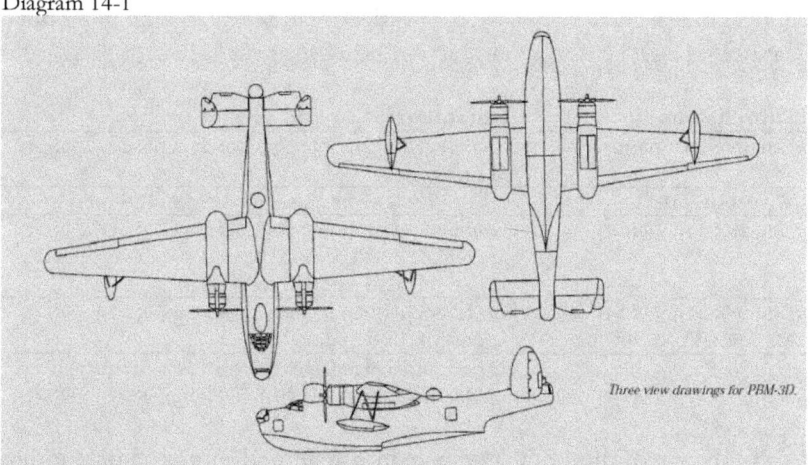

Three view drawings for PBM-3D.

Views of the Martin PBM-3D Mariner patrol bomber
Dictionary of American Naval Aviation Squadrons, Volume 2, Appendix 1

In 1944, the PBM-5 Mariner entered service with upgraded radios, turrets, and radar, and new R-2800 engines that were more reliable than the R-2600s in the PBM-3. Significantly, these aircraft were fitted with Jet Assisted Take-off (JATO) bottles to assist them in making short take-offs in the heavy Pacific sea conditions frequently encountered on Dumbo missions flown for rescue of USAAF B-29 bomber crews that had to bail out or ditch.[7]

Photo 14-3

PBM Mariner takes off with JATO assistance, circa 1944-45.
National Archives photograph #80-G-K-15887

Aircraft characteristics and armament of the PBM-5 and PBM-5A (amphibian version) twin-engine patrol bombers, the most capable of the Mariners, is provided in the table. The aircraft had a range of 2,420 nautical miles, and a crew size that varied from seven to nine personnel. The height listed is for an aircraft on beaching gear.[8]

Dimensions	Armament
Length: 79 ft. 10 in.	Two .50-caliber machine guns each in nose, dorsal and tail turrets
Wingspan: 118 ft.	Single .50-caliber machine gun at waist positions
Height: 24 ft. 10 in.	Combinations of bombs or mines carried in the bomb bay: Eight 1,600lb bombs, eight 1,000lb bombs, twelve 100lb bombs, or eight 325lb bombs
Empty weight: 32,840 lbs. Max. weight: 46,500 lbs.	Eight Mk 26-1 mines, four Mk 13 mines, or four Mk 13-5 mines
	Carried under wing racks: Two Mk 13-3 torpedoes; two 2,000lb mines, or two 1,600lb mines[9]

Only one of the 1,366 Mariners manufactured survives intact today. The Navy received this PBM-5A in 1948 and assigned it to Patrol Squadron VP-33, and later to an aviation indoctrination program for Naval Academy Midshipmen. It was retired in 1953, sold as government surplus, and passed through the hands of several civilian owners who attempted to turn it into everything from a firefighting water tanker to a fresh fish hauler. The Mariner proved uneconomical to operate as a civil aircraft and the Smithsonian Institution accepted its donation in 1973. The aircraft is today on loan and displayed at the Pima Aerospace Museum in Tucson, Arizona.[10]

15

Duty at Saipan, Marianas

Photo 15-1

View of Japanese ships in Tanapag Harbor, Saipan, taken by USS *Wasp* (CV-18) aircraft during Task Force 58 air strikes in early June 1944.
National Archives photograph #80-G-349815

Four days after arriving at Saipan on 8 September 1944, and embarking VP-18 flight crews, *Coos Bay* assumed the additional temporary duty of seadrome control at Tanapag Harbor. The harbor was located on the west side of the island. Three patrol squadrons—VP-17, VP-18, and VP-102—and transient aircraft operated from this drome. On 9 October, the USS *Kenneth Whiting* (AV-14) assumed seadrome control.[1]

At this time, Patrol Squadron VPB-18 (formerly designated VP-18) terminated conducting sector searches, and began a more diverse schedule. In addition to Dumbo duty, and daily mail runs to Ulithi Atoll in the Carolinas and Kossol Passage, Palau, squadron aircraft were also in Dumbo standby and hunter-killer standby status for related tasking. This continued for three weeks, and then VPB-18 assumed day searches and night anti-submarine patrols. On 15 November, the squadron moved ashore from *Coos Bay* to Naval Air Base Tanapag, Saipan.²

Coos Bay then served as a terminal for Kossol Passage-based seaplanes of VPB-17 (which had relieved VPB-18 of its daily mail runs), when calling at Saipan. She also serviced Naval Air Transport Service (NATS) seaplanes carrying mail. This service was a branch of the Navy from 1941 to 1948. Transport Squadron VR-2, based at Alameda, California, serviced the Pacific. The seaplane squadron initially flew converted PBYs and the PB2Y Coronado, and later a transport version of the PBM Mariner.³

RESCUE SQUADRON VH-1

On 16 November, six aircrews of Rescue Squadron VH-1 moved aboard *Coos Bay*, with the other six crews aboard the *Kenneth Whiting*. The squadron was ship-based aboard the two vessels for about two weeks, then moved ashore at Tanapag. VH-1, the first of the Navy's six rescue squadrons, had been established on 1 February 1944 at Naval Air Station Alameda, California. Employing PBM-3D and PB2Y seaplanes, it made nineteen direct rescues via open sea landings during the war, and assisted in the rescue of at least another 119 airmen by directing surface vessels and submarines to their downing locations.⁴

Photo 15-2

PB2Y-3 Coronado at Naval Air Station Honolulu, Hawaii, with mechanics at work. National Archives photograph #80-G-343058

The squadron lost a PB2Y-3 (7186) Coronado at its mooring in Saipan Harbor (just to the south of Tanapag Harbor) on 10 September 1944, when it was boarded by enemy infiltrators in early morning darkness at 0430, and sunk with grenades. There were no personnel casualties, although Ensign Gresham suffered a few abrasions from a grenade that exploded in the bilges beneath him. An entry in VP-18's war diary on 2 August 1944, regarding its planes taking flight from Ebeye, Kwajalein Atoll in the Marshalls, for Saipan, noted, "The Army had declared all organized resistance on the island wiped out by that time, and it was made available for operations by our forces."[5]

PRECEDING INVASION OF SAIPAN, MARIANAS

Map 15-1

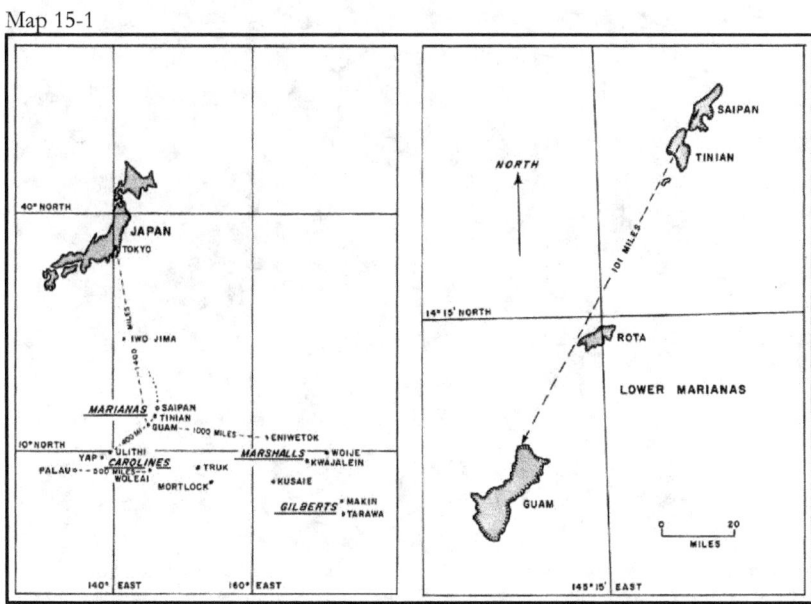

Side by side charts showing the relationships of Central Pacific Island chains to one another, and details of the Lower Mariana Islands
George C. Dyer, *The Amphibians Came to Conquer: The Story of Admiral Richmond Kelley Turner*

The Battle of Saipan (15 June to 9 July) was launched on 15 June, when the 2nd and 4th Marine Divisions and the 27th Army Infantry Division began landing on the island. The objective of the operation, codenamed FORAGER, was to neutralize Japanese bases in the Central Pacific, support the Allied drive to retake the Philippines, and provide bases for a strategic bombing campaign against Japan.[6]

While the Allied assault forces and the enemy ashore fought the Battle of Saipan, the Japanese Combined Fleet sortied to engage the U.S. Navy Fleet supporting the landings. In the ensuing 19-21 June Battle of the Philippine Sea, the American battle force defeated the Japanese in what would be the greatest carrier battle of the war. Following the loss of two carriers, the *Shokaku* and *Taiho*, 346 planes, and their air crews, Japanese naval air was unable to engage Allied forces with parity. This reality resulted in Japanese leadership turning to the use of suicide planes.[7]

Photo 15-3

Aboard USS *Lexington* (CV-16), LTJG Alexander Vraciu, USNR (Fighting Squadron VF-16), holds up six fingers to signify his "kills" achieved on 19 June 1944, during the "Great Marianas Turkey Shoot" (aerial part of the Battle of The Philippine Sea). National Archives photograph #80-G-236841

VP-18 PATROLS FROM SAIPAN

Following the arrival of Patrol Squadron VP-18 at Saipan in early August, its planes took up a five-legged patrol pattern around the entire Saipan-to-Guam area, to protect Allied shipping from Japanese submarines while Isley Field on the southern tip of Saipan, and the airstrip on Tinian, were being reconstructed for Allied use. The Japanese Fleet, having retired in defeat, was not expected to attempt an attack on the heavy surface forces operating out of the Saipan base.

However, with Garapan Harbor (a part of Saipan Harbor) being used as a fleet anchorage, a vigilant anti-submarine program was necessary as the Japanese underwater fleet was still active.[8]

The ASG-9 radar-equipped PBM-3, providing target interception up to ninety miles on land targets and seventy miles on ships, proved to be very effective in adequately maintaining an air reconnaissance screen around Saipan. The biggest impediment to operations were unmarked dangerous coral reefs in the harbor where the flying boats had to land. Following two accidents, one in which a plane was eventually salvaged and returned to service after survey, and another in which the damage to the plane was never completely repaired, squadron personnel took on the task of marking the reefs themselves. This was done intermittently during periods when time away from other pressing duties could be found. The markers were placed accordingly to pilots' reports of reefs as sighted from the air.[9]

In September, 695 hours of day patrols and 385 hours of night patrols comprised the bulk of VP-18 missions. The day patrols were 10-degree sectors in the northeast quadrant, fanning out to a distance of 600 miles. The night anti-submarine sector was a 40-mile square, centered forty miles due west of Saipan. The squadron made one mail run from Saipan to Kossol Passage and return via Ulithi, and one special passenger trip to Ulithi carrying Rear Admiral Smith.[10]

Patrols were decreased over the course of September to allow for aircraft engine changeouts by the Patrol Aircraft Service Unit (PATSU) ashore as the squadron was scheduled to move forward after the campaign to take Palau. However, a change was made and VP-18 (now VPB-18) remained at Saipan. As part of this change Squadrons VPB-16, VPB-202, and VPB-216 were on tenders in Kossol Passage at Palau, and VPB-17 moved into Ulithi after it was taken.[11]

VH-1 ACTIVITIES IN NOVEMBER 1944

Rescue Squadron VH-1 was based aboard the seaplane tender *Kenneth Whiting* from 1-16 November. On the 16th, the detachment was split with six crews moving aboard the *Coos Bay*, and six aboard the *Shelikof*. During this time, PBM Mariners accompanied bomber and fighters on strikes to the Japanese Volcano (Kazan) Islands, which lay between the Bonin Islands to the north and the Mariana Islands to the south. The flying boats also accompanied bombers to Truk in the Carolinas. Truk Lagoon was the Japanese Navy's main base in the South Pacific and home to its Combined Fleet.[12]

The number of each type VH-1 mission in November are summarized in the following table. In addition to these missions there were seven that were called back after being airborne and well on their way to station.

Rescue Squadron VH-1 Missions in November 1944

No.	Missions
12	With B-24s or B-29s to Kazan Islands
12	With B-24s to Bonin Islands
10	With P-47s to Pagan Islands (northern Marianas)
3	With B-29s toward Tokyo
3	With B-29s to Truk
3	With Corsairs, Hellcats, and Avengers from Guam to Ulithi
2	With P-38s toward the Kazan Islands[13]

The U.S. Navy's island-hopping campaign through the Pacific had earlier brought it to Ulithi Atoll in the Carolinas, which offered a large deep lagoon that was perfect for a fleet base. From March to September 1944, aircraft bombed the Japanese forces stationed at the atoll until they eventually withdrew. Vice Admiral Theodore Wilkinson landed a Regimental Combat Team of the U.S. Army 81st Infantry Division at Ulithi on 23 September. The forces met no opposition, and the abandoned atoll was promptly put under development as an important base with its deep-water anchorage.[14]

The rescue squadron operated three different types of aircraft in November: six PB2Y-3 Coronados, five PBM-3D Mariners, and two PBY-5 Catalinas temporarily assigned to them from VPB-100. Three Dumbo missions that month were successful in finding downed aviators. On 2 November, Lieutenant Higgins in a PBM, assisted in spotting and providing cover for a PBM of VPB-18 located on the water forty-five miles west of Saipan. Lieutenant (junior grade) Doheny, while flying a night anti-submarine patrol, had been forced to make a dead stick landing, at night, without flaps, when both his engines cut out because of an air lock in the fuel system. The damaged Mariner was towed back to Tanapag Harbor by the USS *Ellet* (DD-398), a *Benham*-class destroyer named after five members of the Ellet family of Pennsylvania who had rendered valuable service to the Union Army during the Civil War. No injuries were sustained by any of the crew in the emergency landing.[15]

Photo 15-4

USS *Ellet* (DD-398) going astern in February 1939, probably in the New York City area. National Archives photograph #19-N-20352

On 9 November, searching for a ditched B-29 bomber, Lieutenant (junior grade) Hall, in a PBY, sighted its tail gunner and navigator in a raft, and guided a destroyer to them resulting in a safe rescue. Unfortunately, no other survivors of the aircrew were found. A week later, Hall was sent out in his PBY to search for "men in rafts" reported by a B-25. Successfully sighting and locating four men in rafts, the Catalina dropped flares and guided a destroyer to them. A search for other members of the crew in Mae Wests (inflatable life jackets) was made upwind of the rafts, with negative results.[16]

AIR RAIDS ON SAIPAN / *COOS BAY*'S DEPARTURE

During *Coos Bay*'s duty at Saipan, Japanese aircraft conducted several attacks on the airfields and dispersal areas on Saipan and Tinian where B-29 bombers were staged. No attacks were made on shipping in the harbor. The first of these raids for which *Coos Bay* went to General Quarters occurred on 3 November. Her crew remained at battle stations from 0045 to 0235. The last raid endured by her crew at Saipan was on 7 December 1944.[17]

Coos Bay concluded what later proved to be her last seaplane tending operations of the war, when she stood out of Tanapag Harbor on 9 December, bound for San Pedro, California. She was proceeding back to America's West Coast for a badly needed shipyard overhaul, having operated in the forward area for about sixteen months.[18]

Upon completion of the yard period at San Pedro, she sailed for Pearl Harbor, on 11 March 1945, arriving there six days later. She left Pearl on 21 March in company with the aviation supply issue ship USS *Allioth* (IX-204), bound for Ulithi. The two ships sighted and passed Eniwetok on 30 March. A day later, misfortune overtook the *Coos Bay*. On 31 March, her surface radar picked up an unidentified surface contact at 1948, near where an enemy submarine had been reported surfaced.[19]

COLLISON WITH WSA TUG *MATAGORDA*

That morning, *Coos Bay* had received a merchant ship report of a sighting of a surfaced submarine at 12°00'N, 159° 06'E. The seaplane tender then commenced an active anti-submarine patrol. While passing through the submarine's 8-knot position circle (the area in which the sub could then be, if making 8 knots or less), the seaplane tender's crew was at General Quarters.[20]

The previously mentioned radar contact was 28,900 yards distant, almost dead ahead, with both vessels closing one another in the dark. At 2025, after a visual sighting of the other ship only moments before, and with insufficient time to maneuver clear, *Coos Bay* was rammed on her starboard side. The radar contact turned out to be the 185-foot tug *Matagorda*, now colliding with *Coos Bay*. Built in 1943, the V4-M-A1-class tug was operated by the Moran Towing Company for the War Shipping Administration. (It is unclear to the author, based on *Coos Bay*'s war diary, which ship was at fault regarding the collision.)[21]

As *Coos Bay*, damaged and in danger of sinking, was unable to proceed under her own power, *Matagorda* took her under tow to Eniwetok for emergency repairs. Following these, she then towed the tender to Pearl Harbor, for docking and temporary repairs to her hull. In early evening on 27 April, the seaplane tender departed Pearl en route to San Pedro, California, for permanent repairs.[22]

On 16 August 1945, *Coos Bay* departed San Pedro for Pearl Harbor. From there, she carried out some inter-island transport duties until 1 September when she left for Ominato, Honshu, Japan, for post-war occupational duty.[23]

16

Activities of Two Other Future U.S. Presidents

One Avenger torpedo plane, piloted by Lieutenant (jg) G. H. W. Bush, with Lieutenant W. G. White and J. L. Delaney, ARM2c as aircrewmen, was hit by an anti-aircraft burst just as it went into its dive. Bombs were released and the flaming plane levelled by the pilot just long enough for the pilot and one crewman (which one is not known) to bail out. The crewman's parachute failed to open but the pilot landed safely and was rescued by lifeguard submarine, after covering fighters had driven off and sunk two enemy small craft which put out from shore to attempt capture. Lieutenant (jg) Bush was taken on a war patrol by the submarine and eventually returned to the ship two months later. Lieutenant (jg) WHITE and DELANEY are missing in action.

—A Short History of the USS *San Jacinto*, 3 May 1944-15 August 1945, describing the loss on 2 September 1944 by Torpedo Squadron VT-51 of one of its planes, and LTJG William White and Radioman Second John Delaney. Future American president George H. W. Bush, was rescued at sea about 500 miles south-southeast of Tokyo, by the submarine USS *Finback* (SS-230).

On 1 September 1944, the USS *Jacinto* (CVL-30) was under way in the vicinity of the Bonin and Volcano Islands with Task Group 38.4, commanded by RADM Ralph Davidson, USN. The task group was conducting a series of air strikes against enemy shipping, and installation on Chichi Jima, Iwo Jima, and Haha Jima. Embarked aboard the 600-foot light aircraft carrier *Jacinto* were Fighting Squadron VF-51 and Torpedo Squadron VT-51. The former flew F6F Hellcat fighter aircraft, and the latter, TBF Avenger torpedo bombers. The fighter squadron's complement of planes was twenty-four Hellcats, and the torpedo squadron's, nine Avengers. These numbers varied based on aircraft losses and the availability of replacement planes and personnel.[1]

The loss of a VT-51 torpedo plane piloted by Bush and two aircrewmen to combat action occurred the next day during bombing strikes against the Bonin (Ogasawara) Islands. This archipelago lay about six hundred miles south of Tokyo. In a series of thirty-one sorties against installations on the two largest islands—Chichi Jima (father island) and Haha Jima (mother island), LTJG George H. W. Bush, USNR, and his two aircrewmen were shot down.[2]

Photo 16-1

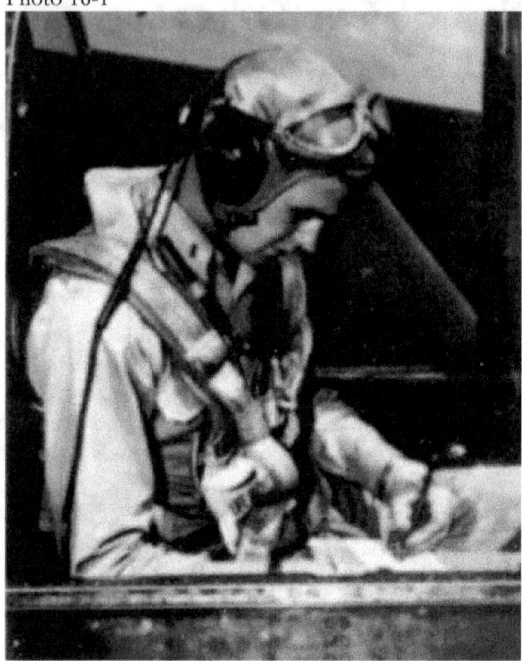

George H. W. Bush as a pilot, sitting in the cockpit of an aircraft.
U.S. Navy Photograph, now in the collections of the National Archives

Bush was then just 20 years old. Two years earlier on 12 June 1942, he had graduated from high school (Phillips Academy in Andover, Massachusetts) and joined the Navy as a seaman, second class. In less than a year, he completed flight training at Naval Air Station Corpus Christi, Texas, and was commissioned an ensign; being for a time, the youngest pilot in Naval Aviation.[3]

LTJG GEORGE BUSH'S STRIKES ON CHICHI JIMA

Bush and his Torpedo Squadron VT-51 squadron mates were assigned to destroy Japanese installations on Chichi Jima in support of the impending 15 September, U.S. invasion of the Palau Islands of Peleliu and Angaur in the Western Caroline Islands. This operation, carried out by VADM Theodore S. Wilkinson's 3rd Amphibious Force, was undertaken because the islands of Morotai, Peleliu, and Ulithi were considered necessary as advance bases before the invasion of the Philippines. The Palaus would later prove useful as staging-points for aircraft and ships en route to Leyte. In retrospect, it was considered hardly necessary, considering that the capture of Peleliu and nearby Angaur cost almost as many American lives as the assault on Omaha Beach, during the Normandy invasion—the gains did not seem to justify the sacrifice.[4]

Photo 16-2

Seaplane base and town on Chichi Jima Island under attack by carrier aircraft on 2 September 1944. Photographed by a USS *San Jacinto* plane.
U.S. National Archives photograph #80-G-248844

On 2 September, Bush awoke aboard *San Jacinto*. It was a sunny morning as he prepared to conduct one of the fifty-eight attack missions he would fly during the war. However, this one, unlike the others which were successful, would end in disaster. The target was a Japanese radio station on Chichi Jima in the Bonin Islands. The enemy on the island had been intercepting U.S. military radio transmissions and warning Japan and occupied enemy islands of impending American air strikes. The communications facility had to be destroyed. Bush, Radioman Second John Delaney, and substitute gunner LTJG William White (replacing Ordnanceman Second Leo W. Nadeau) catapulted off *San Jacinto* in torpedo plane *T-3*. Three other bomb-laden VT-51 aircraft, as well as four of VF-51's Hellcat fighters joined the mission.[5]

Photo 16-3

TBM Avenger of VT-51, taking off from *San Jacinto* for a raid on Guam, 28 June 1944. National Archives photograph #80-G-238783

Nadeau flew as Bush's gunner on all but two of his attack missions. An intelligence officer, White wanted to go along that day to observe the island and had obtained permission to replace Nadeau. Bush, Delaney, and Nadeau had flown into Chichi Jima the previous day and destroyed an enemy gun emplacement. Nadeau later said about the mission, "The anti-aircraft fire on that island was the worst we had seen. I don't think the AA fire in the Philippines was as bad as that." Stanley Butchart, a former VT-51 pilot, later echoed these sentiments, "Chichi was a real feisty place to fly into. As I remember, it had gun emplacements hidden in the mountain areas. In order to get down to the radio facility, you had to fly past the AA batteries, which was risky business."[6]

As expected, the enemy's AA batteries opened fire as soon as Bush and his squadron mates were over the island. Tiny black puffs of smoke thickened around his plane as he put the torpedo-bomber into a steep dive over Chichi Jima. The plane was hit over the northeast side of the island before he reached the radio facility. Bush felt the Avenger "lift" from the impact, but continued his dive toward the target facility and dropped his four 500-pound bombs, scoring damaging hits.[7]

Photo 16-4

TBM-1C Avenger torpedo bomber, of VT-51 embarked aboard the *San Jacinto*, flying over Majuro Atoll, Marshall Islands, on 30 May 1944.
National Archives photograph #80-G-238779

He maneuvered the Avenger, riddled by gunfire hits and aflame, over the water with the hope he could make it back to the *San Jacinto*, but clouds of smoke soon enveloped the cockpit. Choking and gasping for air, Bush and one (not identified) aircrewmen bailed out from about 1,500 feet. The remaining crewman went down with the plane. Bush parachuted safely into the ocean, but dangerously close to the shore, and within sight of a garrison notorious for its brutal treatment of prisoners. Tragically, the crewman who was able to escape the aircraft fell to his death because his parachute failed to open properly.[8]

Once in the water, Bush unleashed his inflatable yellow lifeboat, dragged himself into it, and quickly paddled farther out to sea. The Japanese sent out a boat to capture him. Fortunately, LTJG Douglas West, a fellow VT-51 Avenger pilot, who had observed the ditching strafed the boat. Circling fighter planes transmitted Bush's plight and position to the submarine *Finback* (SS-230) patrolling fifteen to twenty miles offshore from the island. Her patrol mission was two-fold: to conduct lifeguard operations to rescue Allied airmen, and to destroy Japanese combatants and merchant shipping. (Eighty-six different U.S.

submarines would rescue 520 American and Allied airmen during the war.)⁹

FINBACK RESCUES FUTURE PRESIDENT

Photo 16-5

Launching of the submarine USS *Finback* (SS-230) at the Portsmouth Navy Yard, New Hampshire, on 25 August 1941.
Naval History and Heritage Command #NH 53901

A few hours after Bush parachuted from his burning aircraft, CDR Robert R. Williams Jr., the skipper of the submarine, sighted him through the periscope about seven miles from Chichi, surfaced *Finback*, and retrieved the downed pilot. Bush later recounted his reaction upon sighting the submarine emerging from the depths, "I saw this thing coming out of the water and I said to myself, Jeez, I hope it's one of ours." In preparation for the rescue, the sub's bow planes had been rigged out to provide a platform. Crewmembers stepped down from the hull, and plucked Bush from the sea. He was taken inside *Finback*—joining ENS Thomas R. Keene, a TBF Avenger pilot from the carrier *Franklin* (CV-13), and his two aircrewmen, AOM3c James T. Stovall and ARM3c John R. Doherty—then the sub submerged. One minute, Bush was all alone in hostile waters, and the next he was being served food in the red-lite wardroom with music playing on a record player.¹⁰

The following day, *Finback* retrieved LTJG James Beckman, a fighter pilot from the carrier *Enterprise* (CV-6) who was shot down over Haha Jima. The five airmen were put to work as lookouts, standing standard watch rotation, four hours on, and eight hours off. They helped ensure that enemy planes and submarines didn't sneak up on *Finback* during daylight or at night. The submarine often patrolled on the surface in the daytime and always at night when she recharged her batteries. The sub's executive officer, LCDR Dean Spratlin, later remarked about the assimilation of the men, "Bush and the other aviators really got into the submarine experience. Every time an enemy plane would force us down, they'd curse it just like we did."[11]

Many years later, Bush recalled about standing the watches, "I'll never forget the beauty of the Pacific—the flying fish, the stark wonder of the sea, the waves breaking across the bow." The thirty days aboard *Finback* weren't all sanguine, however; as they included being depth-charged and bombed by Japanese ships and planes. Reflecting upon this experience, Bush realized that he much preferred an aviator's lot, over that of submariners:

> I thought I was scared at times flying into combat, but in a submarine you couldn't do anything, except sit there. The submariners were saying that it must be scary to be shot at by anti-aircraft fire and I was saying to myself, 'Listen brother, it is not really as bad as what you go through. The tension, adrenaline and the fear factor were about the same [getting shot at by anti-aircraft fire as opposed to being depth-charged]. When we were getting depth charged, the submariners did not seem overly concerned, but the other pilots and I didn't like it a bit. There was a certain helpless feeling when the depth charges went off that I didn't experience when flying my plane against AA.[12]

While they were aboard, *Finback* sank the Japanese freighters *Hakuun Maru* and *Hassho Maru*, on 11 September. The two cargo ships had been trying to get supplies into Iwo Jima a few months before U.S. forces invaded it.[13]

The rescued airmen remained aboard the submarine until she arrived at Midway Atoll the morning of 30 September. The aviators were transferred to the Naval Operating Base for air transport to Pearl Harbor. "We were supposed to stay at Hawaii for two weeks R&R," Keene recalled, but Bush was concerned about what had happened to his crewmen, and he wanted to get back out to *San Jacinto*. So, we got a ride in a DC-3 and ended up at Guam. We stayed there a few days until we found out where the fleet was." Once back aboard *San Jacinto*, there were few people as happy to see Bush return as his gunner, Aviation Ordnanceman Second Leo Nadeau.[14]

BUSH RECEIVES DISTINGUISHED FLYING CROSS

Lieutenant (jg) George H. W. Bush received the Distinguished Flying Cross for his actions in combat on 2 September 1944:

> The President of the United States of America takes pleasure in presenting the Distinguished Flying Cross to Lieutenant, Junior Grade, George Herbert Walker Bush, United States Naval Reserve, for heroism and extraordinary achievement while participating in aerial flights in line of his profession as pilot of a Torpedo Plane in Torpedo Squadron FIFTY-ONE (VT-51), during the attacks by United States Naval forces against Japanese installations in the vicinity of the Bonin Islands on 2 September 1944. Lieutenant, Junior Grade, Bush led one section of a four plane division which attacked a radio station. Opposed by intense anti-aircraft fire his plane was hit and set afire as he commenced his dive. In spite of smoke and flames from the fire in his plane he continued in his dive and scored damaging bomb hits on the radio station, before bailing out of his plane. His courage and complete disregard for his own safety, both in pressing home his attack in the face of intense and accurate anti-aircraft fire, and in continuing in his dive on the target after being hit and his plane on fire, were at all times in keeping with the highest traditions of the United States Naval Service.[15]

Bush, who would receive three Air Medals in addition to the Distinguished Flying Cross by the time he was discharged in 1945, said in an interview about his service, "There is no question that having been involved in combat has affected my way of looking at problems. The overall experience was the most maturing in my life. Even now, I look back and think about the dramatic ways in which the three years in the Navy shaped my life—the friendships, the common purpose, my first experience with seeing friends die…"[16]

SAN JACINTO CONTINUES COMBAT OPERATIONS

It's not clear when LTJG George Bush was able to catch a hop out to the *San Jacinto* after he arrived at Guam. During the three months since the raids on the Bonin Islands, the ship and her Air Group participated in several combat operations. Actions included strikes against islands in the Western Carolines: Yap Island; Asor in Ulithi Atoll; and Peleliu, Angaur, Ngesebus, and Babelthuap in the Palau Islands—preparatory to amphibious assault operations. Following these operations, came others against Okinawa; Formosa; and Luzon and Visayas, Philippine Islands, in support of the planned occupation of Leyte.[17]

Bush is not cited in Torpedo Squadron VT-51's war diary as having been a participant in the Second Battle of the Philippine Sea, which took place on 25 October 1944. He is included in the list of pilots that took part in successive raids on shipping in Manila Bay and Subic Bay, Philippines, on 13 November 1944. During these strikes, a floating drydock was torpedoed and sunk, two bomb hits were made on a medium size cargo ship, four torpedo hits were made on two Japanese cargo ships, and gunfire was directed at a destroyer escort.[18]

WAR DUTY OF LT GERALD R. FORD JR., USNR

Photo 16-6

LT Gerald Ford taking a sextant reading aboard the light aircraft carrier USS *Monterey*, 1944; and a formal portrait of him taken during World War II.
National Archives Photograph 6923713, and Naval History and Heritage Command Photograph #1043698

While LTJG George Bush was likely still ashore at Guam, *San Jacinto* took part in the Formosa Air Battle (12–16 October 1944), a series of large-scale aerial engagements between Task Force 38 (the Navy' Fast Carrier Force), and Japanese Navy and Army land-based air forces. On 13 October, during fierce combat with a wave of attacking Japanese bombers, *San Jacinto* evaded two aircraft-launched torpedoes, as her gun crews shot down the offending planes.[19]

During attacks on nearby American task groups, the light carrier *Monterey* (CVL-26)—a sister ship of the *San Jacinto*—claimed one enemy plane shot down and assists in downing three others. Aboard her was LT Gerald R. Ford Jr., USNR, the future 38th President of the United States, assigned to the ship as navigation officer, athletic officer, and anti-aircraft battery officer. He had been aboard her since the ship's commissioning on 17 June 1943.[20]

Over the course of the war, Ford would earn the Asiatic-Pacific Campaign Medal with eight battle stars for actions in the Gilbert Islands, Bismarck Archipelago, Marshall Islands, Asiatic and Pacific carrier raids, Hollandia, Marianas, Western Carolines, Western New Guinea, and the Leyte Operation. He also received the Philippine Liberation Medal with two bronze stars for Leyte and Mindoro, and the American Campaign and World War II Victory Medals.[21]

17

Liberation of the Philippines

Should we lose the Philippines operations, even though the fleet should be left, the shipping lane to the south would be completely cut off so that the fleet, if it should come back to Japanese waters, could not obtain its fuel supply. If it should remain in southern waters, it could not receive supplies of ammunition and arms. There would be no sense in saving the fleet at the expense of the loss of the Philippines.

—Adm. Soemu Toyoda, Imperial Japanese Navy, discussing Vice Adm. Takeo Kurita's mission to destroy completely the transports in Leyte Bay following the American invasion of the Philippines, and why there were no restrictions as to the damage that his force might take.[1]

In the summer of 1944, the area separating Allied forces in the Central and Southwest Pacific was growing smaller as MacArthur was concluding his New Guinea campaign and Spruance was continuing his advance through the Central Pacific. Simultaneous landings at Palau and Morotai in September brought them within 500 miles of one another, and made possible a common advance into the Philippines. The Allies wanted Morotai Island (a member of the Molucca Islands lying 300 miles northwest of Sansapor, New Guinea) as a stepping stone for the invasion of the Philippines, and as a base from which nearby Halmahera—a large island hosting a large Japanese garrison and eight air fields—could be neutralized. The Palau Islands, part of the Carolines, were in the eastern approaches to the southern Philippines.[2]

Covering and diversionary operations by Central Pacific forces had begun on 31 August, when a fast-carrier group hit the Bonin and Volcano Islands. This action was followed by further airstrikes and cruiser and destroyer bombardment of the islands on 1-2 September. The entire task force then carried out air raids on Palau and Yap, following which three task groups carried out a series of attacks on Mindanao in the Philippines. As the Japanese forces on Mindanao were unexpectedly weak, the planned attacks were cut short allowing the carriers to move north to fuel and prepare for raids on the Visayan Islands. Located in the central Philippines, the major islands of the Visayas are Panay, Negros, Cebu, Bohol, Leyte, and Samar.³

Map 17-1

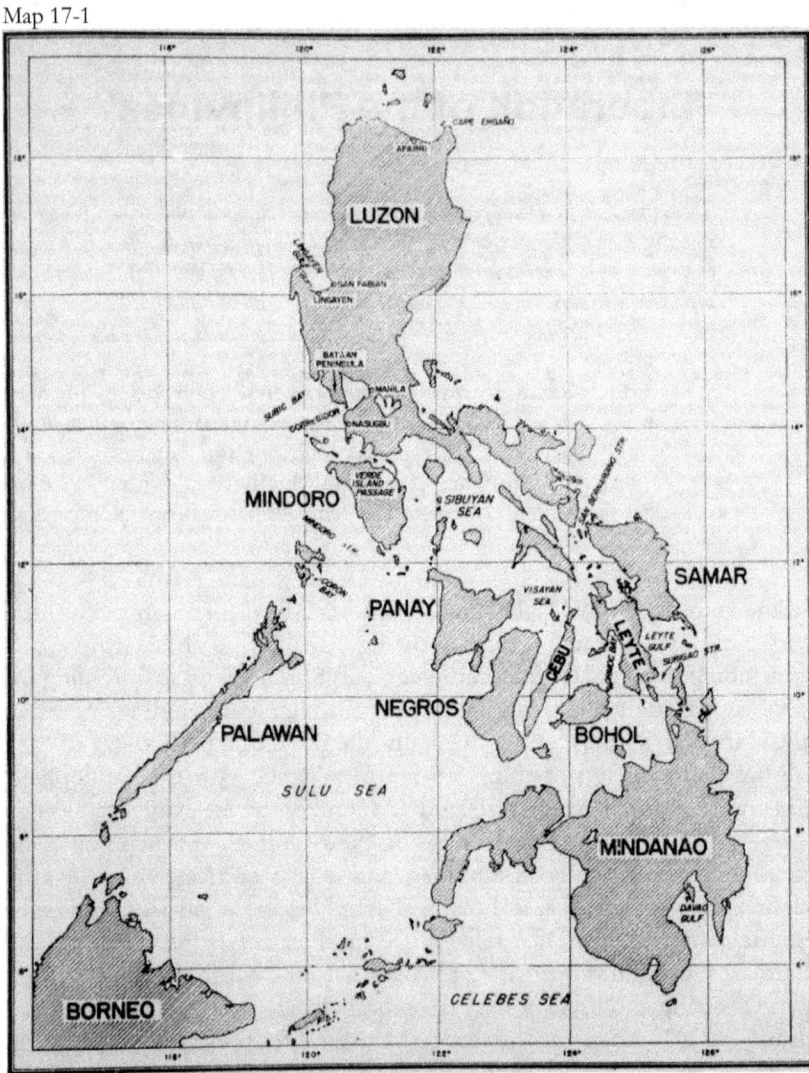

Philippine Islands
United States Navy at War, Second Official Report to the Secretary of the Navy, covering combat operations March 1, 1944, to March 1, 1945 by Fleet Admiral Ernest J. King

On the conclusion of these strikes, one carrier group went south to cover the landings on Morotai, and one east to Palau, while the third replenished fuel, supplies, and arms preparatory to planned air strikes on Luzon. On 15 September, the 1st Marine Division landed on Peleliu in the Palau archipelago, and on the 17th, the 81st Army Infantry Division on Angaur, a two-mile-long island south of Peleliu. Direct air support was furnished by escort carriers, augmented by fast-carrier groups.[4]

Kossol Passage, sixty miles north of Angaur, became a patrol and search base for three squadrons of long-range PBM Mariners including five equipped for air-sea rescue. They arrived on 16-17 September with four tenders: *Pocomoke* (AV-9), *Mackinac* (AVP-13), *Onslow* (AVP-48), and *Yakutat* (AVP-32). By 24 September, captured airstrips were in use by Marine shore-based aircraft, and a heavy bomber runway was operational by 16 October. Opposition to landings by the Army's 31st Infantry Division at Morotai was light.[5]

Photo 17-1

Destroyer USS *Buchanan* (DD-484) in Kossol Passage on 8 October 1944, after rescuing eight of eleven crewmen of a PBM patrol seaplane that had suffered a forced landing and crack-up at sea. Photograph taken from the USS *Chandeleur* (AV-10). Naval History and Heritage Command photograph #NH 97771

On 17 September, forty-two PBM-3D Mariner aircraft of Patrol Bombing Squadrons VPB-16, -202, and -216, and a detachment of VH-1 arrived at Kossol Passage from Saipan. The officers and men of VH-1 initially went aboard the *Onslow*. Following the subsequent arrival of *Chandeleur* (AV-10), they moved aboard her on 27 September. The bombing squadrons conducted long-range searches and anti-submarine patrols. The crews of these planes were able to glimpse the Philippine coast. There was no call for the services of VH-1 during its stay at Kossol. Such rescues for aircraft that were forced down while providing support to operations on Peleliu and Angaur were handled by nearby surface vessels, assisted by scout and observation float planes.[6]

From 21-24 September, the fast-carrier task forces returned to the Philippines to strike airfields on Luzon and the harbor of Manila. After two days of success, the carriers returned to the Visayas. During the month of September, carrier operations in the Philippines destroyed over 800 Japanese aircraft and sank over 150 vessels without damage to U.S. ships and with relatively minor losses in planes. Because of the successes of the fast-carrier strikes and intelligence gathering that indicated the weakness of enemy forces in the Visayas, "top brass" decided to move against them as soon as possible. A plan to capture Yap in the Carolines was dropped, and amphibious landings in the Leyte-Samar area were scheduled for 20 October 1944.[7]

Photo 17-2

Japanese shipping under attack by carrier aircraft off the port of Manila, Luzon, on 21 September 1944. Photographed from a USS *Hornet* (CV-12) plane.
National Archives photograph #80-G-342915

The naval force (comprised of units of the American Third and Seventh Fleets) assembled for the invasion of Leyte was not quite as large as the one that had taken part in June in the invasion of Normandy, but it had more striking power. Embarked aboard the assault vessels were the Sixth Army's Tenth and Fourteenth Corps. The weather at the entrance to Leyte Gulf at daybreak on 20 October was cloudy with altostratus and partial swelling cumulus, a visibility to seaward of twelve miles, and light winds from the southeast. Planners had been concerned that a typhoon might pass through the area and necessitate retirement or diversion of the forces en route from New Guinea. However, the conditions on "A-Day" (Assault Day) were perfect as described by commander, Third Amphibious Force:

> The assault proceeded on schedule following the preliminary bombardment by ships' gunfire and aircraft, a slight onshore tendency of the almost imperceptible wind conveniently drifting the smoke and dust of the bombardment off the beaches and into the interior. The airborne beach observer had made his required report earlier, but the report was unnecessary in this case due to the almost complete absence of surf.[8]

Photo 17-3

Smoke rising from the beachhead on Leyte, during the invasion on 20 October 1944. National Archives photograph #80-G-258354

The landings at Tacloban, located in northeast Leyte on an inlet of the Leyte Gulf, and at Dulag, twenty-five miles to the southward, were made against very little opposition. Naval historian Samuel Eliot Morison noted the following about the operation: "The Leyte landings were easy, compared with most amphibious operations in World War II—perfect weather, no surf, no mines or underwater obstacles, slight enemy resistance, mostly mortar fire." With this beginning, the liberation of the Philippines was off to a good start.[9]

Photo 17-4

Gen. Douglas Macarthur and Philippine president Sergio Osmena in a landing craft on their way to ceremonies proclaiming the liberation of Leyte, 23 October 1944.
National Archives photograph #80-G-289537

The capture of Leyte was part of a strategy to isolate Japan from the countries it had occupied in Southeast Asia, and in particular, to deprive its forces and industries of vital oil supplies. Leyte Gulf was about forty miles wide east to west, and about fifty miles long, and the southern part of the gulf was, in effect, a part of the Surigao Strait which formed a wide waterway between the Pacific and the Sulu Sea. Allied control of the Leyte Gulf area and San Bernardino Strait only 100 miles to the north would deny the Japanese all water routes between the Pacific and the South China Sea except via the northern end of Luzon Island (the economic and political center of the Philippines, being home to the country's capital, Manila) or the southern end of Mindoro Island.[10]

BATTLE OF LEYTE GULF

Photo 17-5

CDR David McCampbell, USN (commander, Air Group Fifteen) in the cockpit of his F6F Hellcat fighter, showing flags denoting 30 Japanese planes he has shot down. The F6F is on board the carrier USS *Essex* (CV-9), 29 October 1944.
National Archives photograph #80-G-258198

The lull following the lightly opposed Leyte landings was short-lived. After receiving word, on 17 October, of the presence of the advance American minesweeping and hydrographic group in the entrances to the Leyte Gulf, Adm. Soemu Toyoda, the commander in chief of the Japanese Combined Fleet, issued the alert for *Sho Ichi Go*

(Victory Operation One), a plan associated with defense of the Philippines. Thus, as amphibious ships were unloading on assault beaches and the Sixth Army was extending to the beachhead, Japanese naval forces were en route to Leyte Gulf to give battle.[11]

U.S. Navy forces during the Battle of Leyte Gulf were comprised of the U.S. Seventh and Third Fleets. The Seventh Fleet (including 6 pre-war battleships and 18 escort carriers, with about 450 Navy aircraft) commanded by VADM Thomas C. Kinkaid was responsible for direct support of the Leyte landings. Third Fleet (including 17 fast aircraft carriers and 6 fast battleships) under ADM William F. Halsey was responsible for providing covering support to the Leyte landings. However, Halsey had the authority to pursue and engage major Japanese fleet forces if the opportunity arose.[12]

In execution of Operation SHO, the majority of Japan's remaining battleship, cruiser, and carrier forces came together in a desperate, multi-pronged attempt to interdict and destroy Allied landing forces off of Leyte and inflict crippling damage on U.S. naval forces. The objective of the Japanese was to destroy the troop transports and supply ships of MacArthur's landing force. The minimal strike capabilities of the enemy's severely depleted carrier air groups were to be augmented by land-based naval and army aircraft from Formosa (Taiwan) and the Philippines.[13]

The Center Force of five battleships and seven heavy cruisers under Vice Adm. Takeo Kurita would reach the central Philippines via passage southward across the Sibuyen Sea and through the San Bernardino Strait, with the intent to attack the U.S. landing force in the Leyte Gulf from the north. A second force of two battleships and a heavy cruiser, commanded by Vice Adm. Shoji Nishimura, would attack Leyte Gulf from the south via passage through the Surigao Strait with intent to draw Seventh Fleet forces away from Kurita's main effort. A smaller force under Vice Adm. Kiyohide Shima, including two heavy cruisers, would join Nishimura to form the Southern Force. A Northern Force of four carriers and two hybrid battleship-carriers, under Vice Adm. Jisaburo Ozawa, would operate northeast of Luzon with the intent of drawing the Third Fleet forces away from Leyte Gulf to further aid Kurita.[14]

Ozawa's Northern Force approached the Philippines from the northeast. It was believed that, if successful in drawing Halsey's Third Fleet away from the Leyte Gulf area, this action would expose the northern flank of Kinkaid's Seventh Fleet invasion force. Possessing superior naval air, Halsey's carrier air wings largely destroyed the Northern Force during the Battle of Cape Engano on 24 October.[15]

U.S. NAVY'S GREATEST ACE AND MOH RECIPIENT

Aboard the USS *Essex* (CV-9), a unit of Halsey's Third Fleet, was CDR David McCampbell, USN, commander Air Group Fifteen. He was responsible for the air group's fighters, bombers, and torpedo bombers aboard the *Essex*. On the morning of 24 October, a large force of Japanese land-based aircraft attempted to attack Halsey's carriers off the east coast of Luzon. Many of the U.S. carriers' aircraft were, at the time, already en route to attack airfields on Luzon and Kurita's battleship/heavy cruiser force, leaving the carriers with a weakened defensive capability.[16]

Taking to the air in a Hellcat fighter, despite the long odds against success and personal survival, and assisted by only one other plane, CDR McCampbell and his wingman, LTJG Roy W. Rushing, intercepted and daringly attacked a formation of sixty enemy land-based aircraft approaching Third Fleet forces. McCampbell shot down nine Japanese aircraft (setting a U.S. single mission aerial combat record), and his wingman shot down six. McCampbell finished the war as the highest-scoring U.S. Navy "ace" of all time with 34 kills in the air. For his actions, on 24 October during the Battle of Leyte Gulf, he was awarded the Medal of Honor. Rushing received the Navy Cross. (Copies of these award citations may be found in Appendix D).[17]

Photo 17-6

Crewmembers of the sinking Japanese carrier *Zuikaku* give a final Banzi cheer after her Naval Ensign was lowered on 25 October.
Naval History and Heritage Command photograph #NH 73070

The Order of Battle (composition of the unit) of Ozawa's Northern Force is identified in the table, including the Imperial Japanese Navy ships sunk in the Battle off Cape Engano, or immediately after. This same convention is used regarding the Order of Battle of the other Japanese forces involved in the Battle of Leyte Gulf.

To preserve space in a book devoted primarily to air-sea rescue, there is little elaboration about the fierce fighting and heroic actions that took place in the skies and on the sea during the Battle of Leyte Gulf. For the same reason, Orders of Battle for the Third and Seventh Fleet's forces have been omitted. Significantly, air strikes that began on 10 October by the four carrier groups of Halsey's Third Fleet against Okinawa and other Ryuku Islands, then Formosa, and the massive air battles on 24 October during the Battle of Leyte Gulf, resulted in approximately 500 Japanese aircraft being shot down or destroyed on the ground. This was more than in the famous "Marianas Turkey Shoot." The loss of these aircraft had a profound effect on the Battle of Leyte Gulf, as Japanese surface forces were deprived of air cover during the battle.[18]

Vice Adm. Jisaburo Ozawa's Northern Force
Battle off Cape Engano (25 October 1944)

Four Aircraft Carriers (Four Sunk)	
Chitose: sunk by aircraft bombing attack	*Zuiho*: sunk by aircraft attack
Chiyoda: sunk by aircraft bombing attack and naval gunfire	*Zuikaku*: sunk by aircraft torpedo attack
Two Battleship-Carriers	
Hyuga	*Ise*: damaged by aircraft bombing attack
Three Light Cruisers (One Sunk)	
Isuzu	*Tama*: sunk by sub torpedo attack
Oyodo (damaged by aircraft attack)	
Nine Destroyers (Two Sunk)	
Akikaze	*Maki*: damaged
Akizuki: sunk by aircraft	*Shimotsuki*
Hatsuzuki: sunk by naval gunfire	*Sugi*
Kiri	*Wakatsuki*
Kuwa	
Supporting Supply Force	
Two oilers	Six escort vessels[19]

The Center Force was attacked by U.S. submarines in the Palawan Strait on 23 October and by U.S. naval air attacks the following day, as it transited the Sibuyan Sea in the center of the Philippine archipelago. Exiting the San Bernardino Strait north of Samar, the force focused on destroying U.S. amphibious shipping to the south in Leyte Gulf.[20]

The following Order of Battle identifies the ships of Kurita's Center Force and total losses, without distinction of which losses resulted from the submarine attack, and two ensuing separate battles.

Vice Adm. Takeo Kurita's Center Force
Submarine attack in Palawan Passage (23 October 1944)
Battle of the Sibuyan Sea (24 October 1944)
Battle off Samar (25 October 1944)

Five Battleships (One Sunk)

Haruna	*Musashi*: sunk by air attack
Kongo: slightly damaged by near misses	*Yamato*
Nagato: lightly damaged	

Ten Heavy Cruisers (Five Sunk)

Atago: sunk by sub torpedo attack	*Maya*: sunk by sub torpedo attack
Chikuma: sunk by air attack	*Myoko*
Chokai: sunk by air attack	*Suzuya*: sunk by air attack
Haguro: severely damaged	*Takao*: disabled by sub torpedo attack
Kumano	*Tone*: lightly damaged

Two Light Cruisers (One Sunk)

Noshiro (sunk by air attack)	*Yahagi*

Fifteen Destroyers (Three Sunk)

Akishimo	*Kiyoshimo*
Asashimo	*Naganami*
Fujinami: sunk by air attack	*Nowaki*: sunk by air attack
Hamakaze	*Okinami*
Hamanami	*Shimakaze*
Hayashimo: sunk by air attack	*Urakaze*
Isokaze	*Yukikaze*[21]
Kishinami	

Owing to the Northern Force's diversion of the U.S. Third Fleet, the Center Force was faced by only three U.S. Seventh Fleet escort carrier task units when it emerged from the San Bernardino Strait in the early morning hours of 25 October. These task units—Taffy 3 off Samar, Taffy 1 southward of Mindanao, and Taffy 2 off the entrance to the Leyte Gulf—had been providing close air support and an anti-submarine screen for the Leyte landings. The resulting clash between the utterly mismatched forces in the Battle off Samar, would prove to be the most dramatic naval engagement of the Leyte campaign.[22]

The Battle off Samar, the culminating action of the Battle of Leyte Gulf, was decided by the courage and resolve of the officers and men of "Taffy 3," the radio callsign of a small U.S. force of six escort carriers, three destroyers, and four destroyer escorts. Facing an overwhelmingly superior Japanese force, Taffy 3 turned certain defeat into a hard-fought victory, with assistance from carrier aircraft of Taffy 2.[23]

Photo 17-7

American survivors of the Battle off Samar, which took place on 25 October 1944, are rescued by a U.S. Navy ship the following day. Some 1,200 survivors of escort carrier USS *Gambier Bay* (CVE-73), and destroyers USS *Hoel* (DD-533), USS *Johnston* (DD-557) and USS *Samuel B. Roberts* (DE-413) were rescued in the days after the action. National Archives photograph #SC 278010

BATTLE OF THE SURIGAO STRAIT

Meanwhile, Vice Adm. Shoji Nishimura's Brunei-based First Attack Force, and Vice Adm. Kiyohide Shima's Kure-based Second Attack Force, combined to form the "Southern Force." On 24-25 October, this force was soundly defeated by the U.S. Seventh Fleet's Bombardment and Fire-Support Group in the Battle of Surigao Strait as it attempted to force its way into Leyte Gulf from the south.[24]

This battle was the last naval battle in WWII in which air power did not play a significant role (except in pursuit of withdrawing Japanese vessels), and was thus the U.S. Navy's last battle-line engagement. Ironically, of the six U.S. Navy battleships participating—*Mississippi* (BB-41), *Maryland* (BB-36), *West Virginia* (BB-48), *Tennessee* (BB-43),

California (BB-44), and *Pennsylvania* (BB-38)—all except *Mississippi* were veterans of the attack on Pearl Harbor in December 1941—they had been damaged during the attack but had been raised/repaired to avenge that fateful day. The battle also marked the last use of the "Crossing the T" maneuver (which resulted in the sinking of the battleships *Fuso* and *Yamashiro*).[25]

Crossing the T was a classic naval warfare tactic used from the late 19th to mid-20th centuries, in which a line of warships endeavored to cross in front of a line of enemy ships, allowing the crossing line to bring all their guns to bear (to either port or starboard) while receiving fire from only the forward guns of the enemy.

Vice Adm. Shoji Nishimura's Southern Force
Battle of Surigao Strait (24-25 October 1944)

Two Battleships (Both Sunk)	
Fuso: sunk by destroyer torpedo attack	*Yamashiro*: sunk by destroyer torpedo attack, and naval gunfire
One Heavy Cruiser (Sunk)	
Mogami: sunk by naval gunfire and destroyer torpedo attack	
Four Destroyers (Three Sunk)	
Asagumo: sunk by destroyer torpedo attack, and naval gunfire	*Shigure*
Michishio: sunk by destroyer torpedo attack	*Yamagumo*: sunk by destroyer torpedo attack

A Japanese force of two heavy cruisers, one light cruiser, and seven destroyers under Vice Admiral Shima, trailing behind Nishimura's force, turned back after the light cruiser *Abukuma* was badly damaged by U.S. PT boats. With his flagship *Nachi*, damaged in a collision, Shima decided that forcing the strait to enter the Leyte Gulf would be futile. Shima's force, part of the Southern Force, is often referred to as the Second Striking Force.[26]

Vice Adm. Kiyohide Shima's Second Striking Force

Two Heavy Cruisers (One Slightly Damaged)	
Ashigara	*Nachi*: damaged in collision with *Mogami*
One Light Cruiser (Sunk)	
Abukuma: torpedoed by PT boat, sunk by USAAF bombers	
Seven Destroyers (Two Sunk)	
Akebono	*Shiranuhi*: sunk by aircraft from USS *Enterprise* on 27 October
Hatsuharu	*Ushio*
Hatsushimo	*Wakaba*: sunk by aircraft from USS *Franklin* on 24 October
Kasumi[27]	

By 26 October 1944, U.S. naval forces had thwarted Operation SHO and destroyed the strategic threat posed by the Imperial Japanese Navy. Despite hard battles ahead, which would include attacks by waves of kamikaze aircraft, the American offensive in the Philippines, and beyond, was to continue unabated.[28]

Photo 17-8

Personal flag of RADM Jesse B. Oldendorf, flown from the mast of the heavy cruiser USS *Louisville* (CA-28), during the Battle of Surigao Strait, 25 October 1944.
Naval History and Heritage Command photograph #NHHC 2003-58-16

COMPARISON OF NAVAL SHIP LOSSES

The Imperial Japanese Navy lost 26 warships during the Battle of Leyte Gulf—Northern Force (7), Central Force (10), and Southern Force (9). These ships are identified in preceding tables. The U.S. Navy lost seven ships. Of note, *Gambier Bay* was the only carrier sunk by naval gunfire during the war, and *St. Lo* was the first major naval combatant to be sunk by a kamikaze attack.

Battle of the Sibuyan Sea

Light carrier USS *Princeton* (CVL-23) sunk by aircraft bombing attack

Battle off Samar

Escort carrier USS *Gambier Bay* (CVE-73) sunk by naval gunfire
Escort carrier USS *St. Lo* (CVE-63) sunk by kamikaze attack
Destroyer *Hoel* (DD-533) sunk by naval gunfire
Destroyer *Johnston* (DD-557) sunk by naval gunfire
Destroyer escort *Samuel B. Roberts* (DE-413) sunk by naval gunfire

After the Battle of Leyte Gulf (28 October)

Destroyer escort USS *Eversole* (DE-404) sunk by submarine torpedo attack[29]

18

Philippines Campaign Air-Sea Rescue

> *The actual rescue of the* COOPER*'s survivors was accomplished in two days. All but Lieutenant Orr and his party were picked up on the first day by five or six PBY trips. A world's record for weight carrying was established in the first two. One plane with Lieutenant Joe Ball had a total of 64 on board, which included 56 survivors. Another with Lieutenant (jg) Essary as pilot had 45 survivors. No one knows how they all got in the plane. However, the plane was approximately 3,000 pounds heavier than the designers said it would fly.*
>
> —CDR Mell A. Peterson, commanding officer of the
> USS *Cooper* (DD-695), sunk by a Japanese torpedo
> in Ormoc Bay, Philippine Islands,
> the night of 3 December 1944.[1]

The seaplane tender *Orca* (AVP-49) arrived in San Pedro Bay the morning of 4 November 1944 and anchored off Jinamoc Island, before receiving aboard, later that day, VPB-34 and two Army Air-Sea Rescue crews. On 18 August 1943 the Army Air Force had drafted plans to organize seven air-sea rescue squadrons, equipped with PBYs for rescue operations, Stinson L-5 Sentinels for liaison, and AT-7 or AT-11 trainer aircraft for utility purposes. The plan called for completion of the program by the spring of 1944, with assignment of most of the new units to the Pacific air forces.[2]

U.S. ARMY AIR FORCE RESCUE SQUADRONS

This plan proved too ambitious. There were only two air-sea rescue units, which the Army termed Emergency Rescue Squadrons (ERS), in operation by summer 1944. Another became operational by the end of 1944, in the Southwest Pacific, but the others did not achieve that status until 1945, and some only at the close of the war. However, despite this tardiness, between July 1943 and April 1945, Army emergency rescue squadrons working with the U.S. Army Fifth Air Force saved 1,841 persons, 360 of them in the month of January 1945, when the Fifth had its own emergency rescue squadron. The Fifth Air Force had received its first emergency rescue squadron (2d ERS) in July 1944. Meanwhile, units of the Seventh Fleet and Navy PBYs had supplemented the resources of the Fifth.[3]

In the Southwest Pacific, air-sea rescue was handled, until summer of 1942, on an emergency basis with whatever equipment was available. Fortunately, the Royal Australian Air Force possessed a few PBYs before creation, in late 1944/early 1945, of its own dedicated Air-Sea Rescue Flights; and in August 1942, the Fifth Air Force had received four PBYs of its own. After the U.S. Army Thirteenth Air Force completed its successful campaigns of the Solomon Islands and joined forces with the Fifth, under the newly created Far East Air Forces (FEAF), the 2d ERS was assigned to the Thirteenth in October 1944, when the Fifth received the newly arrived 3d ERS.[4]

Despite the new self-sufficiency made possible by the presence of two AAF rescue units, FEAF continued to receive the assistance of Navy Dumbo squadrons. On 24 November 1944, the Twenty-First Bomber Command began long-range operations against Japan from Saipan, and the 4th ERS was ordered to the central Pacific for support. But it was not until April 1945 that the 4th ERS began operations with three of its PBYs at Peleliu. Meanwhile, rescue service for the B-29s—which had to fly from the Marianas across 1,400 miles of open sea to reach their targets on Honshu—was largely shouldered by the Navy.[5]

Vice Admiral John H. Hoover, USN, served as commander, Forward Area, Pacific Ocean Areas, during the buildup of B-29 bomber units in 1944-45 in the Mariana Islands. He established a supporting Air-Sea Rescue Task Group (TG 94.11) under CDR H. R. Horney, with units at Saipan, Guam, Peleliu, Ulithi, and, after February 1945, on Iwo Jima. Subordinate units were designated as follows:

- Guam Unit (94.7.1)
- Saipan Unit (94.7.2)
- Iwo Jima Unit (94.7.3)[6]

OPERATIONS FROM *ORCA*

Army air-sea rescue planes sometimes supplemented the efforts of Navy patrol squadrons aboard a few seaplane tenders for short periods, but did not replace the Navy PBYs or PBMs. Information about the Army planes is essentially non-existent in host seaplane tender war diaries. VPB-34 remained only three days before being transferred to the *Tangier* at Morotai Island for temporary duty. A replacement squadron, VPB-11, reported aboard *Orca* two days later.[7]

The *Orca* moved to a new anchorage in San Juanico Strait (separating Samar and Leyte) on 12 November. That afternoon, one of her Army rescue planes departed at 1230 on a mission. Escorted by P38 fighter cover, it arrived an hour later over Ponson Island, which lay to the east of Cebu and west of Leyte. After making a water landing as a prelude to rescuing American servicemen sighted in the native village of Pilar, a native boat came alongside and reported that it was too dangerous to bring the men out because there were three boats with Japanese in them cruising around the bay.[8]

Undeterred, the pilot obtained identification about which boats contained the enemy and then taxiing over to them, strafed them with .50-caliber machine gun fire. The fighter cover took over the attack and eliminated the threat, allowing the Army plane to rescue five men. It then proceeded to Baybay on the west coast of Leyte, picked up four more men, and returned to the *Orca* in the early evening. As air-sea rescue operations from the seaplane tender continued, VPB-11 transferred to U.S. Seaplane Base, Mios Woendi (sited on a small island near Biak off northwestern New Guinea) on 22 November, and VPB-34 returned aboard the *Orca* for duty.[9]

COOPER SURVIVORS RESCUED BY PBY-5 PLANES

Photo 18-1

Destroyer USS *Cooper* (DD-695) when first completed, circa March 1944.
Naval History and Heritage Command photograph #NH 55382

Around midnight on 2 December, the destroyers *Cooper* (DD-695), *Allen M. Sumner* (DD-692), and *Moale* (DD-693) were patrolling Ormoc Bay on the west side of Leyte Island, under orders to seek out and destroy

Japanese transports attempting to reinforce Leyte through the port of Ormoc. Shortly before midnight, a group of three enemy planes, using land background to mask its approach, attacked the three ships of Destroyer Division 120. Two of the attackers were shot down.[10]

On 3 December, two minutes into the new day, the *Cooper* opened fire at a range of 12,000 yards on an enemy destroyer with troops aboard, located close to the beach in Ormoc Bay. Shelled by the *Sumner* as well, the target caught fire and began to sink. The *Cooper*'s 5-inch battery then shifted to a second target in the same vicinity, a ship the size of a destroyer escort, five miles distant. The first salvo hit, and fire was continued until an immediate halt was necessary to avoid hitting the *Moale*, which had crossed her line of fire while taking a ship under fire northward of the *Cooper*'s target. At least one of these targets was sunk, but tragically at 0017, a torpedo struck the *Cooper* on her starboard side and blew a large hole in her hull. The destroyer rolled 45-degrees to starboard, continued over onto her side, and broke in two and sank in less than thirty seconds.[11]

A majority of her survivors spent most of the next fifteen hours in the water. Only one-quarter to one-third could be on a raft, in a floater net, or in a rubber boat at a time. Some of the survivors made it to shore on northwest Leyte, and on Ponson Island, and were aided by Philippine guerillas. CDR Mell A. Peterson, the destroyer's commanding officer, expressed in an interview his gratitude for the assistance provided:

> The Philippine guerillas apparently have an organization to take individuals in the water or groups of survivors to the PBYs so that they do not have to taxi all over the area and pick up small groups.... On northwest Leyte, in the mountains, Lieutenant Orr of the *Cooper* and 22 men were kept the night of December 3rd. These men were outfitted from a small store of Japanese clothing which somehow had been collected by the natives.... The performance of the Philippines guerillas was outstanding and they cared for my people in a very fine fashion.[12]

On 3 December, all the survivors except for LTJG John Irwin Orr Jr. and his party were picked up by VPB-34 over the course of five or six PBY trips. It's unclear how Orr and the others made it back to the fleet, but at least some were probably rescued in ensuing days by squadron aircraft. (Lieutenant Orr later served aboard the heavy cruiser USS *Indianapolis*, and tragically was one of the hundreds of men lost when she was torpedoed by a Japanese submarine, the subject of Chapter One.)[13]

Photo 18-2

Philippine guerillas pose with a captured Japanese flag, a 75mm light field gun and other equipment taken during a raid on Dimasalang, Philippines, 1-4 March 1945. National Archives photograph #80-G-259134.

The two seaplanes rescuing the first two groups set a record for weight taken aloft by Catalinas. An account by the *Orca*'s navigation officer, indicated one of the fifty-seven survivors picked up by the first PBY proved to be a Japanese sailor—presumably from one of the enemy ships sunk in battle—who seemingly adopted an "any port in a storm" philosophy about escaping from the sea. The account provided other details about the rescue effort as well:

> One plane had 57 survivors on board and the other had 45.... All these in addition to the regular 9 men in each plane's flight crew. About this time the Japs on shore, only a few hundred yards away, apparently woke up to what was going on, and they let go with quite a fusillade at the two Catalinas. The pilot of each was positive that the planes could not take off with such a load – but they had to "Get the Hell out of there" – and quickly!! They poured on full throttle and away they went taxiing for dear life for the open sea. After a three-mile run, to the utter and complete astonishment of all hands, the old Cats gave a mighty grunt and heaved themselves into the air and flew.[14]

Photo 18-3

A PBY Catalina of VPB-34, skims the waters of San Pedro Bay, Philippines. Naval History and Heritage Command photograph #L01-11.04.02

Between 3 and 5 December 1944, PBY-5 Catalinas of VPB-34 tended by the *Orca* were able, by sheer daring, to save 166 survivors of the *Cooper*, with 191 men lost. LT Joe Ball, the pilot of the plane that retrieved the first fifty-six Cooper survivors (and one Japanese), many of them severely injured, received the Navy Cross. The pilot of the plane that lifted the second group, LTJG Melvin S. Essary, USN, was awarded the Distinguished Flying Cross. Lieutenant Ball's medal citation reads:

> The President of the United States of America takes pleasure in presenting the Navy Cross to Lieutenant Joe Frederick Ball, United States Naval Reserve, for extraordinary heroism in operations against the enemy while serving as Commander of a Navy PBY-5 Patrol Plane in Patrol-Bombing Squadron THIRTY-FOUR (VPB-34). On 3 December 1944, Lieutenant Ball, as Patrol Plane Commander of a Navy Catalina aircraft landed his plane on the waters of Ormoc Bay, Leyte, and picked up a total of 56 survivors from the U.S.S. *COOPER* which had been sunk during the previous night. He carried out the entire rescue with consummate skill and with total and repeated disregard for his personal safety, remaining on the water for almost an hour with many enemy planes in the vicinity, and repeatedly taxiing his plane well within point-blank range of guns on the enemy-held coastline and of two enemy warships, in his effort to pick up survivors. When his plane could hold no more, he was forced to make a run of three miles in order to get off the water. Upon becoming airborne, he elected to fly his

plane home unescorted in order to provide the quickest possible medical treatment for his passengers, many of whom were wounded, and succeeded in returning his plane and passengers safely to base. His courage and heroic conduct throughout were in keeping with the highest traditions of the United States Naval Service.[15]

PATROL SQUADRON VPB-34 ACCOMPLISHMENTS

In addition to the accomplishments previously described, Patrol Bombing Squadron VPB-34 conducted a variety of missions from San Pedro Bay between 23 November (after returning aboard the seaplane tender *Orca*) and 22 December 1944. These included: air-sea rescue missions; evacuation missions; and transport of scouts, irregulars, and supplies. During these rescue and related operations, a total of 204 Army and Navy personnel were picked up—all from the vicinity of enemy bases or in waters regularly patrolled by enemy aircraft. Rescues were made near Cebu in the Camotes Sea; near Ormoc town, off the Bondoc Peninsula on Luzon; and in various other areas in the Philippine Archipelago. All of these operations were carried out without the loss of a single plane, and no loss or injuries to personnel.[16]

In December 1944, squadron aircraft recovered a total of 183 personnel, one of them Japanese. The latter individual was listed in a war diary entry as having been rescued on 4 December; he was apparently logged aboard *Orca* after being left in the aircraft overnight, following his retrieval from the sea the previous day.

VPB-34 Rescues in December 1944 (183 total)

Date	Pilot/Plane	#Survivors	Location Recovered
3	LTJG Ball (#33)	56	Ormoc Bay
	LTJG Harrison (#32)	7	Ormoc Bay
	LTJG Gillard (#81)	9	Ormoc Bay
	LTJG Essary (#32)	44	Ormoc Bay
	LTJG Day (#21)	26	Ormoc Bay
4	LTJG Ball (#33)	1 Japanese	Ormoc Bay
5	LTJG Gillard (#32)	12	Maltong, Ormoc Peninsula
10	LTJG Utgoff (#6)	7	Bondoc Peninsula, Luzon
11	LTJG Day (#6)	1	Ormoc Bay
	LTJG Kelly (#32)	1	Calugan Grande Island
12	LTJG Pfleeger (#87)	1	11°20'N, 124°10'E
16	LTJG Gillard (#89)	12	Northern shore of Mindanao
21	LTJG Comery (#81)	1	Valencia, Bohol
22	LTJG Ball (#81)	5	Bantayan, Cebu[17]

On 23 December 1944, Patrol Bombing Squadron VPB-34 was relieved of its duties by VPB-54, and proceeded to Manus, Admiralty Islands, via Mios Woendi, New Guinea, for further transit to the United States. Squadron aircraft then flew from Manus to Kaneohe, Hawaii, via stops at Tulagi, Funafuti, Canton, and Palmyra islands. On 10 January, all personnel embarked aboard the escort carrier USS *Hollandia* (CVE-97) for transportation to San Diego.[18]

The 512-foot *Hollandia*, like the escort carriers that fought so valiantly in the Battle of Leyte Gulf, was relatively small and slow. Escort carriers were former merchant vessels, or hulls laid down as such, to which a flight deck was affixed atop them. *Hollandia* had been converted from a Maritime Commission hull by Kaiser Co., Inc., of Vancouver, Washington. Following arrival of the escort carrier at Naval Air Station, San Diego, on 16 January, squadron personnel were reassigned and reclassified by commander, Fleet Air West Coast. Six days later, all personnel were detached on the 22nd and the squadron was reduced to caretaker status. Patrol Bombing Squadron VPB-34 was disestablished on 7 April 1945.[19]

Photo 18-4

USS *Hollandia* (CVE-97) arriving at Naval Air Station Alameda, California, on 29 November 1945. She was transporting 1,100 servicemen home from Eniwetok and Kwajalein, in the Marshall Islands.
Naval History and Heritage Command photograph #NH 106721

SUPPORT FOR STRIKES ON BYPASSED AREAS

During the Philippines Campaign, some of the Patrol Squadrons in the Philippines were temporarily reassigned where needed elsewhere, before returning to the Philippines. One such was Squadron VPB-33, which left the Leyte area on 29 November, bound for Los Negros Island in the Admiralty Group. On 1 December, seven planes were sent to Emirau, Green, and Treasury islands, from which they covered strikes on the Gayelle Peninsula of New Britain, on New Ireland, and on Bougainville Island. Dumbo planes from Green Island also landed Australian scouts and coast watchers on Lihir, Tanga, and Feni islands, which lay north, northeast, and east of New Ireland, respectively. During this period, two air-sea rescues were made, both on Christmas Day.[20]

Map 18-1

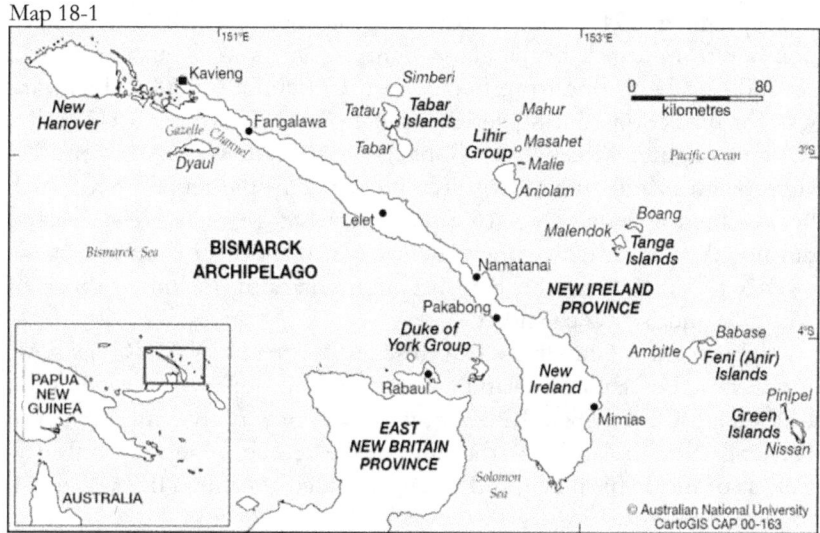

Bismarck Archipelago
Courtesy of Australian National University

RESCUE OF TWO RNZAF F4U-1D CORSAIR PILOTS

United States DFC   **United Kingdom DFC**   **French Croix de Guerre**

LTJG Jack L. Thurman, USNR, when based at Green Island, rescued Squadron Leader Mortimer Tuke Vanderpump, RNZAF, in

Ataliklikun Bay, New Britain, off Keravat Air Strip on Christmas Day, 1944. No shore fire was received during the rescue, but anti-aircraft fire was encountered at Cape St. George en route to the bay. Vanderpump, piloting an F4U Corsair with RNZAF No. 24 Squadron, had been shot down by AA fire earlier that day while making a strafing run and had to ditch in the bay.[21]

It is interesting to note that Vanderpump, the commanding officer of No. 24 Squadron, was the holder of two Distinguished Flying Crosses from earlier in the war. He received the first one, an American DFC on 17 September 1943, for his part in protecting a disabled American bomber from attack by Japanese fighters. He was "Gazetted" for a second DFC (this one from the UK) on 15 February 1944. The award was for his actions during aerial combat on 17 December 1943 when, during a fighter sweep over Rabaul, he shot down an unidentified aircraft and a Zero, two of five aircraft claimed by the squadron to which he was then assigned (No. 16 Squadron, RNZAF). As New Zealand did not then have its own awards system, the British Monarch made awards to subjects of the British Commonwealth, including those of Australia and New Zealand.[22]

The second Christmas Day rescue was made by LTJG Jack W. Jones, based at Emirau Island. He picked up Squadron Leader John Rutherland Clark Kilian, RNZAF, twenty to twenty-five miles north of Kavieng, New Ireland. At the time of the rescue, Jones had only 150 gallons of fuel in his plane, and when he landed at Emirau this was down to just 70 gallons.[23]

Kilian had parachuted into the sea earlier that day, after his Corsair was hit by flak. Then the commanding officer of No. 19 Squadron, he was one of New Zealand's fighter 'aces' during the war. Kilian joined the Royal New Zealand Air Force in 1937. His first combat duty was in Europe flying Spitfires with No. 485 (NZ) Squadron, and Nos. 122, 222, and 504 RAF Squadrons. During this service, he commanded No. 122 Squadron (in 1942) and No. 504 Squadron (1943). Kilian earned the French Croix de Guerre for his part in Operation JUBILEE, an Allied assault on the German-occupied port of Dieppe, France, on 19 August 1942. After returning to New Zealand in November 1943, he commanded No. 19 Squadron and late in the war, No. 14 Squadron.[24]

The VPB-33 planes and crews at Green and Emirau were recalled to Los Negros on 6 January 1945, in preparation for the squadron's return to the Philippines and Leyte Gulf. The planes at Mios Woendi preceded them, on 20 December, in returning to the Philippines to conduct anti-submarine patrols, tended by and with crews embarked aboard the *San Carlos*. By 10 January 1945, the entire squadron had returned to the Philippines to operate again in a Dumbo capacity; with their personnel being stationed aboard the seaplane tender *Tangier*, together with those aboard the *San Carlos* continuing to run anti-submarine patrols.[25]

ACTIVITY ABOARD *TANGIER*

Lying at anchor in San Pedro Bay, *Tangier*, was under the operational control of the USAAF 5th Air Force's Air-Sea Rescue Group 5276. Her duties included routine maintenance, upkeep, and repair of PBY and PBM aircraft by the Patrol Aircraft Service Unit (PATSU 17-3-1) attached to the ship. The PBYs and PBMs conducted rescue and cover missions for Army bombing strikes against enemy installations in the central Philippines. The Air-Sea Rescue Group assigned the missions, and fighter cover was provided for each one.[26]

The seaplane tender was overcrowded in supporting Patrol Bombing Squadrons VPB-54 and VPB-33, which were aboard for duty, and Squadrons VPB-20, VPB-25, and VPB-71 embarked for temporary berthing and subsistence. Over the course of the month, January 1945, until *Tangier* departed for Lingayen Gulf on 24 January, rescues were made on nine of those days, resulting in the recovery of eighty-two servicemen and three Japanese prisoners. Summary information about these operations, which follows, provides an overview of Allied offensive operations in progress in the Philippines. *Tangier* war diary entries do not identify which squadrons were responsible for individual rescues.[27]

AIR-SEA RESCUE OPERATIONS

On 3 January 1945, four planes covering strikes on Davao and Cebu also searched water areas for survivors. These efforts resulted in the recovery of six men from Panay Island who were attached to the 500th Bombing Squadron, 345 Bombing Group. The following day, three planes covering Manila strikes searched for survivors off Samar and southern Luzon recovered eleven men of VPB-104 off Samar, and also picked up an Army pilot from a life raft off southeastern Luzon.[28]

LTJG Woodford W. Sutherland, piloting a PB4Y-1 Liberator bomber on the return leg of a patrol along the China coast, had sighted an enemy barge-gunboat (Type C) anchored close inshore at San Fernando, Luzon. As he made a low strafing run, shore batteries opened up with heavy, medium, and light anti-aircraft fire, which burst all around the plane. Using violent evasive action, he avoided the flak, and reached the vicinity of his base at Tacloban, Leyte. Clouds obstructed the field, so the searchlight beam was not visible. As navigation equipment in the plane was out of commission, and with only 40 gallons of gas left, he had been forced to ditch off the east coast of Samar. All of the flight crew were rescued by friendly natives, and returned to base by the Dumbo two days later.[29]

On 6 January, *Currituck* (AV-7), flagship for commander Aircraft, Seventh Fleet, stood out of San Pedro Bay to rendezvous with Task Force 79 (Philippine Invasion Force) en route to Lingayen Gulf. While making the passage, she left Squadron VPB-20 temporarily with *Tangier*, which was now tending 40 planes with 225 officers and approximately 1,200 men aboard ship. On the day of *Currituck*'s departure, three separate missions to the central Philippines had been made and a total of eight survivors had been returned to *Tangier*. Three had been badly injured when their ship burned and sank as a result of enemy action. The others were from downed Army Air Force planes.[30]

The next rescues were made on 10 January, when five men, attached to various Navy carrier squadrons, were recovered from the central Philippines. Overcrowding aboard *Tangier* diminished somewhat with the departure of Squadron VPB-20 for the Lingayen Gulf. All missions on 11 and 12 January were cancelled owing to strong winds and heavy seas.[31]

On 15 January, seventeen parachute infantry men were rescued from Palawan Island and between it and Mindoro. A separate landing on Fonhead Island recovered an Army fighter pilot and information from guerrillas on others who would be ready for evacuation the following day. Unfortunately, two missions made to the central Philippines the next day, with the hope of recovering these individuals, produced negative results.[32]

Three missions over the central Philippines on 17 January rescued eleven members of VPB-20, who had crashed off Mindoro the day after Christmas. They were part of a mission on the night of 26 December, when four PBM-3D Mariners of the squadron had attacked a Japanese naval task force engaged in shelling Allied beachheads at Mangarin Bay, Mindoro. This force had consisted of a heavy cruiser, a light cruiser, and six destroyers. One destroyer was damaged by the attack. Two of VPB-20's planes were shot down, but their crews were rescued (some during the mission of 17 January). One man, fatally injured, died after rescue. A third plane of the squadron while returning, ran out of fuel short of base and had to be refueled in enemy waters by a crash boat with motor torpedo boat escort.[33]

Also recovered on a 17 January mission, were two carrier pilots and three Japanese prisoners. The prisoners were turned over to Army authorities.[34]

The last rescues before *Tangier* left San Pedro Bay for Lingayen Gulf occurred on 20 and 21 January. Five Navy carrier pilots were rescued from southern Luzon on the 20th. Three missions over the central Philippines the following day recovered five Army Air Force B-25 officers and men. What was unusual about these missions and very heart-warming, was that they included two U.S. Marines and one Navy radioman who were on Corregidor and Cavite in 1941 and had escaped from a Palawan Japanese prison camp on 14 December 1944. Of the 146 enlisted men and four officers held in the camp, they were among only eleven survivors of a massacre that occurred that day.[35]

MASSACRE OF AMERICAN POW AT PALAWAN

> *During an attack on a branch camp by the Allies, the main force shall keep strict guard over POWs, and if there is any fear that the POWs would be retaken due to the tide of battle turning against us, decisive measures must be taken without returning a single POW.*
>
> —Written order sent to each Japanese branch camp commander in May 1944, which was introduced at the Palawan Massacre trial which began on 2 August 1948 in Yokohama, Japan.[36]

Japanese Prisoner Camp 10-A was located near Puerto Princesa, the capital city of the island of Palawan. The POWs had been brought there to provide the labor necessary to build an airfield for their captors. The 131st Airfield Battalion under Capt. Nagayoshi Kojima, was in charge of the prisoners and airfield at Palawan. General MacArthur had previously sent correspondence to Field Marshal Count Hisaichi Terauchi, the Japanese commander in chief in the Philippines, warning him that his command would be held responsible for any abuse of prisoners, internees, and noncombatants. This directive failed to serve as a deterrent to the murder of every POW, except for the eleven who were able to escape on 14 December 1944.[37]

On 14 December, Japanese aircraft sighted an American convoy, which Japanese commanders believed was destined for Palawan but which was actually headed for Mindoro. Afterward, following a sighting by Japanese at the prison camp of two P-38 Lightning fighter aircraft, the POWs were ordered into air raid shelters. After the prisoners emerged from their shelters, a second alarm in early afternoon sent them back into the shelters, where they remained closely guarded.[38]

Suddenly, in a planned act, Japanese soldiers acting under the orders of 1st Lt. Yoshikazu Sato doused the wooden shelters with buckets of gasoline and set them afire with torches, followed by attacks with hand grenades. As men engulfed in flames broke out of their deathtraps, the guards machine gunned, bayoneted and clubbed them to death. Thirty to forty Americans escaped from the massacre area, either through the barbed-wire fence perimeter of the camp, or under it, where some concealed tunnels had been prepared by the prisoners for use.[39]

Map 18-2

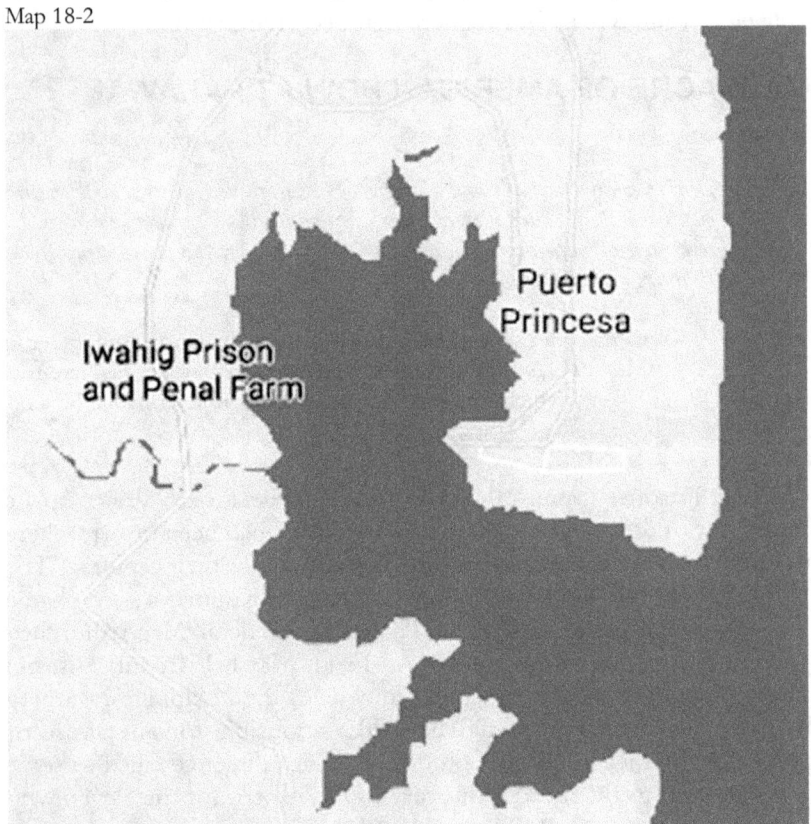

Central east coast of Palawan Island, Philippines

The men fell and/or jumped down the cliff above the beach area, where most sought hiding places among the rocks and foliage. Several immediately attempted to swim across Puerto Princesa's bay, but were shot in the water. Japanese guards used dynamite to force prisoners from their hiding places of cracks and crevices along the beach, and shot them when found. Finally, a call for dinner diverted them from their deadly task. Without the abandonment of the search for surviving prisoners in favor of hot food, there may well have been no survivors.[40]

Most of the eleven survivors swam across the bay and were rescued by inmates of Palawan's Iwahig Penal Colony, where several of the guarding officials were involved with the local resistance movement. Guerrillas escorted three of the Americans—Marine Sergeant Douglas Bogue, Marine PFC Glen McDole, and Navy Radioman First Class Joseph Barta—down the coast to Brooke's Point on southeast Palawan. The seaplane that picked them up, evacuated the men to Leyte where they told their story to U.S. military authorities. (It's unknown to the author how, or when the other survivors were recovered.)[41]

LINGAYEN GULF LANDINGS

> *The enemy's large scale employment of suicide planes give rise to highly concentrated and desperate anti-aircraft fire.... About forty-three planes reached their targets [Allied ships], many after flying through the heaviest concentration of gun fire. A plane diving at speed in excess of four hundred fifty knots becomes in effect a low velocity projectile; its momentum is so great that it may not be deflected by volume of hits alone. The loss of a tail or the total disintegration of a wing generally, but not always, causes the plane to swerve from its general course. Killing the pilot may or may not deflect the plane.*
>
> —VADM Thomas C. Kinkaid, commander, Luzon Attack Force, highlighting the grave danger that kamikaze aircraft presented to Allied shipping during Lingayen Gulf landings in January 1945.[42]

The climax of the Philippines campaign was the invasion of Lingayen Gulf in western Luzon. At 0930 on 9 January 1945, assault forces of the U.S. Sixth Army under Lt. Gen. Walter Krueger commenced landing on Lingayen Gulf beachheads in the Lingayen-Dagupan-San Fabian-Rabon area of Luzon. The objectives of the operation were the seizure of the central Luzon plain and Manila area, with the subsequent denial to the Japanese of the northern entrance to the South China Sea. As the attack force came ashore, the enemy offered little opposition to landing craft approaching the beach or soldiers disembarking, with the only resistance coming on the northeastern flank. This reflected the new tactics adopted by the Japanese to withdraw from established beach defenses and to instead fight inland in the jungle.[43]

The Philippines campaign revealed the poor condition of the Japanese air force. Although the Empire's production of planes had increased in 1943 and 1944, and the quality of aircraft had improved, the ratio of losses in combat accelerated dramatically. Having lost many superb pilots earlier in the war, the Japanese were left with untrained pilots, but adequate numbers of planes. In desperation they resorted to kamikaze suicide tactics.[44]

ROYAL AUSTRALIAN AIR FORCE AIR-SEA RESCUE

Before progressing on to the invasions of Iwo Jima and Okinawa, as part of the drive through the Central Pacific to the Japanese home islands, the next three chapters will visit Australian air-sea rescue efforts. Three RAAF Air-Sea Rescue Flights (No. 111, 112, and 113) were then operating, or soon would be from Madang, New Guinea; Darwin, Australia; and Morotai, Molucca Islands, respectively.

19

Air-Sea Rescues from Madang, New Guinea

Photo 19-1

Madang Harbour, New Guinea, in October 1945 with two ships at anchor and, in the background on the left, an airstrip.
Australian War Memorial photograph P02257.025

Madang, on the north coast of New Guinea, was home to RAAF No. 111 Air-Sea Rescue Flight, which was formed there on 13 December 1944. The Flight was one of the units assigned to Northern Command RAAF, that coincidentally was headquartered at Madang. Northern Command was responsible for conducting operations in New Guinea, New Britain, and Bougainville. A chronology of the Flight's activities is related in the following paragraphs.[1]

Photo 19-2

Sign on the entrance of RAAF Northern Command Headquarters, at Madang, New Guinea, showing the whereabouts of RAAF radio station, "The voice of the islands," circa October 1944.
Australian War Memorial photograph OG1956

Flight Lieutenant Ian James Lock Wood (DFC) and his Catalina crew were the first to report to the new unit. Wood had just received an immediate award of the Distinguished Flying Cross, on 6 December 1944, for heroic action on 10 November 1944. While assigned to No. 8 Communications Unit based at Madang, Wood had been the pilot of a Catalina which rescued five American airmen from a raft off Cape Wom near Wewak, New Guinea. They were survivors of a downed B-25 aircraft of Fifth Bomber Command that had crash-landed in the sea in enemy-controlled waters. Wood was well acquainted with New Guinea, having been a patrol officer there before the war.[2]

Photo 19-3

Jacquinot Bay, New Britain, 21 November 1944. Flight Lieutenant Ian J. L. Wood is pictured on the left, with Warrant Officer A. V. Dawson at the right of the photograph. Australian War Memorial photograph 077157

Over the remainder of the month, three additional aircrews, the adjutant, and twenty-two other personnel joined them. Catalinas *A24-98* and *A24-104* were received in late December 1944 from RAAF No. 8 Communication Unit. All members of the Flight were billeted and rationed (berthed and fed) by the communication unit stationed there until their own camp was built. On 6 January, the officer ordered to command, Flight Lieutenant C. W. Miller (DFM) arrived, and Wood relinquished the temporary position he had assumed on his arrival to the more senior officer. (Wood had been an acting flight lieutenant when he received the DFC.) Among the personnel that were gradually reporting to the squadron, were three airmen who had been previously decorated for bravery—recipients of the Distinguished Flying Cross or Flying Medal.

- Flight Officer George William McMaster (DFM)
- Pilot Officer N. M. Hall (DFM)
- Pilot Officer J. R. Riddle (DFM)[3]

In January 1945, the Flight received Catalina *A24-91*. By month's end, it had over one-half of its allowance of personnel, and three of its four aircraft. The summary information in the table provides insight into the composition of the 4-Catalina air-rescue flight at this time.

Personnel	Allowance/Strength	Aircraft	Allowance/Strength
Officers	17/10	Catalina	4/3
Warrant Officers (WO)	2/5		
Flight Sergeants (Sgt)	1/3		
Sergeants (Sgt)	21/10		
Corporals (Cpl)	18/5		
Air Crews	41/25		
Total allowance versus strength[4]	100/58		4/3

On 9 February, Flight Lieutenant Donlan and aircrew arrived with the Flight's final aircraft, Catalina *A24-107*. Miller took her out on night bombing missions over the Wewak area on 15 and 17 February, searching for targets of opportunity. Results of the unit's bombing and strafing were indeterminate. After that, the Flight's work involved Dumbo covering of RAAF Beaufort strikes on Adler Bay, Jacquinot Bay, and Wide Bay in New Britain, and responding to a handful of air-sea rescue missions on which no sightings of any survivors were made. On 19 April, a Catalina evacuated wounded personnel from Annanberg, a village in the Ramu Valley in northeast New Guinea. (Beaufort was a reference to the twin-engined RAAF Bristol Beaufort torpedo bomber; "covering" meant being in position to rescue aviators downed by combat, or aircraft malfunction.)[5]

Photo 19-4

Astrolabe Bay, Madang, New Guinea, 1 October 1945. A group of officers aboard RAAF Crash Launch *08-40* of No. 111 Air-Sea Rescue Flight. The man standing (extreme left) is Flying Officer Allen Robert Griffiths, while the man standing next to him may be Gordon Thomas. The boat is flying the RAAF ensign (flag) from its mast. Australian War Memorial photograph P02257.082

On 24 April, Flight Lieutenant Cuthbertson and crew in Catalina *A24-104* practiced air-sea rescue and homing and bearings in company with an air-sea rescue launch. The following day, they proceeded to Nightingale Bay on the north coast of New Guinea, to attempt to locate the crew of a downed U.S. Army Air Force Douglas A-20 Havoc light bomber. One survivor was rescued.[6]

On 2 May, No. 111 ASR Flight personnel began building a slipway and engineering sections at the base. This was necessary as the Flight was to receive PBY-5A Catalinas (the amphibious version of the aircraft, which had retractable tricycle landing gear) to replace its PBY-5s which were pure flying boats. On 10 May, Catalina *A24-104* practiced dinghy dropping in cooperation with an air-sea rescue launch. The following day, the same aircraft proceeded over Wewak to cover airstrikes by Beaufort aircraft at amphibious landings. None of the aircraft were forced to ditch in the sea.[7]

Replacement PBY-5A Catalina *A24-57* arrived at Madang on 12 May; followed by Catalinas *A24-44*, *A24-55*, and *A24-59* in June. On the last day of the month, Flight Lieutenant Miller lifted off in *A24-44* to fly air-sea rescue cover for HRH Prince Henry (Duke of Gloucester, and Governor-General of Australia) travelling from Madang to Jacquinot Bay aboard the *Endeavor*, an Avro York transport aircraft. As July broke, No. 111 Air-Sea Rescue Flight had 76 of its 111 authorized personnel, and four PBY-5As had replaced the older PBY-5 Catalinas.[8]

Photo 19-5

Jacquinot Bay, New Britain, 3 July 1945. The *Endeavor*, an Avro York transport aircraft used by the Governor-General of Australia, stands ready to take him to Bougainville. Australian War Memorial photograph 093621

Photo 19-6

Torokina, Bougainville, 5 July 1945. Note Vice-Regal crest on the *Endeavor*.
Australian War Memorial photograph 093688

EPIC RESCUE OF A USAAF FIGHTER PILOT

Photo 19-7

Wewak area, northeast New Guinea, 8 July 1945. A Supermarine Walrus amphibian aircraft of a RAAF Air-Sea Rescue Flight lands on the swift running waters of the Karawari River to rescue a downed U.S. flyer.
Australian War Memorial photograph OG3067

The most dramatic air-sea rescue of the Flight's wartime service, came the first week in July. While Miller in *A24-44* was escorting the *Endeavor* to Jacquinot Bay, and from there to Torokina, Flying Officer Bergmann in Catalina *A24-59* conducted a search for a C-47 military transport aircraft missing from Port Morseby. Unsuccessful in locating the transport he landed at Port Morseby and the following day, 5 July, escorted the Duke of Gloucester from Port Morseby to Cairns, Australia. Bergmann, taking over Miller's assignment, freed the Flight commander to search for the pilot of a downed P-38 Lightning.[9]

That same day, 2nd Lt John P. Carter of the US Army Air Corps took off from the Nadzab Airfield (26 miles outside Lae) on a training mission in a P-38 Lightning fighter aircraft. While passing over the Sepik River, both engines failed. Carter bailed out at 5,000 feet, and landed unhurt in kunai grass (an aggressive, green-leaved plant which could grow to seven feet in height)—about 250 yards from the Karawari River, and near the Sepik River, in northern New Guinea. His aircraft crashed into a swamp nearby.[10]

Carter had only a pint of water and a D-ration chocolate bar to sustain him in the sweltering heat of the tropical lowland rainforest. The next day, two B-25 Mitchell bombers dropped emergency rations and supplies, which landed too far away to be retrieved by Carter. The same day, Flight Lieutenant Miller in *A24-44* located Carter in the dense grass, but the nearest area of the river suitable for alighting was about two miles away. The Catalina was too large to land in the Karawari near his position, but a smaller RAAF Walrus seaplane was suitable for the purpose. Carter was only one-quarter mile from a point where a Walrus could land, but it was too late in the afternoon for one to arrive before dusk. The Catalina led the pilot toward the desired river location, then departed.[11]

On 7 July, a RAAF Walrus piloted by Flying Officer Neil M. E. Agnew circled Carter, but was unable to locate him due to a low cloud base, and departed. It was not possible to carry out another attempt before dark. Miller, in *A24-44*, did locate the downed flyer and dropped messages, keeping him apprised of the situation. Miller also dropped emergency supplies of food and water, which Carter was able to retrieve. After satisfying hunger and thirst, Carter fell asleep from exhaustion. One of the messages instructed him to wait until the next day for rescue, and to make his way to the river. Carter was unable to do so, owing to the thick grass and swamp.[12]

On 8 July, the Walrus escorted by the Catalina arrived back on scene. Agnew piloted the rescue plane with Lt. Johnston from the Australian New Guinea Administrative Unit and two native police boys aboard. He successfully landed in the Karawari despite floating debris and a prevailing four to six knot current.[13]

While the Walrus waited on the river, the rescue party deployed an inflatable rubber dinghy, and used machetes to cut a trail to reach Carter and bring him back to the river. Two Beauforts from No. 100 (RAAF) Squadron covered the rescue by strafing suspect Japanese positions along the Karawari, keeping them away from the rescue site. Once the rescue party and recovered pilot were aboard, the Walrus took off and landed at Tadji Airfield, located in a sago swamp area inland from the north coast of New Guinea.[14]

Photo 19-8

Two native police boys landed by a RAAF Walrus on the Karawari River, make for the shore in a dinghy. They spent an hour hacking their way with machetes through the kunai grass to 2nd Lt John Carter, 250 yards away, then bringing him back to the plane. Australian War Memorial photograph OG3072

Photo 19-9

Group photograph of those involved in the downed pilot's epic rescue. From left: Lieutenant Johnston; 2nd Lt John Carter; the two native police boys, for whom the rescued pilot had a special word of thanks; and Flying Officer Neil Agnew, RAAF. Australian War Memorial photograph OG3076

20

Two RAAF Bombers and a Catalina lost in Dutch East Indies

Photo 20-1

Near Flores Island, Netherlands East Indies, 6 April 1945. A PBY Catalina piloted by Flight Lieutenant Robin M. Corrie, lands to pick up survivors from two RAAF B-24 Liberator bombers, and another Catalina destroyed by enemy aircraft after retrieving aircrewmen from the sea. Corrie's aircraft was itself attacked while attempting to take off, but successfully returned to base with its crew and survivors intact.
Australian War Memorial photograph NWA0802

On 6 April 1945, two PBY Catalinas of No. 112 Air-Sea Rescue Flight attempted a daring rescue of the survivors of two crews of RAAF B-24 Liberator bombers shot down by Japanese Zeros in the Netherlands East Indies. The seaplanes, based at Darwin, Australia, specialized in long-range rescues of downed aircrews in Japanese-held waters. In carrying out their mission, one of the Catalinas was destroyed on the water by an enemy fighter. The other Catalina, piloted by Flight

Lieutenant Robin Corrie, evaded two Irvings (twin-engined Nakajima J1N fighters) and, heavily laden with three survivors of the downed bombers and all but one of the crew of her sister flying boat, returned safely to Darwin.

The series of events leading to the loss of the bombers and Catalina began two days earlier on 4 April, when the Japanese cruiser *Isuzu* left Surabaya, Netherlands East Indies. Escorted by the torpedo boat *Kari* and minesweepers *W-12* and *W-34*, she was bound for Kupang on Timor to transport an Army detachment to Sumbawa Island. Three American submarines operating together—*Charr* (SS-328), *Besugo* (SS-321), and *Gabilan* (SS-252)—sighted the group off Paternoster Island, but aircraft forced the pack to dive and they were unable to attack.[1]

The following day, 5 April, RAAF twin-engine de Havilland "Mosquito" fighter-bombers from No. 87 Squadron (based at Coomalie Creek Airfield in Australia's Northern Territory) shadowed *Isuzu* and the smaller warships. In early morning on 6 April, an intelligence message from ComSubPac (commander, Submarine Force, Pacific Fleet, at Pearl Harbor) provided the submarines with location information about the enemy ships, and the submarines took up position to intercept the enemy near Bima Bay, on the northeast coast of Sumbawa. The US submarines were joined there by the British submarine HMS *Spark*.[2]

Map 20-1

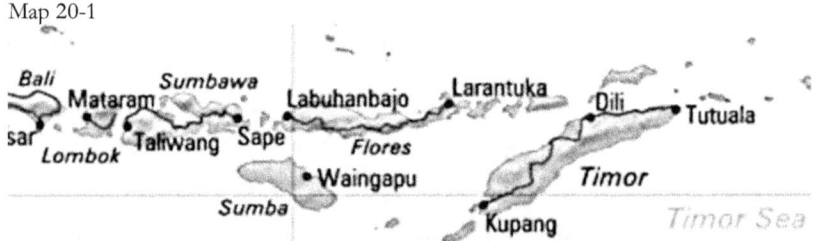

Lesser Sunda Islands (Sumbawa, upper left, is one of the Lesser Sunda Islands, a string of volcanic islands stretching eastward from Java toward northwest Australia. Flores Island lies eastward.)

Isuzu embarked the soldiers at Kupang Harbor at dawn on 6 April, and departed for Sumbawa. North of her destination, she came under attack by ten B-25 Mitchell bombers of the Dutch 18th Squadron based at Darwin. She was slightly damaged by near misses of 300kg bombs that exploded off her starboard bow, but was able to land the troops at Bima Bay. Later, following her departure eastward, B-24 Liberator bombers of RAAF No. 21 and No. 24 Squadrons, based in Northern Australia, attacked the group near Flores Island. *Isuzu* was further damaged by bomb hits in the bow, but Mitsubishi A6M Zeros shot down two of the B-24s.[3]

While making a second attack run on the *Isuzu*, B-24 bomber *A72-77* (of No. 24 Squadron) piloted by Flight Lieutenant Eric Valentine Ford, was attacked by a Zero head on and slightly below. Ford was flying in the number 3 position of a three-plane formation. His plane's No. 3 engine caught fire and flames were seen from near the nose wheel compartment. The starboard wing was then alight and Ford motioned to Warrant Officer Colin George Vickers, 2nd Pilot, to abandon the aircraft. The back hatch was jettisoned and four members of the aircrew jumped from 12,000 feet. The plane had been holding a level course, but suddenly rolled over to port and plunged toward the sea, exploding on impact. Meanwhile Bomber *A72-81* was also shot down with at least one survivor reaching the sea surface.[4]

Photo 20-2

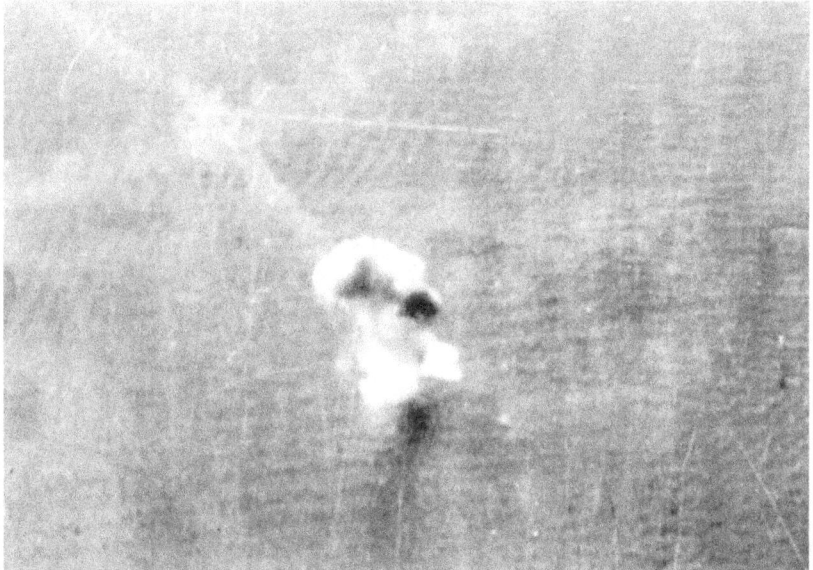

A RAAF B-24 bomber crashes into the sea near Flores Island, and explodes on impact. Australian War Memorial photograph NWA0803

Flight Lieutenant Charles Ralph Bulman set Catalina *A24-54* down on the water, and had picked up three survivors of bombers *A72-77* and *A72-81*, when strafed by an enemy fighter. The Catalina caught fire, was abandoned, and sank. All occupants of the flying boat managed to reach dinghies dropped by circling Liberators.[5]

Photo 20-3

Two aircrew members of a Fairey Fulmar aircraft shot down in the Mediterranean cling to a dinghy supplied by HMAS *Quiberon*.
Australian War Memorial photograph P06165.014

About an hour later, Catalina *A24-58*, with Flight Lieutenant Robin Corrie at the controls, landed. Warrant Officer Vickers and Flt Sgt Walter W. T. Sayer (*A72-77*), Warrant Officer K. R. Shilling (*A72-81*), and Bulman and all his crew, with the exception of Flt Sgt A. E. Jones, were taken aboard.[6]

A search was made for further survivors, but had to be abandoned when the Catalina was attacked by two Irvings. The PBY blister-gunner returned fire while Corrie made an emergency take-off, with water cascading through the blisters (opened for the firing of machine guns). Pursued by the enemy fighters, the overweight Catalina twisted and turned only a few feet above the sea, in an effort to evade them.[7]

Photo 20-4

Images of Japanese Irving 11 reconnaissance aircraft; one of several variants.
Source: Japanese Operational Aircraft "Know Your Enemy!"
CinCPac - CinCPOA Bulletin 105-45

Photo 20-5

Left: Aerial gunner manning his .30-caliber machine gun, in the waist blister of a PBY Catalina. Right: A Catalina receives repairs to a gun blister at a repair base, April 1945. National Archives and NARA photographs #80-G-K-13509 and #80-G-K-4907

When a covering Liberator occupied the attention of the Japanese fighters, Corrie made a labored climb up to 3,500 feet and the safety of a cloud bank. The flying boat, escorted by the Liberator, eventually arrived safely back at Darwin.[8]

JAPANESE CRUISER *ISUZU* SUNK BY SUBMARINES

In late afternoon that same day, 6 April, the submarine pack intercepted the enemy cruiser. USS *Besugo* fired nine torpedoes. All the "fish" missed *Isuzu*, but one sank the minesweeper *W-12*. The following morning at 0605, *Isuzu* was struck by one of five torpedoes fired by USS *Gabilan*. Hit portside below the bridge, the cruiser's speed fell below 10 knots, she took on a list and was down by the bow. While her crew was performing emergency repairs, USS *Charr* fired four torpedoes at her from 1,200 yards. Two hit her portside near the aft engine room. *Charr* fired two more torpedoes and obtained another hit. At 0843, *Isuzu*'s bow broke off. Three minutes later, she rolled to port and disappeared into the deep. *Isuzu*'s remaining escorts counter-attacked *Charr*, but she went deep and evaded them.[9]

Photo 20-6

Japanese light cruiser *Isuzu* at Yokohama in 1944.
Mikasa Memorial Museum, Yokosuka, Japan

CORRIE AND BULMAN DECORATED FOR BRAVERY
Flight Lieutenants Robin Morton Corrie and Charles Ralph Bulman, Royal Australian Air Force, and both members of No. 112 Squadron, were awarded the Distinguished Flying Cross.

Photo 20-7

Flight Lieutenant Robin Morton Corrie, at Darwin, 7 April 1945. Australian War Memorial photograph NWA0827

21

No. 113 Air-Sea Rescue Flight

Photo 21-1

Labuan Island, North Borneo, 15 August 1945. Flight Lieutenant Wally Mills, DFC, famed pilot of No. 113 Air-Sea Rescue Flight, flew the last RAAF aircraft to return from a mission in the war. His pet monkey, Tojo, was his constant companion.
Australian War Memorial photograph OG3227

Of the RAAF's five Air-Sea Rescue Flights, No. 113 saw the most action, and Walter (Wally) Raymond Mills was its most decorated member. He began his service with the RAAF, when he enlisted in Sydney, on 22 June 1941, as an aircrew trainee. After initial training at Lindfield, New South Wales (NSW), he joined No 8 Elementary Flying Training School at Narrandera, NSW, in August 1941. He completed his flying training at 7 Service Flying Training School, RAAF Station

Deniliquin, NSW, from October 1941 to March 1942. Mills was promoted to Pilot Officer on 9 March 1942 and to Flying Officer on 9 September. After further training, including Catalina second pilot qualification, he reported to No 20 Catalina Squadron at Bowen Queensland in December 1942.[1]

From Bowen, 20 Squadron conducted seaward reconnaissance throughout New Guinea, the Solomons, and New Caledonia. Squadron crews were also increasingly ordered to perform night attacks on enemy bases and the frequency of these increased while he was at Bowen. In mid-1943, the Squadron moved to Cairns, Australia, to undertake offensive mining operations with the flying boats ranging from Manus Island, through the Dutch East Indies, and as far north as Manila, Philippines. During 1943 the squadron made regular combined attacks with other aircraft across New Britain, in support of Allied attempts to recapture the island. Mills qualified as a Catalina Captain in April 1943, and was promoted to Flight Lieutenant on 9 March 1944. In April 1944, he took up ferry pilot duties, flying Catalinas from the United States to Australia.[2]

Mills was posted to No 113 Air-Sea Rescue Flight on 12 January 1945 shortly after it had formed in Cairns. The flight was equipped with PBY Catalinas and was planned to deploy to Morotai in the Maluka Islands group in Netherlands East Indies. Mills was made OIC of the advanced party and departed Cairns on 19 January, overnighting at the strategic airstrip at Merauke on the south coast of Dutch New Guinea, arriving Morotai on 20 January.[3]

The Flight conducted its first operational missions in February, providing support for Allied air strikes and making supply drops. On 7 February, while investigating reports of dinghy sightings, Mills' aircraft was attacked by a Japanese 'Rufe' float-plane. Despite the Rufe fighter bomber (based on the A6M Zero) making several attacks at his plane, the Catalina emerged unscathed.[4]

An Operations Record Book entry describes the air combat:

> Attacked by one Rufe with very dark brown camouflage; the Rufe was sighted from the Catalina at approximately 4 o'clock 500 feet above the Catalina. The Rufe made a diving attack from 5 o'clock and cannon fire was seen to strike the water about 100 yards short of the Catalina. The Rufe then climbed and carried out a second attack from about 7 o'clock. Then breaking away from behind the Catalina climbed to 500 feet and tried to roll off the top and to a vertical attack. The Rufe maneuvered like a spin and broke off to a beam position about 1000 feet above the Catalina and attempted to attack out of the sun. The Rufe climbed and made extreme range

pass with the land as a background, broke off and climbed on a course of 090° at 1130 hours.[5]

Photo 21-2

Rufe II navy fighter seaplane.
Source: Japanese Operational Aircraft "Know Your Enemy!"
CinCPac - CinCPOA Bulletin 105-45

At 1133 with the Catalina unscathed, Mills set course for Padjang Island. A sighting was made a little over an hour later of anti-aircraft fire in the air, dead ahead, from an enemy destroyer at position 2°00' N, 119°10'E. While shadowing the destroyer until 1325, Mills was fired on several times. After the warship turned south, the Catalina continued surveillance operations and later returned to base at Morotai, Molucca Islands, that evening.[6]

COMBAT WITH ENEMY FORCES AT AMBON ISLAND

On 31 March 1945, Mills and his flight crew took off from Morotai at 0545, to cover a Liberator strike on Southeast Borneo. At 0800, he received a report of a downed Bristol Beaufighter night fighter from No. 31 Squadron, and was ordered to divert and pick up the two survivors who were in a dinghy. At 1045, a sighting was made of Liang airstrip on Ambon, Netherlands East Indies (occupied by the Japanese through the remainder of the war) and off it, the dinghy in Haroekoe Strait—Dutch spelling of Haruku, as shown on the map on the following page. Ambon was one of the famous Spice Islands. Mills came in at near zero elevation, slightly above the water, and was subject to heavy fire from anti-aircraft sites on both sides of the strait. The survivors, while taking fire from Waai Village on the shoreline, were swimming alongside their dinghy.[7]

Mills landed the Catalina under heavy AA fire from nearby enemy gun positions, stopped his engines, and lowered his wheels to keep the aircraft stationary to pick up the two downed airmen. The PBY was now protected from heavy gunfire (sitting on the water, it was slightly below the angle to which the AA guns could depress), but was met with a heavy hail of rifle and machine gun bullets. The enemy fire was returned from the unprotected blister of the Catalina by Flying Officer Norman William Hastie. During this exchange, Hastie was severely wounded in the abdomen, but continued to carry out his duties, until relieved by Sgt. Robert Charles Ballingall.[8]

Map 21-1

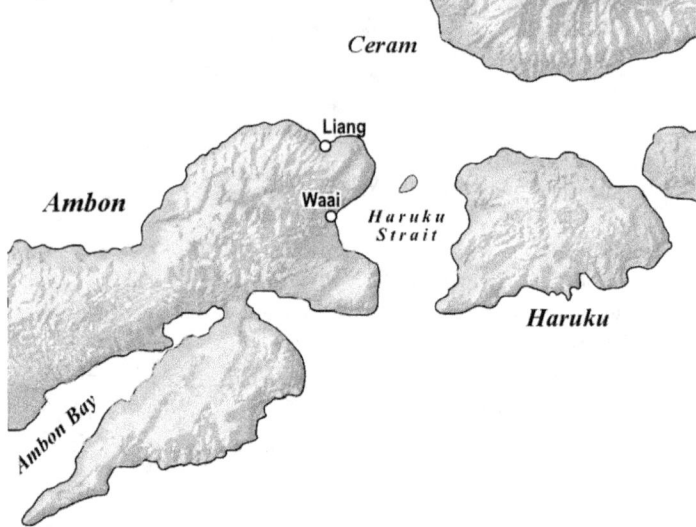

Ambon, Moluccas ("Spice Islands")

Photo 21-3

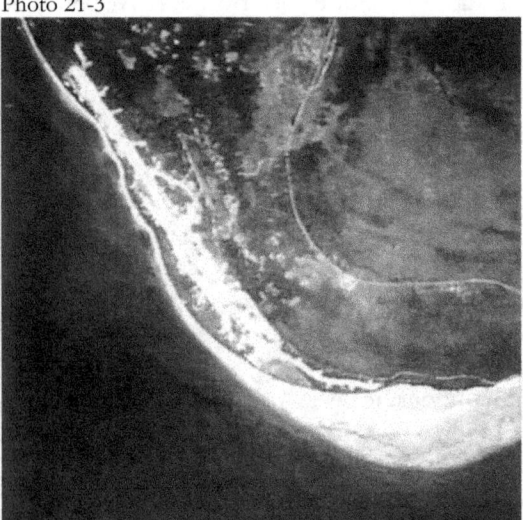

Japanese airstrip, believed to be at Liang, Ambon Island.
Australian War Memorial photograph P00956.045

Photo 21-4

Members of No. 31 (Beaufighter) Squadron RAAF standing by, ready for the signal to board their aircraft at their base at Coomalie Creek, Northern Territories, for action against the enemy, December 1943.
Australian War Memorial photograph NWA0510

Sergeant Arvie Maberry, an American medical orderly aboard the aircraft, attended Hastie under fire. The flying boat was badly holed by rifle and machine gun fire but Mills was able to take off at 1100 in rough water, even though the heavy swell caused some additional structural damage to the aircraft. The action, however, was not yet over. At 1120, Mills received a report from one of the beaufighters which had been covering the recovery on the water, that it had lost one of its two engines through enemy action. Mills covered (accompanied) it during the return flight to Morotai. He landed there at 1355. Hastie was rushed to 155th American General Hospital for surgery; following the operation, his condition was reported as serious. The rescued men were Warrant Officer R. K. Roberts and Flight Sergeant B. W. Phillips, the pilot and navigator of the downed beaufighter.[9]

MILLS, HASTIE, OTHER FLIGHT CREW DECORATED

Flight Lieutenant Walter Mills and Flying Officer Norman Hastie were awarded the Distinguished Flying Cross. (Their award citations may be found in Appendix G.) The other members of Mill's flight crew were honoured with a Mention in Despatches.

Name	Crew Position	Award
FL Walter Raymond Mills	Pilot/Plane Captain	DFC
FO Vernon John Boyd Morris	2nd Pilot	MID
T/WO Reginald Charles Grilley	Navigator	MID

FO Robert William Thompson Francis	WAG	MID
FO Norman William Hastie	WAG	DFC
T/F/Sgt. Arthur Frederick Taylor	Engineer	MID
Sgt. Robert Charles Ballingall	Gunner	MID
Sgt. Arvie Maberry*	Medical Orderly	MID[10]

* Member of 2nd American Emergency Rescue Squadron

Months after being wounded, and when sufficiently recovered from his extensive injury, Hastie was discharged from the Royal Australian Air Force. He had joined as an air gunner in October 1940, and had been commissioned as an officer in 1943 after completing a Gunnery Leader's Course.[11]

Wireless Operator/Air Gunner Brevet Patch

No. 113 Air-Sea Rescue Flight relocated from Morotai beginning in July, to become operational at Labuan on 29 September. Mills took over as commanding officer on 17 September 1945. During his combat duty, he flew 160 missions, mainly covering fighter and bomber strikes, and recovered fifteen aviators who had been shot down.

22

Assault of Iwo Jima

At great cost, you'd take a hill to find then the same enemy suddenly on your flank or rear. The Japanese were not on Iwo Jima. They were in it! I'd known combat in the Solomons with its sly ambushes and jungle firefights, but Iwo was another kind of war. On Iwo by the 8th day, only two officers of my second battalion (26th Marines, 5th Marine Division) were standing.... We had one prisoner—unconscious, his clothes blown off.

—Col. Thomas M. Fields, USMC (Ret.)[1]

The battles of Iwo Jima and Okinawa, fought by Allied forces between February and August 1945, were operations preliminary to securing island bases for a final B-29 bomber assault on Japan. The Iwo Jima operation was conducted first because it was expected to be easier than an assault on Okinawa. Because of the enemy's prolonged and bitter defense of Leyte and Luzon, the planned dates for both actions had slipped. Thus, the Fifth Fleet had to cover and support both invasions, while the Seventh Fleet and its amphibious forces were concurrently engaged in liberating the Philippines.[2]

Preliminary bombings of Iwo and the minor airbase at Chichi Jima (over which LTJG George Bush and his aircrew had been shot down) were conducted by shore-based aircraft from the Marianas. Supporting operations for the invasion were begun on 16 and 17 February when fast carrier raids were made on the Tokyo area of the Japanese home islands. During these raids, and one, ensuing on the 25th, 420 enemy planes were shot down, 228 were destroyed on the ground, and some effort was directed against aircraft engine plants and airplane factories.[3]

On 19 February, as naval gunfire pounded the island, more than 450 ships massed off Iwo Jima. Marines of the 4th and 5th Divisions hit the four assault beaches shortly after 0900, initially finding little enemy resistance. Coarse volcanic sand hampered their movement as they struggled to move up the beach from the surf zone. As the protective naval gunfire subsided to allow for troop advancement, the

Japanese emerged from fortified underground positions to begin a heavy barrage of fire against the invading force. The 4th Marines continued to push forward against heavy opposition to take the Quarry, a Japanese strong point, while the 5th Marine Division's 28th Marines isolated Mount Suribachi that same day.[4]

The 3rd Marine Division joined the fighting on the fifth day, charged with securing the center sector of the island. The fortified enemy defenses linked miles of interlocking caves, concrete blockhouses and pillboxes, which required frontal assaults to gain nearly every inch of ground. Maj. Gen. Harry Schmidt, commanding the Fifth Amphibious Corps—of which the 3rd, 4th, and 5th Marines were a part—declared Iwo Jima secured on 16 March. Ground fighting, however, continued between then and the official completion of the operation on 26 March 1945.[5]

Photo 22-1

A few of the hundreds of landing craft staging for assault landings on Iwo Jima. Naval History and Heritage Command photograph #NH 104203

SEAPLANE OPERATIONS

Patrol plane tenders anchored in the lee of Iwo Jima to service PBM's of Task Group 50.5 at such times as sea conditions would permit the operation of seaplanes. These PBM's were used for dumbo missions and for the extension of the more important search sectors toward Japan, commencing on 28 February. The tenders and PBM's were withdrawn when activation of the airfield ashore permitted their replacement by land planes and amphibians. The performance of the PBM's, with jet assisted take-off, in the bad sea conditions which normally existed around Iwo Jima, was a remarkable tribute to the ability of their pilots.

—From a report titled, Amphibious Operations – Capture of Iwo Jima - 16 February to 16 March 1945[6]

Two seaplane tenders—*Hamlin* (AV-15) and *Chincoteague* (AVP-24)—had left Guam on 19 February for Iwo Jima. Commodore Dixwell Ketcham, commander Fleet Air Wing One, and his staff were embarked aboard *Hamlin*. His orders were to establish a floating operating base for seaplanes on the south side of Iwo Jima; to conduct seaplane searches as ordered by commander, Fifth Fleet; to conduct air-sea rescue missions; and to refuel battleship and cruiser scout planes (Vought OS2U Kingfishers and Curtiss SOC Seagulls) as required.[7]

Photo 22-2

Commodore Dixwell Ketcham, commander, Fleet Air Wing One, (front row, center) with his staff, on board his flagship, which may be USS *Hamlin* (AV-15). Naval History and Heritage Command photograph #NH 95728

Commissioned on 26 June 1944, the *Hamlin* was a relatively new ship. Her first experience of being in the front line was a thrilling one for many aboard. Night retirement groups of ships moving out to sea caused some ticklish moments for *Hamlin* as she approached the island, and her radar proved its worth many times. Her commanding officer, CAPT Gordon Alexander McLean, USN, was in charge of the seaplane base under operational control of commander, Group 1, Amphibious Force, Pacific (RADM William H. P. Blandy). Aircraft of Patrol Bombing Squadron VPB-19 and Rescue Squadron VH-2 were assigned to the task group for the Iwo Jima operation.[8]

Arriving at 0800 on 20 February, the two tenders received orders to lay off the southeast side of the island in the transport area. Later in the day, the *Hamlin* and *Chincoteague* proceeded to Area William to await orders to establish a seadrome. This did not occur immediately because of bad sea conditions and heavy gunfire from the beach which lasted for four days. Finally, on the morning of 24 February, *Hamlin* anchored 600 yards from the shore, and began laying seaplane moorings, while working in very close to the beach with a heavy sea still running. Prior to receiving planes, and after their arrival, her boats swept the area continuously removing as much floating debris as possible. Flotsam presented one of the major obstacles to seaplane operations at Iwo Jima. There was so much detritus from the amphibious landings that only the larger pieces could be removed, including rafts, timbers, ammunition cases, and other incidentals, which drifted continuously through the seadrome.[9]

In addition to serving as a seaplane tender, *Hamlin* was designated a rescue center to receive survivors. Such personnel were treated and berthed aboard her until arrangements could be made for their further relocation. While at Iwo Jima, she provided care for nineteen aviators, and survivors of the sea; including four men that her crew had rescued from a sinking LCM landing craft; and twenty-three Marines.[10]

From her anchorage under the shadow of Mount Suribachi, *Chincoteague* tended part of VPB-19 and two PB2B-2 (PBY Catalinas manufactured by Boeing in Canada) from Rescue Squadron VH-2. The three planes of VPB-19 left Saipan on 21 February for Iwo Jima, but had to be recalled because, at the time, the seadrome had not been established. Six days after the recall, the same three planes were ordered forward on 27 February, and landed at Iwo without further incident. The squadron's remaining twelve PBM-3Ds arrived between 28 February and 3 March. During the period 28 February-5 March, VPB-19 planes flew up to seven searches daily, reaching within 100 miles of Honshu, despite having to land and take off under essentially open sea conditions. Nineteen sightings of enemy vessels were made and reported, and as part of these sightings, a Japanese picket boat was sunk by strafing.[11]

Photo 22-3

Shells explode ashore during the bombardment of Iwo Jima on 17 February 1945. This view of Mount Suribachi's west side was probably taken during pre-invasion minesweeping activities, as the ship in the right foreground appears to be a YMS.
Naval History and Heritage Command photograph #NH 104142

To lighten their loads, the PBMs did not carry any bombs to allow take off from the unprotected roadstead. Despite heavy seas, the patrol aircraft with a load of 1,800-2,000 gallons of gasoline, and 250 rounds per gun, were able to make good all attempted take-offs with JATO assistance.[12]

On 28 February, two PB2B-2 planes of Rescue Squadron VH-2 reported to commander, Task Group 50.5 for operation from the *Chincoteague*. The plane commanders of these aircraft were squadron commander LCDR Harold A. Wells (#44244) and Ensign Gargas (#44240). In spite of the unsatisfactory seaplane operating conditions, the status of "immediate readiness" was maintained for any air-sea rescue missions demanded by operations at Iwo during daylight hours, until 7 March. After the South Air Field on the island was secured and made usable, three land-based PBY-5As, and four crews of VPB-23, were substituted for the VH-2 aircraft and personnel, which returned to Naval Air Base, Tanapag, Saipan, on 8 March 1945.[13]

The PBM-3Ds of VPB-19 also returned to Saipan, and from there, were ferried to Parry Island, Eniwetok Atoll, during the period 6-17 March, reporting to commander, Marshalls-Gilberts Area for duty assignment. *Hamlin* and *Chincoteague* left Iwo Jima on 8 March, bound for Saipan. Because *Chincoteague*'s next assignment was to tend a squadron that flew PBM-5 planes, a change over from her PBM-3D allowance of repair parts and maintenance items was necessary.[14]

HEAVY CASUALTIES SUFFERED AT IWO JIMA

The acquisition of Iwo Jima—strategically important as an air base for fighters escorting B29s flying long-range bombing missions against mainland Japan, and as an emergency landing strip for crippled B29s unable to make it back to their base in the Mariana Islands—came at a high cost. The vital link in the chain of bases was gained through the individual and collective courage of Marines over a thirty-six-day period. The brutal fighting resulted in 26,000 American casualties, including 6,800 dead. Only 1,083 of the 20,000 Japanese defenders of Iwo Jima survived.[15]

23

Invasion of Okinawa

AFTER many years of derision, the patrol seaplane was finally, during the war in the Pacific, recognized by all pilots as a most useful and valuable creature. This realization came most forcefully to many as they or their friends were picked out of an inhospitable sea by a flying boat which landed for that purpose. This phase of the airplane's activity was, however, only one of many and by no means the most important of an extensive repertoire.

The seaplane came into its own during the invasion of Okinawa. In the words of Commander Fleet Air Wing One, "Never before had search planes and tenders attempted so much under such difficult combat, weather, and base conditions." And the attempt was successful beyond all but the wildest hopes! Long before fields were available for the first land-based aircraft, the seaplane organization had established a large, protected, and well supplied floating base from which operations were established days before the initial landing on Okinawa had taken place. For a period of weeks the only long-range planes in the area were the Martin Mariners from Kerama Retto, the group of small islands about 15 miles west of Okinawa proper.

—LCDR R. R. Boettcher, USN, in his article
"Low and Slow – Making Seaplane History,"
Proceedings, September 1949.

The eighty-two-day-long Battle of Okinawa was fought on Okinawa in the Ryukyu Islands. It was the largest amphibious assault in the Pacific. The Ryukyus, a chain of Japanese islands stretching southwest from Kyushu to Formosa (now Taiwan), includes the Osumi, Tokara, Amami, Okinawa, and Sakishima islands. The larger islands are mostly volcanic, the biggest being Okinawa, and the smaller ones are mostly coral. Four divisions of the U.S. Tenth Army (the 7th, 27th, 77th, and 96th) and two Marine divisions (the 1st and 6th) fought on Okinawa, supported by naval, amphibious, and air forces. The purpose of the operation, lasting from 1 April through mid-June 1945, was to capture the island for use as a base for air operations for the planned invasion of Japan.[1]

Some Japanese accounts of the battle refer to it as *tetsu no ame* or *kou no kaze*, "iron rain" or "steel wind," respectively, due to the ferocity of the fighting, the intensity of kamikaze attacks from the Japanese defenders, and the sheer numbers of ships and armored vehicles that assaulted the island. The battle produced the highest casualty numbers in the Pacific Theater during World War II. The Tenth Army suffered 7,613 killed or missing in action, and 31,800 wounded, while Marine Corps casualties overall—ground, air, and ships' detachments— exceeded 19,500. Navy losses were 34 vessels and craft sunk and 368 damaged, with over 4,900 sailors killed or missing in action, and over 4,800 wounded. Japan lost over 100,000 soldiers, either killed, captured or victims of suicide.[2]

ARRIVAL OF THE ADVANCE SEAPLANE GROUP

As a prelude to the invasion of Okinawa Gunto (meaning group of Okinawa islands), the Advance Seaplane Group of seaplane tenders— *Hamlin* (AV-15), *St. George* (AV-16), *Chandeleur* (AV-10), *Yakutat* (AVP-32), *Onslow* AVP-48), *Shelikof* (AVP-52), and *Bering Strait* (AVP-34)—left Saipan for Kerama Retto on 25 March 1945. Following an uneventful transit and arrival on the 26th, the Kerama Retto Sea Plane base under the command of Captain McLean in *Hamlin* (CTU 51.2) was established. It was announced ready for operations at noon the next day.[3]

Patrol Bombing Squadrons of seaplanes VPB-21, -208, -18, and -27, and Rescue Squadron VH-1 were then ordered forward from the Marianas and the Carolines on 28 March. PBM Mariners of VPB-27, one of the first squadrons to arrive, took up anti-submarine patrols (ASP) around Kerama and Okinawa on the 29th. Their initial assignment of three inner ASP sectors by day, and three inner and outer sectors at night proved too much for them. Accordingly, VPB-21 and -18 were assigned limited ASP duties until VPB-208 arrived to fill in the gap. These squadrons were not specially trained in anti-submarine warfare, nor equipped with special anti-submarine warfare (ASW) armaments (torpedoes or depth charges). Considering these shortcomings, the performance of LTJG Edgar C. Newkirk of VPB-21, which contributed to the sinking of a Japanese submarine on the night of 30/31 March, is remarkable.[4]

The crews of Squadron VPB-21 flying their PBM-5 Mariners arrived from Saipan at Kerama Retto in the early morning hours of 29 March, joining their maintenance crew personnel carried there aboard the *Chandeleur*. As their aircraft made their way to moorings, shadowed by rocky cliffs of the small islands of the group, their movements were accompanied by heavy gunfire and the rattle of small arms fire, as

American forces ashore dynamited caves and cleared out snipers. The Kerama Retto Islands, fifteen miles to the west of Okinawa, had been seized only three days earlier, concurrent with the arrival of the seaplane tenders, in order to provide a protected anchorage and a base for logistic support of the planned invasion of Okinawa. The Group did not require level ground for an airfield, but it could not tolerate enemy fire from positions ashore.[5]

JAPANESE SUBMARINE SUNK BY DD *MORRISON*

> *ASP plane 14-V-461 was circling overhead to help direct the attacks and to provide illumination. Throughout the attack the ASP rendered excellent assistance, particularly in well-placed illumination.*
>
> —USS *Morrison*'s war diary entry describing the assistance lent by the PBM-5 Mariner piloted by LTJG Edgar Newkirk, during the destruction of a Japanese submarine.[6]

VPB-21 Crew Seven, headed by LTJG Newkirk, made the first enemy encounter for the Group while covering the night ASP sector 114V461 southeast of Okinawa on 30 March. LTJG John R. Peacock, USNR, detected a surface contact on their Mariner's radar, and Newkirk made a beeline toward it. At 2045, a visual sighting was made of a large, type I-5 Japanese fleet submarine on the surface. The submarine crash dived, and "hold down" tactics (involving remaining overhead to help deter it from surfacing to recharge batteries) were employed by the aircraft until destroyer USS *Stockton* (DD-646) arrived. Newkirk then adopted hunter-killer tactics: illuminating the scene with flares, relaying communications, and guiding in *Stockton* and later a second destroyer, *Morrison*, on contacts.[6]

Stockton, while proceeding with the other vessels of Task Unit 5018.42—fleet oiler *Tomahawk* (AO-88), cargo ship *Las Vegas Victory* (AK-229), and destroyer *Gillespie* (DD-609)—to Kerama Retto, had made a radar contact 12,800 yards off her starboard bow at 2306. The task unit was then ninety miles southeast of Okinawa. Ordered to investigate, *Stockton* left her screening station to do so. At 2332, the contact disappeared from radar, but the destroyer immediately made a sonar contact.[8]

The task group proceeded on its way, leaving *Stockton* at the scene of the contact. After midnight, the destroyer began making depth charge attacks on the submarine. At 0300 on 31 March, the destroyer *Morrison* (DD-560) arrived at a rendezvous position, marked by a lighted buoy dropped by the PBM Mariner, and relieved *Stockton*, which had lost contact and was then carrying out a search plan based on the submarine's probable location. *Morrison* had been at Okinawa, shelling a beach to prevent enemy forces ashore from firing at Underwater Demolition Team personnel clearing obstacles from the shallows, in advance of planned D-Day assault landings at Okinawa on 1 April.[9]

Stockton reported that she had made eight depth-charge attacks, and that the sub's last course was 130°T, speed 6 knots, at a probable depth of 300 feet. A noticeable odor of oil in the area suggested the submarine may have been damaged by the *Stockton*'s attacks, or the sub may have discharged oil as a ruse to convince attackers that she had been sunk.[10]

Gaining a sound contact at 0302, *Morrison* prepared to make her own depth-charge attack. However, due to the short range and insufficient bearings and ranges to determine a good attack course, no charges were dropped. As sonar is ineffective at close range, contact was lost at 50 yards, indicating the sub was very close. At 0310, a contact was regained, but it proved to be the submarine's wake not the vessel. (A standard submarine escape tactic was to get inside the turning circle of an attacker, so as to force sound operators to take ranges on its wake rather than its hull, or on "knuckles" in the water caused by a sudden increase in propeller speed.)[11]

At 0324, the destroyer regained contact with the sub and initiated an attack, dropping an eleven-charge pattern with shallow settings for the depth charges. *Morrison* then opened range to the submarine to set up for another attack, when the sound operator, SoM2c Eugene V. Suaggi, heard accelerated propeller noises, and the submarine's ballast tanks being blown, indicating that it was surfacing. The sub broached the surface, 45-degrees on the destroyer's starboard bow, at a distance of 900 yards. *Morrison* took her under fire with 20mm, 40mm, and 5-inch batteries, and attempted to ram her, but the submarine was turning hard right, apparently in an effort to bring her forward torpedo tubes to bear on the destroyer.[12]

Morrison passed down the enemy's starboard side, astern, then up her portside, keeping the submarine under fire. This prevented her crewmen from manning the deck gun. With the submarine 200 yards away, the destroyer fired depth charges from her starboard "K" guns with shallow settings. At 0351, following a loud explosion, the submarine settled by the stern, with only her bow and conning tower above the water. Resurfacing on an even trim, she came under continuous fire from *Morrison*, and at 0412 sank stern first at 25°29'N, 128°35'E. The destroyer then took up search plan "Operation Observant" (the same one employed earlier by *Stockton*) to await sunrise for further evaluation and confirmation of sinking.[13]

At 0630, one survivor and two mangled bodies were sighted off the ship's port bow in considerable debris—mostly planking. The Japanese survivor was recovered by boat, and given medical treatment aboard ship for multiple shrapnel wounds. The prisoner, later determined to be PO2C Mukai Takamasa, was transferred to the custody of commander, Western Islands Attack Group (CTG 51.1) at Kerama Retto for interrogation. When asked by his questioner to identify his submarine, he replied, "If your ship was sunk, would you release the name?" The submarine was later determined to be the *I-8* under the command of Lt. Comdr. Shinohara Shigeowas, who perished with the remainder of his crew.[14]

LTJG Edgar C. Newkirk was awarded the Distinguished Flying Cross for his display of skill and tenacity during the long action.[15]

BATTLESHIP *YAMATO* AND ESCORT SHIPS SUNK BY WAVES OF U.S. CARRIER AIRCRAFT ATTACKS

> *All cadets prepare to leave the ship.*
> *Distribute sake to all divisions.*
> *Open the ship's store.*
>
> —Orders broadcast over the battleship *Yamato*'s public address system on the evening of 5 April 1945, the night before the Japanese Special Sea Attack Force departed Tokuyama Bay, on the Inland Sea, on a suicidal mission to cripple the American fleet at Okinawa.[16]

The primary mission of the seaplane tender and land-based patrol aircraft of Fleet Air Wing One during the first thirty days of the Okinawa operation was anti-submarine patrol, but long-range search of the waters between the Empire and Okinawa was almost as important. The function of Wing aircraft engaged in patrol was to give early warning of enemy surface units which might proceed toward Okinawa to resupply the Japanese garrison, or to attack the large concentration of U.S. Navy shipping off the beachheads.[17]

Diagram 23-1

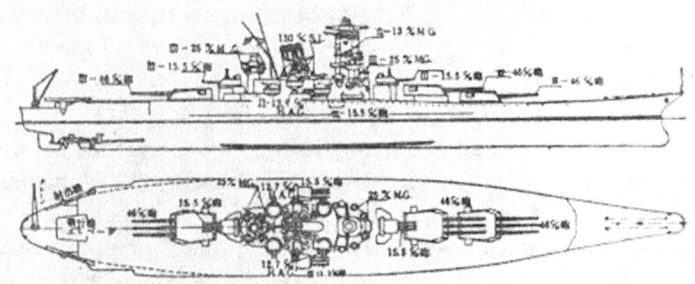

Builders plans for IJN Battleship *Yamato*
Naval History and Heritage Command photograph #NH 111711

On the evening of 6 April, the submarine USS *Threadfin* (SS-410) detected the Japanese battleship *Yamato*, and accompanying escort ships proceeding south through Bungo Suido, a strait separating the Japanese islands of Kyushu and Shikoku, and connecting the Inland Sea to the Pacific. *Threadfin* was on patrol near Fukashima, a tiny island at the mouth of the strait. She reported the movement of the enemy force to commander, Submarine Forces, Pacific, at Guam. Later *Hackleback* (SS-295) also sighted the *Yamato* and reported her position.[18]

Fleet Air Wing One (FAW-1) planes were assigned 24-hour tracking missions of the enemy force, surface units at Okinawa were notified and suitably dispositioned to meet a possible attack, and the fast carrier force (Task Force 58) prepared to intercept *Yamato* and her accompanying escorts.[19]

The Japanese ships sailing together were titled the "Special Sea Attack Force (SSAF)," an ordinary-sounding designation, except for the fact that "special" was a euphemism for a much more sinister word. These cobbled together remnants of the once mighty Imperial Japanese Navy (IJN) were a "Suicide Sea Attack Force" intended to be sacrificed in an attempt to cripple the American fleet at Okinawa. *Yamato* (the world's largest battleship and pride of the IJN) was to be escorted to her death by the light cruiser *Yahagi* and the eight destroyers of the Second Destroyer Squadron. After the ensuing actions, only four of the ten members of the SSAF would survive.

Japanese Special Sea Attack Force (SSAF)		
battleship *Yamato* (S)	destroyer *Hamakaze* (S)	destroyer *Kasumi* (L)
light cruiser *Yahagi* (S)	destroyer *Hatsushimo*	destroyer *Suzutsuki* (D)
destroyer *Asashimo* (S)	destroyer *Isokaze* (L)	destroyer *Yukikaze*
destroyer *Fuyutsuki*[20]		

(S): Sunk   (D): Damaged   (L): Scuttled as a result of bomb damage

On 3 April 1945, in a desperate attempt to contest the U.S. military invasion of Okinawa, the commander in chief of the Japanese Combined Fleet, Adm. Soemu Toyoda, informed the officers and men of his just-formed Special (Suicidal) Sea Attack Force: "The fate of our Empire depends upon this one action. I order the Special Sea Attack Force to carry out on Okinawa the most tragic and heroic act of the war." This planned action involved sending all of the SSAF's sailors on a mission to fight "to the last man." Two days later, Toyoda issued more specific amplifying guidance:

> The Surface Special Attack Unit is ordered to proceed via Bungo Suido Channel at dawn on Y-1 Day [7 April] to reach the prescribed holding position for a high-speed run-in to the area west of Okinawa at dawn on Y-Day. Your mission is to attack the enemy fleet and supply train and destroy them. Y-Day is April 8th.[21]

On 6 April, waves of Japanese planes dove in suicidal attacks into Fifth Fleet ships as part of Operation KIKUSUI. First Japanese pilots, and then the sailors of the SSAF had been ordered to end their lives in the same supposedly heroic manner. The purpose of their mission, named Ten Ichi-Go ("Heaven Number One"), was grimly simple: sail directly into the American ships and transports supporting the Okinawa landing and inflict as much punishment on them as possible. Following this action, *Yamato* was to run aground and employ her 18.1-inch main batteries and smaller guns in support of Okinawa's land defense forces.[22]

All crewmembers aboard the battleship, except those assigned to gun crews, were to go ashore and fight alongside the soldiers of Okinawa's defense garrison. Believing it likely that these men and the sailors of the escort ships would all die at sea or ashore, the Japanese Naval General Staff instructed that each ship be given only sufficient fuel for a one-way trip to Okinawa, but harbor officials disobeyed this order risking execution and fueled the ships to capacity, enabling them to return home if they somehow survived.[23]

On 7 April at 1000, following notification of the SSAF's path south-southwest toward Okinawa, VADM Marc A. Mitscher (the commander of Fast Carrier Force 58) ordered separate flights of 280 and 106 planes aloft to attack the Japanese task force in waves. The first enemy casualty was the destroyer *Asashimo*, which had fallen behind the group owing to an engine failure. As *Yamato*'s radar detected increasing numbers of American planes in the skies overhead, she received a radio transmission from *Asashimo* at 1210 that the destroyer was engaging enemy planes;

then the radio went silent. *Asashimo* had been sunk, and her entire crew perished going down with the ship.[24]

Attacked by increasing numbers of U.S. Navy F6F Hellcat fighters, F4U Corsair fighters, SB2C Helldiver dive-bombers, and TBF Avenger torpedo bombers, the Japanese Special Sea Attack Force soon suffered other ship casualties. The destroyer *Hamakaze* was damaged by a bomb hit at around 1247 followed by a torpedo strike, and sank. About the same time, a third destroyer, the *Suzutsuki*, received a bomb and caught fire. Hit again, she somehow managed to make it back to Japan—thanks in part to the harbor officials who had defied orders and provided her sufficient fuel to do so.[25]

Photo 23-1

Japanese battleship *Yamato* blowing up, following massive attacks by waves of Task Force 58 carrier aircraft, on 7 April 1945. An escorting Japanese destroyer is at the left. Naval History and Heritage Command photograph #NH 62582

With the battleship's sheer size making her a tempting target, she continued to be pummeled by bombs, bullets, and torpedoes. Immediately after the first wave of attacking U.S. Navy planes, a second wave began a coordinated strike. As dive-bombers from high overhead began their attacks, torpedo bombers came in from varying directions, just above the wave tops. The second attack lasted about a half-hour, during which *Yamato*, already damaged by two bombs in the first strike, was hit by at least two more bombs and a minimum of four torpedoes. At 1405 during strikes by a third wave of aircraft, in which she was hit by additional bombs and torpedoes, she blew up and sank.[26]

The light cruiser *Yahagi* was hit by a bomb at 1330 and, after continued strikes by aircraft, sank. In addition to the loss of the battleship, light cruiser, and two destroyers sunk in battle, the destroyers *Kasumi* and *Isokaze* were sunk in scuttling actions by the *Fuyutsuki* and *Yukikaze*, in late afternoon and early evening, respectively, after survivors were taken off, owing to extensive bomb damage to the ships.[27]

FOUR IJN DESTROYERS ESCAPE DESTRUCTION; AMERICAN AIRCRAFT LOSSES RELATIVELY LIGHT

Following three waves of carrier aircraft strikes on the Japanese Special Sea Attack Force, only the destroyers *Fuyutsuki*, *Hatsushimo*, *Suzutsuki*, and *Yukikaze* survived the failed mission to attack shipping at Okinawa. Suffering varying degrees of damage, they retired to their base at Sasebo, Japan. Task Force 58 lost ten planes and fourteen personnel from enemy anti-aircraft fire. In addition, three airmen were injured.[28]

SUPPORTING ACTIONS OF PBM-5 MARINERS

Fleet Air Wing One patrol aircraft had tracked the enemy force to a point fifty miles southeast of Kyushu, before planes of the fast carrier forces began their assault on the battleship and escorts. Lieutenants Richard L. Simms and J. R. Young (piloting two PBM-5 Mariners of Patrol Bombing Squadron VPB-21) had detected the Japanese ships in early morning on 7 April. They then shadowed the enemy group for five hours, sending in reports and homing (guiding) in a portion of the carrier aircraft which sank the *Yamato*, cruiser *Yahagi*, and four destroyers. However, the gallantry of the Mariners pilots and crews were not confined to the long, tiring tracking mission through the persistent variable weather encountered.[29]

In the course of the torpedo and bombing runs, LTJG W. E. Delaney piloting a TBM Avenger from the carrier *Belleau Wood* (CVL-24), was shot down and crash landed near many Japanese survivors in the water. During the air strikes, the Mariners had been fired on continuously by enemy ships. Disregarding their fire, because of the possibility of the downed pilot's capture, Young made an open sea landing and picked him up. While the rescue was in progress, Simms circled to distract enemy gun crews and draw fire that might otherwise be directed at Young's Mariner alighted helplessly on the water.[30]

CHANGE IN LEADERSHIP OF FLEET AIR WING ONE
On 23 April 1945 as part of Navy restructuring, Commodore Dixwell Ketcham, USN, was relieved by RADM John D. Price, USN, as commander Fleet Air Wing One. At the same time, operational control of FAW-1 units in the Marianas, Carolines, and Iwo Jima passed to VADM John H. Hoover, USN, commander, Forward Area, Central Pacific (CTF 94). This change reduced the units assigned to Fleet Air Wing One to only those at Kerama Retto, or en route there.[31]

FLEET AIR WING ONE OPERATIONS AT KERAMA
Throughout April, seaplane operations at Kerama Retto were carried out under extraordinarily difficult conditions with regard to aircraft maintenance. Overhaul work normally accomplished ashore was done by the tenders, much of it in the water without the planes being hoisted aboard. Additionally, seventy enemy air raids, and associated time spent by tender crews at General Quarters compounded the difficulties encountered by seaplanes at the base. On 29 April, Fleet Air Wing One including the original Advance Seaplane Group consisted of three task units:

- TU 50.5.1: ASW and Air-Sea Rescue Unit at Kerama Retto
- TU 50.5.6: Seaplane Search Unit at Kerama Retto
- TU 50.5.5: Landplane Search Unit at Yontan Airfield[32]

PB4Y-2 Privateer patrol bombers of VPB-118 arrived at Yontan Field (near Sobe on the western coast of Okinawa Island) on 22 April, after the invasion had secured the airfield, to inaugurate the landplane unit of Fleet Air Wing One. These aircraft were assigned offensive searches and anti-shipping sweeps in the East China Sea and Tsushima Strait. The sea passage between Kyushu and Tsushima links the East China Sea and the Sea of Japan in the northwest Pacific. Patrol Bombing Squadron VPB-109, and its PB4Y-2 Privateer patrol aircraft, joined the landplane unit on 14 May.[33]

Photo 23-2

PB4Y-2 Privateer patrol planes over Miami, Florida, July 1949.
National Archives photograph #80-G-440193

AIR-SEA RESCUE COMMAND AND CONTROL

Task Unit 50.5.1 was composed of the seaplane tenders, *Hamlin, Norton Sound, Yakutat, Onslow, Shelikof, Bering Strait*, and *Suisun*, and Squadrons VH-3, VPB-26, VPB-27, and VPB-208. A majority of the missions carried out by patrol aircraft were anti-submarine patrols. Rescue Squadron VH-3, supported by the *Bering Strait*, was responsible for air-sea rescue missions. Commander, Air-Sea Control Unit (Task Group 51.10), aboard the amphibious force command ship USS *Eldorado* (AGC-11), was in control of all air operations in the objective area. (CAPT Richard F Whitehead served in this role until mid-April, when relieved by RADM Alfred M. Pride.) This control, including that of air-sea rescue, lasted from 1 April until 17 May. (On that date, control was passed to commander, Fifth Fleet Amphibious Force.) Supervision of rescue missions tasked to *Bering Strait* and her VH-3 Squadron was the responsibility of commander, Task Group 51.20 Captain McLean, commanding officer of USS *Hamlin*.[34]

SCOPE OF ACTIVITIES/FACILITIES FOR RESCUES

Air-Sea Rescue activities during this operation were on a scale larger than in any previous amphibious operation in the Pacific and extended to areas far beyond the immediate vicinity of the objective. Neutralization strikes carried out regularly against SAKASHIMA GUNTO and the AMAMI-O-SHIMA group accounted for a substantial proportion of the rescue incidents covered by units of Task Force 51 [Joint Expeditionary Force] and the operations of Task Force 58 [Fast Carrier Force] and of the search planes of Task Group 50.5 [Search and Reconnaissance Group] in the vicinity of KYUSHU, resulted in a number of rescue missions at ranges of over 300 miles from Okinawa.

—Excerpt from an Action Report by ADM Richmond K. Turner of 25 July 1945, related to the capture of Okinawa Gunto, Phase I and II, 17 February to 17 May 1945.[35]

Photo 23-3

Vought OS2U Kingfisher floatplane, from the battleship USS *North Carolina* (BB-55) off Truk with nine aviators on board, awaiting rescue by USS *Tang* (SS-306) on 1 May 1944. After landing inside Truk Lagoon and recovering downed airmen, it was unable to take off with such a load, and taxiied out to *Tang*, which was serving as a lifeguard submarine during the 29 April-1 May carrier strikes on Truk.
National Archives photograph #80-G-227990

Prior to the commencement of PBM Dumbo operations from Kerama Seaplane Base on 29 March, facilities for rescue in the Okinawa area had consisted of small ships in the vicinity of the island objective and VOS (the U.S. Navy designation for scouting-observation planes) OS2U Kingfisher and Curtiss SOC Seagull floatplanes carried aboard cruisers and battleships of Task Force 54 (Gunfire and Covering Force). During this period, two of the shipborne aircraft normally employed for scouting and observation purposes, were daily maintained in a standby status for rescue purposes from dawn until dusk.[36]

On several occasions, VOS aircraft engaged in spotting duties were diverted for rescue missions of such urgency that delay involved in launching one of the dedicated "rescue VOS" would have jeopardized the chances of success. (The term "spotting" referred to the use of aircraft to observe the fall of gun rounds of a battleship or cruiser, and report whether they were short, long, left, or right of target. This information allowed the ship to make necessary corrections and to "walk its fire" onto the target.)[37]

After the six PBM-5 Dumbos of VH-3 took up their duties on 29 March, they became the primary means for air-sea rescue in the Okinawa area and beyond into Japanese waters. In rescuing more downed aviators than any other type of air, ship, or submarine assets, they covered wider areas than could have been accomplished through the use of other available means. (A copy of VH-3 instructions to downed airmen may be found in Appendix F.) PBMs of Task Group 51.20 (anti-submarine patrol and search planes) also engaged in search and rescue missions and made numerous rescues.[38]

Significantly, in the waters in and around Okinawa Gunto, the majority of rescues were performed by destroyers, destroyer escorts, and smaller vessels. The stationing of screening, minesweeping, and picket vessels was such that it was unlikely that a plane could go down during daylight hours without being seen by at least one ship.[39]

A summary of the total number of 132 personnel rescued from this area, means of their rescue, and task force or task group to which they were assigned, may be found in the following side-by-side tables. In three instances, personnel forced down in enemy territory made their own way to shore, then swam or rowed out to sea for rescue by seaplanes. One airman parachuted behind enemy lines, but was able to cross over to American front lines, thus requiring no air-sea rescue.

Rescues Associated with the Assault and Occupation of Okinawa

Means of Rescue	No	Assignment of Rescued	No
PBM Dumbo	63	Task Force 58 (Fast Carrier Force)	59
VH-3 Ships/Craft	44	Okinawa Gunto (TG 99.2 and TG 51.20)	42
PBM Search and ASP Aircraft	18	Task Group 52.1 (Support Carrier Group)	23
VOS Aircraft	5	Task Force 57 (British Carrier Force)	3
Land Forces	1	Iwo Jima TF 93 (Strategic Air Force, Pacific Ocean Areas)	3
Made Shore Unassisted	1	Task Force 54 (Gunfire and Covering Force)	2
Total number	132	Total number	132[40]

HEROIC ACHIEVEMENTS OF SQUADRON VH-3

The above table, while demonstrating the breakdown of the rescues for a certain period, understates the full number of rescues made. The report from which the above information was obtained did not specify the beginning and end dates for the 132 rescues by aircraft, ships, submarines, and land forces. It cites the rescue of 63 individuals by VH-3 aircraft, but the squadron's war diary documents 183 survivors picked up by VH-3 aircraft in the last five months of the war. Details relating to these survivors are shown in the following table. Although mostly pilots and aircrew of downed aircraft, those rescued included ten sailors from the destroyer USS *Braine* (DD-630), blown into the sea by the explosion of a kamikaze aircraft crashing their ship.

VH-3 Rescues in April, May, June, July, and August 1945

Month	Number Open-Sea Landings	Number Survivors Recovered	Landings Made in the Face of Hostile Enemy Fire (EF) or with Enemy Aircraft (EA) in the Area
April	27	39	5 (EF), 3 (EA)
May	15	39	3 (EF)
June	14	42	6 (EF)
July	20	62	
August	1	1	
Total	77	183	14 (EF), 3 (EA)[41]

Owing to expanded aerial activity in the approaches to the Japanese homeland, VH-3's number of open sea landings and associated rescues was even higher in July 1945 than in previous months. There is a story involved with every one of these missions. Space limitations permit only a few to be elaborated in this book. Of the rescues in May, one of particular interest occurred on 12 May. It involved a PBM-5 Mariner of Patrol Bombing Squadron VPB-21 (based with VH-3 at Kerama Retto) which was forced to make a dead-stick landing on the open sea. This action was necessitated by damage suffered during prolonged aerial combat between it and another squadron Mariner, and a group of six to seven Japanese fighter aircraft—which highlights dangers PBMs often faced, whether engaged in search or Dumbo duties. Details of this engagement are presented in a later section.

VH-3 Rescues in May 1945 (15 open-sea landings/39 survivors)

Date	Pilot	Location	Survivors
3 May	LCDR Bonvillian	40 miles SW of Yoro Shima (1 survivor)	ENS L. L. Colin, VBF-12
4 May	LTJG Dorton	10 miles SE of Kikai Shima (1 survivor)	LTJG William H. Marr, VBF-85
5 May	LT Palm	3 miles west of Kikai Shima (2 survivors)	LTJG O. Payne and LTJG L. D. Welch VBF-85
10 May	LT Solomon	west of Tokuno Shima (1 survivor)	ENS Robert D. Flodquist, VF-46
11 May	LT Dunn	2 miles NW of Iheya Shima (1 survivor)	LT M. C. Replogle VF-47
	LT Kouns	2 miles NE of Miyako Retto (1 survivor)	ENS A. G. Cooper VF-24
	LT Mansueto	off Tokuno Shima (1 survivor)	LTJG K. A. Moose VF-85
12 May	LT Eddy	80 miles WSW of Kyushu (13 survivors)	LT R. L. Simms; LTJG W. Graves; LTJG R. W. Wolf; AMMF2c E.D. Brown; AOM2c H.E. Callahan; ARM2c J.M. Digulio; AMMF2c W. Rutz; AMM2c D.E. Powell; ARM3c J.L. Ferganis; ARM3c R.W. Hullett; S1c L. Cottrell; ARM2c R.E. Arthur; CPhoM Fuchay VP-21
14 May	LT Dunn	west of Ishigaki Shima (3 survivors)	ENS R. L. Rice; ARM3c J.R. Thacker; AOM3c D. Galilei VT-25

Date				
18 May	LT Eddy	Ishigaki Shima (1 survivor)		LT Robert W. Allen VMF-312
22 May	LT Eddy	SE of Amami Shima (1 survivor)		LT H. W. Sturdevent VF-30
24 May	LCDR Bonvillian	off Yaku Shima, and at entrance to Ariake Bay, Kyushu (2 survivors)		1st Lt. R. J. MacInnis, VMF-123; LT R. D. Cougar, VF-17
27 May	LT Kouns	50 miles east of Okinawa (10 crewmen from USS *Braine* (DD-630), blown overboard by the explosion of a kamikaze plane crashing the ship and exploding)		TM2c Earl F. Biens; S1c Thomas J. Burke; S1c Arthur L. Quinn; RDM3c Brengle S. Johnson; S1c Thurman C. Lord; Cox Lawrence B. Armour; RM1c Ernest Y. Thomason; S1c Homer L. Schutz; RM3c Robert J. Cornelius; Ernest W. Teetsel (deceased)
31 May	LT Solomon	off Miyako Retto (1 survivor)		ENS Vincent J. Calo VT-40

A summary of the aircraft carriers (in two cases, a shore station and a seaplane tender), and associated squadrons to which the survivors were assigned provides insight into the scope of air operations at that time during the Okinawa campaign.

Shore Station or Aircraft Carrier and Associated Squadrons of Survivors

Ruby Field, Okinawa: VMF-312	USS *Hornet* (CV-12): VF-17
USS *Battan* (CVL-29): VF-47	USS *Independence* (CVL-22): VF-46
USS *Belleau Wood* (CVL-24): VF-30	USS *Randolph* (CV-15): VBF-12
USS *Bennington* (CV-20): VMF-123	USS *Santee* (CVE-29): VF-24, VF-25
USS *Chandeleur* (AV-10): VP-21	USS *Shangri-La* (CV-38): VBF-85
USS *Chenango* (CVE-28): VT-25	USS *Suewanee* (CVE-27): VT-40[42]

FORTY-FIVE MINUTE RUNNING AIR BATTLE

> *About midway in the fray, the deck, starboard waist and tail guns of Hook's plane caught a Tojo on a medium high side run so that many tracers were observed pouring into his engine. He pulled up and off at once, such heavy black smoke issuing from his engine that the pilots at first believed their own port engine on fire; he crabbed and slipped off on port wing and was last seen falling off to starboard from an altitude of 700 feet when other attacks obliged all hands to direct attention elsewhere.*

> *Early in the battle (about third run) bow, deck and tail guns of Simms' plane caught a Tojo on a level frontal run and recovery astern and he was observed to stream smoke, then blow up violently astern of both planes. Cottrell, L., S1c, on starboard waist gun gets credit for kill; he caught the Tojo broadside during its recovery from a run from 11 o'clock, pouring hits into and about the cockpit. He saw the Tojo pour smoke and Ferganis, J. L., in tail turret saw him blow up astern the PBM.*
>
> —Description of a portion of a prolonged aerial combat on 11 May 1945, between two PBM-5s and a group of six or seven "Jacks" and "Tojos" carrying out a series of attacks on them.[43]

On 12 May, LT Eddy's PBM-5 Mariner of VH-3 rescued at sea the entire crew of a PBM-5 of Patrol Bombing Squadron VPB-21 based aboard the *Chandeleur* at Kerama Retto. Two PBMs, piloted by Naval Reserve Lieutenants Richard L. Simms and John C. Hook, had been searching the southern coast of Korea the previous day. After sighting seven enemy destroyers anchored among coastal islands, they were jumped by a "Rufe," the Allied designation for a Japanese Navy fighter seaplane.[44]

Flying at an altitude of 800 feet while investigating coves along the coast, the patrol aircraft had allowed separation of a mile to develop between their positions. Upon sighting the fighter above them (seven miles distant at 2,000 feet), the VPB-21 aircraft quickly joined up "wing-and-wing" for mutual protection. The Rufe descended to their level and paralleled their course, 1,500 yards away on their starboard side—just out of gun range. Suddenly, the enemy turned in on a rather half-hearted side-pass, fired two bursts (as did Hook's PBM), then swerved to port, and sped back out again at 4 o'clock position. With no apparent damage suffered by either side, the fighter withdrew, apparently not liking the Mariners' firepower, and was not seen again.[45]

Photo 23-4

Above: Two views of a "Rufe" Navy fighter seaplane

Left: A "Jack" Navy interceptor fighter, and at right: a "Tojo" Army fighter. Japanese Operational Aircraft "Know Your Enemy!" CinCPac-CinCPOA Bulletin 105-45

    This brief encounter was a precursor to what soon developed with a group of six or seven "Jacks" and "Tojos" near the northwestern area of Gotto Retto, in the narrow slot between the Retto and the southwestern Kyushu coast. The Army and Navy fighter aircraft, when sighted, appeared to be lazying over two enemy merchant ships headed in the direction of a nearby harbor. On detection, the Japanese planes were at 2,500 feet altitude, twelve miles distant. With a 1,000 yard spacing between them, the PBMs were flying at 1,500 feet, with Simms leading. The patrol aircraft immediately closed in tight with Hook snug up against Simms' port side, and descended to an altitude of 200 feet.[46]

    As the only available cloud cover was above 6,000 feet, the pilots knew that an attempt to reach that altitude would slow down their aircraft, making them too vulnerable to successful enemy attack—it was too risky to consider. Both pilots believed that their best defense was to remain in close formation, fly low to the water, and employ only mild evasive tactics, hoping that all gun stations were in commission and firing. In the event some guns were out of action (which would be the case this day), they believed their salvation would be radical maneuvers and out-guessing the enemy, while remaining in formation as close to the water as possible.[47]

    At first, the enemy remained out of range and to the side of the Mariners. Slightly in front, they were apparently sizing up the situation and planning their attacks. Their formation then split up, pushed over, and its planes came in singly, about one-half mile apart. At least two of the fighters made runs on each Mariner. They released an estimated two phosphorous bombs per plane, about 3,000 feet out in front and 300 feet higher than the patrol aircraft. Each Jack (Mitsubishi J2M Raiden land-based fighter) could carry two 132-pounders. The pilots believed that the bombs were propelled in some manner (in addition to the forward motion of the planes launching them), as trails of smoke were left by the spiraling ordnance. Hook noted racks or rails about four or five feet long, and square in appearance—one under each wing of an enemy aircraft, directly beneath its 20mm guns.[48]

The bursts of the exploding bombs were cherry red in color, with characteristic spidery, white phosphorous fingers reaching out over a large area. Simms saw two bombs burst; while Hook took the brunt of four near him. The first pair burst about 250 feet in front and an equal distance below his aircraft. The second pair below his starboard wing abreast of the galley, did considerable damage to the bottom of the wing and to tail surfaces. After releasing their bombs, the enemy made their recoveries directly over the Mariners in steep, climbing turns. Hook saw many tracers from his deck turret enter a Jack on the bottom of its fuselage in the wing section. These and other hits, which failed to down the enemy, gave the impression that it was well protected by armor.[49]

Following a lull in the action, repeated attacks ensued, as described in a report, "Forty-five Minute Running Battle between Two PBM-5 Mariners and Six or Seven Jacks and Tojos" as told by Lieutenants Simms and Hook:

> Then the enemy paralleled their targets for a while, leisurely chased tails, finally breaking away for attack individually with half a mile between each one. Sometimes they bracketed the flying boats with three on one side, one at least on the other (possibly some still astern), but on head-on attacks made no attempts at coordination. In all nearly fifty individual runs were made, the great majority coming from the forward quarter, largely from twelve o'clock, with the Mariners heading directly into threats when they were not directly from head-on.
>
> These runs were mostly from slightly above or slightly below, but never from high overhead. In the latter half of the running fight the enemy seemed more aggressive, perhaps spurred on by his losses, showing signs of more varied runs. A threat by three from one side and one from the other was broken up by heading directly into the larger force, whereupon the lone enemy showed no desire to come in astern. In fact, the fighters showed no desire to mix it up with the tail turrets, trying only a few unsuccessful runs from 5 and 7 o'clock which were not pressed close.[50]

As revealed in the quoted material at the beginning of this account, PBM gunfire was responsible for the certain destruction of one Tojo, and likely a second one as well. The first was seen to stream smoke and then blow up violently astern of both planes. Black smoke was pouring from another Tojo, as it fell off to starboard from an altitude of 700 feet. Its demise went unseen, the attention of PBM aircrews being devoted to fending off still more attacks. Finally, the remaining enemy drew astern the Mariners after a run, hung there a while, then retired.[51]

The skillful and courageous defense put up by the PBM-5 Mariner patrol aircraft was particularly noteworthy in that Simms' bow turret was out of commission from nearly the beginning of the action, and his deck turret was shot out midway through the 45-minute running battle. One of Hook's bow turret guns went out of commission midway with the azimuth motor burned out, and on several of the enemy passes the other gun had spent its ammunition before it could be reloaded.[52]

During the attacks the Mariners stayed as close as possible to one another, flying between 700 and 50 feet off the water. They pushed down into attacks from lower levels, and up into those from above, and always headed into threats so as to make best employment of their bow and other guns. Once the fighters were committed to runs and could not change, the Mariner pilots skidded and slid slightly off their courses, sufficient to move out of the tracers (streaking past a little out from the aircraft's tail), but not enough to make their maneuvers apparent to the enemy. Few hits were made on the flying boat's hulls, with most damage being absorbed by their wings and tail surfaces. The pilots felt that they could not have done better flying closer to the water, as they risked catching a float or wing tip in the sea while making turns.[53]

By the time the remaining Japanese fighters dropped out of the battle, Simms' aircraft was in bad shape. The main fuel line from hull tanks to wing tanks had been severed by enemy 20mm fire, and fuel was pouring out as from a hydrant. The starboard engine controls were gone, leaving the engine stuck at 2,500 RPMs, rapidly depleting the 400 remaining gallons of fuel available for use. Simms informed Hook that he would have to land and at 1545, he skillfully executed a dead-stick landing in rough seas at 30°20'N, 128°20'E.[54]

It was impossible to avoid some bounce, and the engine-less Mariner took one of 60-100 feet which, after slamming down on the water, left her lying with port float gone, wing submerged, and hull split open. The crew orderly abandoned the aircraft, and took to their rafts. After being covered all night by other Mariners from Kerama Retto, Lieutenant Eddy picked them up at 0845 on 12 May in a skillful open-sea landing and take-off. Lieutenant Simms, LTJG W. I. Graves, and

ARM2c J. M. Digulio had superficial wounds and everyone was suffering from exposure, but apart from that all were in good condition.[55]

Hook's aircraft safely made it back to base, but in a seriously damaged condition.[56]

COLATERAL BENEFIT OF INVASION OF OKINAWA – CHANGE IN TOP JAPANESE LEADERSHIP

The invasion of Okinawa was the largest amphibious operation of the Pacific war. The Allied expeditionary force included 1,213 ships, 564 support aircraft aboard escort carriers, and 451,866 Army and Marine Corps ground troops. Also available to provide air support, and prevent enemy interference and reinforcement were a fast-carrier force (Task Force 58) with 82 ships and 919 planes, and a British carrier force (British Pacific Fleet) with 22 ships and 244 planes. Additionally, for interdiction and neutralization raids against Japanese air bases, there were the USAAF's Twentieth and Far East Air Forces.[57]

On 21 June, organized resistance ceased on Okinawa. As a result of unremitting Allied military pressure by U.S. Navy, Marine Corps, and Army, and British and Commonwealth forces during the invasion of the Okinawa, there had been a change of cabinet in Japan. Prime Minister Koiso Kuniaki resigned on 5 April, four days after U.S. forces landed on Okinawa. His replacement was Kantaro Suzuki, a retired admiral of the Imperial Japanese Navy, a hero of the Battle of Tsushima, and the final leader of the *Taisei Yokusankai* (Japan's wartime organization). Suzuki became the 42nd Prime Minister of Japan on 7 April 1945.[58]

In the final days of the war while serving as prime minister, Suzuki faced great challenges in trying to pursue peace while placating senior army and naval leaders. These were the ones who had long dominated Japanese policy, precipitated the war in the first place, and wanted to continue the struggle to the bitter end. Following the Potsdam Declaration on 26 July by the governments of the United States, the Republic of China, and the United Kingdom (offering Japan an opportunity to end the war), Suzuki declared that his government would *mokusatsu* ("kill with silence"), meaning make no comment regarding the Declaration. This was a compromise with fervent Army leaders who wanted a blunt rejection. Suzuki resigned on 17 August, shortly after the surrender of Japan was announced. This action cleared the way for Prince Higashikuni to become prime minister. It was believed, he would be better able to enforce on military leaders the Emperor's decision to surrender.[59]

After the fall of Okinawa, with the decimation of its fleet and air power and with access to fuel, food and natural resources blocked, the fate of Japan as an expansionist Pacific power was sealed—she could no longer control other lands. However, the fanatic pride and anticipated tenacity of her people in resisting an invasion and occupation of their homeland remained a huge obstacle to an easy ending of the war. Thus the precipitated change in leadership, although not obvious to many, was a huge added benefit of the Okinawa success.

24

The British Pacific Fleet

When a kamikaze hits a US carrier it means six months of repair at Pearl [Harbor]. When a kamikaze hits a Limey carrier it's just a case of 'Sweepers, man your brooms.'

—Comment made by the USN liaison officer aboard
the British fleet aircraft carrier HMS *Indefatigable*.[1]

Photo 24-1

Port side view of the aircraft carrier HMS *Formidable* with a Wildcat fighter about to land, circa 1943.
Australian War Memorial photograph 302388

As the invasion in Europe was well advanced in late 1944 and aircraft carriers were no longer required there in great numbers, the Royal Navy offered to assist the U.S. forces by supplying a fleet of carriers and other vessels to work in battling the Japanese in the Pacific. Initial U.S. opposition to this offer was overcome when it was established that the British force would be largely self-sufficient and would not put additional pressure on an already stretched America supply system. The British Pacific Fleet (BPF) ultimately mobilized to aid in the potential invasion of Japan was the most powerful strike force ever assembled by the Royal Navy. It included 6 fleet carriers and their squadrons, 4 light carriers, 2 maintenance carriers, 9 escort carriers, 4 battleships and dozens of cruisers, destroyers and lesser combat ships as well as a huge train of supply and maintenance ships, oilers and assorted auxiliaries.[2]

Under Royal Navy command, ships and/or personnel of Canada, Australia, New Zealand, and the Netherlands participated in the British Pacific Force that was formed on 22 November 1944 under Adm. Bruce Fraser, based at Sydney, Australia. In January 1945 while still under independent Royal Navy operating command as British Task Force 63, en route to Sydney, it successfully attacked Japanese oil refineries in Sumatra, eliminating the bulk of their supply of aviation fuel. Operation MERIDIAN significantly reduced production at Pladjoe, and that at Soengei Gerong stopped completely until the end of March. The combined production of both refineries was returned to no more than thirty-five percent of their previous capability by the end of May and the Japanese never achieved full production before their surrender. The attacks on the refineries resulted in the enemy being desperately short of oil, and this incalculable effect on the war may well have been the British Pacific Fleet's greatest contribution to Allied victory.[3]

From 26 March to 20 April 1945, while supporting the invasion of Okinawa, the British Pacific Force was responsible for neutralizing Japanese air bases in the Sakishima Islands and on Formosa (Taiwan), which were a constant threat from the southwest. Gunfire and air attack were used against potential kamikaze staging airfields that might otherwise be used to support attacks against U.S. Navy ships at Okinawa. The Sakishimas, southwest of Okinawa, are a part of the same Ryukyu Island chain previously mentioned.[4]

The BPF (Task Force 57) began operations attached to the U.S. Fifth Fleet at the end of March 1945—initially consisting of some twenty-five surface combatants including four fleet carriers and a Fleet Train of some thirty ships designated Task Force 112. The fighting core of the task force was the 1st Aircraft Carrier Squadron, under Rear Adm. Philip Vian, RN, which included HMS *Indomitable*, HMS *Victorious*, HMS *Illustrious*, HMS *Indefatigable*, HMS *Implacable,* and HMS *Formidable* (the latter two were still en route). The British fleet carriers were subjected to heavy and repeated kamikaze attacks off the Sakishimas. However, boasting armored flight decks, they were quite resilient and returned to action relatively quickly.[5]

The Commonwealth contribution was especially important in terms of the aircrews that made up the BPF's naval air squadrons. A significant portion of the Royal Navy's pilots came from Commonwealth countries, either serving in the Royal Navy and its reserves or as members of the Royal New Zealand Navy, Royal Canadian Navy, or Royal Australian Navy and their attached reserves. At least one flyer of the Royal Netherlands Naval Air Service (RNNAS), Bouke K. Swart, also participated. He made the ultimate sacrifice when killed on 24 July 1945 during operations off Tukushima, Shikoku.[6]

Embarked about the six carriers were twenty Royal Navy Fleet Air Arm Squadrons. Comprising their collective 354 British and American aircraft were: 105 Avenger torpedo-bombers, 109 Corsair fighters, 29 Hellcat fighters, 88 Supermarine Seafire fighters, 21 Fairey Firefly fighters, and 2 Walrus air-sea rescue aircraft.

RN Aircraft Carrier	RNFAA Squadrons	RN/USN Aircraft
HMS *Indomitable*	857 squadron	15 Avengers
	1839, 1844 squadrons	29 Hellcats
HMS *Victorious*	849 squadron	14 Avengers
	1834, 1836 squadrons	37 Corsairs, 2 Walrus ASR
HMS *Indefatigable*	820 squadron	20 Avengers
	887, 894 squadrons	40 Seafires
	1770 squadron	9 Fireflies
HMS *Illustrious*	854 squadron	16 Avengers
	1830, 1833 squadrons	36 Corsairs
HMS *Formidable*	848 squadron	19 Avengers
	1841, 1842 squadrons	36 Corsairs
HMS *Implacable*	828 Squadron	21 Avengers
	801, 880 Squadrons	48 Seafires
	1771 Squadron	12 Fireflies[7]

HMS *Formidable* joined the British Pacific Fleet in April 1945, her first major action was strikes against Japanese air bases supporting Japan's defense of Okinawa. Aircrew losses were heavy, the British carriers lost forty-seven aircraft to enemy anti-aircraft fire and operational causes.[8]

BRITISH AND AMERICAN AIR-SEA RESCUE

While the BPF was largely independent in supply, airplane and ship repair and maintenance, they lacked recovery facilities for downed airmen. The only organic resources of the fleet carriers were two small Walrus flying boats carried aboard HMS *Victorious*. They did their best, but could not rescue all the downed flyers. The fleet therefore had to rely on the U.S. Dumbo and submarine recovery system already in place.

Perhaps the saddest incident of air-sea rescue in the Pacific war occurred on 29 January 1945. A Fleet Air Arm Walrus was returning rescued aviators from the Sumatra refinery attacks to HMS *Illustrious*, when the carrier was hit by friendly fire and the plane was destroyed and some of the rescued airmen were killed. A war diary entry of the battleship HMS *King George V* details this incident:

> A Walrus amphibian, with recovered aircrew, had just landed on *ILLUSTRIOUS*, when two Sallies [Mitsubishi Ki-21 Army Type 97 heavy bomber] attacked the *ILLUSTRIOUS*. One dropped a bomb astern of the *ILLUSTRIOUS* that failed to explode and they then flew down the length of the deck. The cruiser *EURYALUS* was shooting at the attacking aircraft and failed to check her fire as the enemy flew over the *ILLUSTRIOUS*. Two of *EURYALUS*'s 5.25in shells struck the *ILLUSTRIOUS* hitting the superstructure and destroying the Walrus and killing some of the rescued aircrew.[9]

During the Okinawa Campaign, some other pilots of downed aircraft were saved by the Walrus air-sea rescue planes, and at least one other by American Dumbo aircraft. On 16 April, during strikes on Sakishima Gunto, one of the TBF Avengers from 848 Squadron aboard the *Formidable* was hit by flak and crashed into the sea just out from Hirara, on the island of Miyako. Only the navigator aboard the torpedo bomber was able to bail out. *Victorious'* ASR Walrus dashed close inshore and, while under small-arms fire from the coast, plucked the airman from the water as the Firefly TarCAP (Target Combat Air Patrol) overhead, peeled off to engage the shore-based guns.[10]

Photo 24-2

Royal Navy Fleet Air Arm Avengers, Seafires and Fireflies line up on the deck of HMS *Implacable* and warm up their engines before taking off for strikes in the Pacific in 1945. Australian War Memorial photograph 019037

The pilots of the two Walrus aircraft (at that time, affectionately known as *Darby* and *Joan*—a proverbial reference to a married couple content to share a quiet life of mutual devotion) were Australians. Flight Lieutenant Dave Howard and Pilot Officers Bruce Ada and Viv Lohmeyer were on loan for service with the British Pacific Fleet. After a crash course at Nowra, New South Wales, on deck landings for carrier air-sea rescue work, they had flown aboard HMS *Illustrious*, off Bondi Beach at Sydney, Australia, on 1 March. They later transferred to HMS *Victorious* at Manus, Admiralty Islands. (The air station at which the Walrus pilots trained was HMS Nabbington. Previously RAAF base Nowra, it had been transferred to the Royal Navy in 1944. Following the war and the RN's determination that the base was surplus to its needs it was commissioned into the RAN in 1948 as HMAS Albatross, the primary base for the RAN Fleet Air Arm.)[11]

Ada carried out the hazardous rescue of Sub Lieutenant John Gass off Miyako Island on 17 April 1945. After parachuting into the sea west of Hirara, Gass was propelled two and one-half miles by a fast current. Ada and Sub Lieutenant A. R. Marshall alighted their bi-plane on the water near some tiny islands to which Gass had drifted. Ada landed in shallow water about half a mile from the survivor and, while under fire from Japanese anti-aircraft guns, wove his way slowly through jagged coral beds. Fortunately, the guns did not depress fully and the rounds passed harmlessly overhead. Marshall hauled the survivor on board. Later that same day, Capt. Michael Denny, RN (commanding officer of the aircraft carrier HMS *Venerable*), sent a signal to HMS *Victorious*, "We are most grateful to the air-sea rescue for such a fine rescue," acknowledging the return of Sub. Lieutenant Gass to the fleet.[12]

Another air-sea rescue was made on 20 April, after four strikes of Avengers and rocket-armed Fireflies were launched against Sakishima Gunto's airfields. The Fireflies were to use their rockets against coastal shipping and other targets of opportunity, while the Avengers, as usual, dropped their ordinance primarily on the airfields. An Avenger of 848 Squadron was hit over Ishigaki Island, southwest of Okinawa, and crashed into the sea. Its aircrew managed to escape and were picked up by an American Dumbo aircraft a day later.[13]

Ada took flight in Walrus W3085 (*Darby*) on 10 August 1945 (the final day of the war) to rescue a New Zealand flyer. He sighted the pilot's raft, but had to ditch before reaching it because he ran out of fuel. Fortunately, the pilot (Derick Morton) had been picked up by the submarine USS *Peto* (SS-265) just before arrival of the aircraft. Ada then became a rescued flyer, after being picked up by an American destroyer. Following his transfer by breeches buoy to several other ships including the battleship HMS *King George V*, he arrived back aboard HMS *Victorious*, which was heading south at 26 knots. (This type of personnel transfer method consisted of a canvas sling, similar in form to a pair of breeches, suspended from a life buoy moved between ships by a rope and pulley system.)[14]

Photo 24-3

ADM Arthur W. Radford, USN, being passed to USS *Saratoga* (CV-3) his flagship, in a breeches buoy, November 1944.
National Archives photograph #80-G-205657

During World War II, U.S. submarines rescued 504 airmen from all services and Allied nations. Eleven of these airmen were recovered by submarines USS *Kingfish* (SS-234) and USS *Bluefish* (SS-222) during the Okinawa Campaign. Identified in the table are eight survivors from the BPF and three from the escort carrier USS *Chenango* (CVE-28).[15]

USS *Kingfish* (SS-234)

Date	Survivor	Ship
27 Mar 45	Lt. Comdr. Fred Charles Nottingham, RNVR	HMS *Illustrious*
31 Mar 45	Lt. Comdr. William Stuart, RNVR	HMS *Indomitable*
	Lt. Ian Joicey, RN	HMS *Indomitable*
	CPO W. Pine, RN	HMS *Indomitable*

	USS *Bluefish* (SS-222)	
30 Apr 45	ENS Victor T. Molinaro, USNR	USS *Chenango*
	AMM3c John A. Heintz	USS *Chenango*
	AMM3c Jack F. Bentley	USS *Chenango*
12 May 45	Sub Lt. T. H. Staniforth, RNVR	HMS *Indomitable*
	Sub Lt. W. Illingworth, RNVR	HMS *Indomitable*
	PO Gunner R. F. Mortimer	HMS *Indomitable*
16 May 45	Lt. D. T. Chute, RNVR	HMS *Victorious*[16]

Lt. Comdr. Fred Charles Nottingham, a TBF Squadron leader from HMS *Illustrious*, was rescued near the Sakishima Islands. Four days later, the submarine picked up three more British airmen from HMS *Indomitable*. With *Kingfish* heading toward the area were a TBM Avenger was reportedly having difficulties, the torpedo bomber crash landed 1,000 yards ahead of the submarine. The plane sank within a minute as its crew cast off in their rubber boat, and were then quickly taken aboard *Kingfish*. (Grumman TBF Avenger torpedo bombers were designated TBM if manufactured by General Motors.)[17]

Airmen who owed their lives to their life jacket, dinghy (rubber raft), etc., could join the Goldfish Club, formed in November 1942 by C. A. Robertson of the United Kingdom's PB Cow & Co.—one of the world's largest manufacturer of air-sea rescue equipment. Money, social position, or power could not gain a man or woman entry to the exclusive Goldfish Club. To become a member, one had to spend time upon the sea with nothing but a Carley Rubber Float between them and a watery death. (A placard advertising the requirements for club membership may be found at Appendix E.)[18]

Bluefish picked up the American crew of a TBM-3 Avenger from VT-25 aboard USS *Chenango* on 30 April, and the British 3-man crew of an Avenger from HMS *Indomitable* (857 Squadron) on 12 May. Four days later, a Corsair pilot from HMS *Victorious* was rescued.[19]

CANADA'S LAST VICTORIA CROSS WINNER

Aboard *Formidable* was Lt. Robert Hampton Gray, RCNVR. Gray, a Corsair fighter pilot assigned to 1841 Squadron, and a member of the Royal Canadian Navy Volunteer Reserve (RCNVR), colloquially known as "The Wavy Navy."[20]

The preceding year, Gray had received a Mention in Dispatches (MID) on 29 August 1944, for his participation in an attack on three German destroyers, during which his plane's rudder was shot off. On 16 January 1945, he received a second MID for "undaunted courage, skill and determination in carrying out daring attacks on the [German battleship] ... *Tirpitz*" in August 1944.[21]

By July 1945, the combined Allied fleets were attacking the Japanese mainland, striking any targets found. On 9 August, while heading a strike against the Japanese naval base at Maisuru (located on an inlet of the Sea of Japan on norther Honshu), Gray made a direct hit on a Japanese destroyer, setting it afire and ultimately sinking it. Admiral Vian sent a congratulatory message to HMS *Formidable*, praising Gray's "resolute and professional manner" and recommended him for the immediate award of the Distinguished Service Cross. However, the valiant Canadian would not live to receive this award.[22]

On the day of Gray's attack, as a second atomic bomb fell on Nagasaki, three days after the first one was dropped on Hiroshima, air strikes against Japanese targets continued unabated. Leading two flights of Corsairs against airfields in the Matsushima area, northern Honshu, Gray found little enemy activity, an earlier strike from *Formidable* having left the targets in ruins. This being the case, he decided to hit the secondary target of naval ships at nearby Onagawa Bay.[23]

Photo 24-4

Japanese escort ship *Etorofu*, May 1943, the same class ship as the *Amakusa*.
写真日本の軍艦第7巻 p. 198 (photographer unknown)

There, the Flight found a number of Japanese ships and dived down to attack. Furious fire was opened on the carrier aircraft from army shore batteries and from warships in the bay. Gray selected for his target a Japanese destroyer. As he leveled out and made straight for it, his Corsair was hit with cannon and machine gun fire, set aflame, and one of his 500lb. bombs was shot off. Gray steadied the aircraft, and released his remaining bomb, which struck *Amakusa* below her after gun turret. The munition detonated the ammunition locker, blowing out the starboard side of the escort vessel, which then rolled over and sank.²⁴

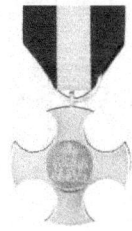

Victoria Cross   Distinguished   Mention in Despatches
                   Service Cross   (oak leaf device worn)

Lt. Robert Hampton Gray, RCNVR, did not return from this mission, having given his life at the very end of his fearless bombing run. One of the last Canadians to die in the war, he was posthumously awarded the Victoria Cross, the highest medal for valor in the British Commonwealth, on 13 November 1945. (This award followed that of the Distinguished Service Cross on 31 August 1945.) This citation for the Victoria Cross (which remains today the last one issued to a Canadian) reads as follows:

> For great bravery in leading an attack to within 50 feet of a Japanese destroyer in the face of intense anti-aircraft fire, thereby sinking the destroyer although he was hit and his own aircraft on fire and finally himself killed. He was one of the gallant company of Naval Airmen who, from December 1944, fought and beat the Japanese from Palembang to Tokyo. The actual incident took place in the Onagawa Wan on the 9th of August 1945. Gray was leader of the attack, which he pressed home in the face of fire from shore batteries and at least eight warships. With his aircraft in flames he nevertheless obtained at least one direct hit which sank its objective.²⁵

Gray's was only the second Victoria Cross earned by the Fleet Air Arm in WWII. The other was earned by Lt. Comdr. Eugene Esmonde RN, posthumously, after leading an attack on German battle cruisers *Scharnhorst* and *Gneisenau*, and the cruiser *Prince Eugen* as they dashed from Brest, France, up the English Channel to the safety of the North German port of Brunsbuttel in 1942.²⁶

RCNVR flier Lt. Arthur William Sutton was killed on 24 January 1945 when his Corsair of 1830 RN Squadron aboard HMS *Illustrious* crashed into a hanger full of Japanese aircraft in Palembang, Sumatra (today Indonesia). He was recommended for the VC, but the award was denied because it could not be determined that he was still alive in the final moments of the attack, and had purposefully, and heroically carried out this action.[27]

GRAY'S SQUADRON MATES CARRY ON ATTACK

When Gray's aircraft rolled inverted and crashed into the waters of Onagawa Bay, his squadron mates took out their anger at the loss of their leader on the other ships in the bay by repeatedly strafing them. Having vented their frustration and reeling from the loss of Gray, the Corsairs returned to *Formidable* off the coast. Once recovered, flight deck crews armed and spotted the planes for another attack on Onagawa while the pilots hurriedly grabbed some food.[28]

Among the pilots of the planes launched for the second strike were Canadians Lt. Charles Edgar Butterworth, RCNVR, and Lt. Gerald Arthur Anderson, RCNVR. As this flight hammered away at the remaining ships in the bay, they were joined by forty Hellcats from nearby American carriers. By the time the Corsairs from *Formidable* were heading back to the carrier mid-afternoon, there was total devastation at Onagawa Bay. Only the auxiliary minesweeper *Kongo Maru No. 2* remained afloat. The day's combat operations had been successful, but the loss of much-liked Robert Gray was a bitter pill to swallow, particularly because the war was winding down.[29]

Anderson's aircraft had been hit by anti-aircraft fire during the second attack and he was leaking fuel rapidly during the 150-mile return flight. He had two options. He could ditch his damaged Corsair or try to make the carrier. Anderson may have been wounded and believed that extraction from the aircraft after ditching might prove difficult. He chose to land aboard and made an approach from the port rear quarter. It looked like he was going to make it when, just feet from the flight deck rounddown (deck edge), Anderson's engine quit on him. The 9,000-pound aircraft slammed hard and flat into the rounddown, angled toward the sky. The aft section of the Corsair behind the pilot then broke away and fell downward into *Formidable*'s turbulent wake, created by her three churning propellers.[30]

For several long seconds, the forward part of the Corsair lay precariously at the back of the ship. The 22-year-old Anderson was slumped forward, unconscious, his head resting against the control panel. Flight deck personnel were momentarily stunned, and before any action could be taken, Anderson and his wrecked plane slid back, pitched up, and toppled over the stern of the ship and disappeared into the deep. One can only hope that Anderson, in his sealed cockpit, did not regain consciousness as his Corsair settled to the sea floor. (It has been stated that a pint more of fuel would have saved the day for Andersen, the last Canadian combat casualty in the war.)[31]

Of the seven Canadian pilots attached to *Formidable*'s air group only two, Lieutenants William Atkinson and Charles Butterworth, survived. Atkinson was the last Canadian to become an "ace" during the war. The seven pilots are identified below, along with the date of their death (if killed during the war), and awards for valor they earned.

Canadian RCNVR Pilots aboard HMS *Formidable*

Name	Killed	Awards
Lt. Gerald Arthur Anderson, RCNVR	9 Aug 45	
Lt. William Bell Asbridge, RCNVR	18 Jul 45	
Lt. William Henry Isaac Atkinson, RCNVR	Survivor	DSC, MID
Lt Charles Edgar Butterworth, RCNVR	Survivor	DSC
Lt Robert Hampton Gray, RCNVR	9 Aug 45	VC, DSC, 2 MID
Lt. James Finlay Ross, RCNVR	30 Jul 45	2 MID
Lt. Robert Ross Sheppard, RCNVR	22 Mar 45[32]	

Robert Sheppard was the brother of Donald John Sheppard, a pilot aboard HMS *Victorious* and the only other Canadian naval flyer to become a Pacific "ace" during the war. He and Atkinson were two of only sixteen WWII Fleet Air Arm pilots to achieve five or more air victories.[33]

The above tabulation is an example of participation for one country aboard only one fleet carrier between 22 March and 10 August 1945. In addition to five Canadians, other Fleet Air Arm squadron losses aboard *Formidable* included ten British (eight officers and two petty officer airmen), and a Dutch Royal Netherlands Naval Service (RNNAS) pilot. These losses resulted from a variety of causes associated with war:

- Commander of 1842 Squadron missing in action (16 April)
- Loss of 848 Squadron Avenger aircraft (17 April)
- Kamikaze attack off Okinawa (4 May); two of the four victims perished that day, the remaining two died of their wounds on 7 and 16 May, respectively
- Loss of aircraft to combat action (18 July)
- Aircraft losses off Tukushima (24 July)
- Aircraft losses during attack on Japanese shipping at Onagawa Wan (9 August)
- A/Sub Lt. Leslie Alan Maitland, RNVR, lost (details unknown)

RN Fleet Air Arm Squadron Losses aboard HMS *Formidable* (22 March-10 August 1945)

Name	Killed	Squad	Awards
Lt. Robert Ross Sheppard, RCNVR	22 Mar 45	1845	
A/Lt. Comdr. Anthony M. Garland, RNVR	16 Apr 45	1842	DSC & bar
T/A/PO Airman Charles W. Irwin	17 Apr 45	848	
Sub Lt. Douglas R. Whitehead, RNVR	17 Apr 45	848	
A/Sub Lt. John F. Bell, RNVR	4 May 45	1842	
A/Lt. Alan D. Burger, RNVR	4 May 45	1842	
A/Sub Lt. Donald G. Jupp, RNVR	16 May 45	848	DSC
Lt. William Bell Asbridge, RCNVR	18 Jul 45	1845	
Sub Lt. Walter Thomas Stradwick, RNVR	18 Jul 45	1842	MID
A/Lt. Alfred Cecil Francis, RNVR	24 Jul 45	848	MID
T/A/PO Airman Gordon C. Rawlinson	24 Jul 45	848	
Sub Lt. Bouke K. Swart, RNNAS	24 Jul 45	1842	
Lt. James Finlay Ross, RCNVR	30 Jul 45	1842	2 MID
Lt. Gerald Arthur Anderson, RCNVR	9 Aug 45	1845	
Lt. Robert Hampton Gray, RCNVR	9 Aug 45	1845	VC, DSC, 2 MID
A/Sub Lt. Leslie Alan Maitland, RNVR	10 Aug 45	1841	2 MID[34]

JAPANESE ATROCITY AGAINST BPF PRISONERS

During the Operation MERIDIAN attacks on Sumatran oil refineries in January 1945, nine flyers and aircrewmen who bailed out of planes hit by anti-aircraft fire, were captured by the Japanese. The prisoners were held in Outram Road jail in Singapore until on or after VJ Day, then transported to a beach at Changi (located in the eastern area of Singapore) and beheaded. Following this act, their bodies were placed in a boat, which was sunk at sea to hide evidence of the crime.[35]

After the war, an Allied investigation of the reported murder found that Captains Toshio Kataoka and Ikeda were primarily responsible for the atrocity. Both men killed themselves on 26 December 1945. Kataoka, in a will made before taking his life, wrote:

> We took nine prisoners from Outram Road in a lorry to the beach at the northernmost end of Changi and executed them with Japanese swords. The bodies were put in a boat prepared beforehand and sunk in the sea with weights attached. Now that the responsibility must be borne out publicly, I hereby pay for my deeds with suicide.[36]

25

Air-Sea Rescue, Ryukyus

Our Fleet had operated, from the very start, on the theory that the men who risked their lives to rocket, bomb and strafe the enemy wherever and whenever possible should, under no circumstances, be left to fend for themselves when disaster struck them. Admiral HALSEY and Admiral SPRUANCE made it clear that those airmen who were downed were to be rescued at all costs. This attitude on the part of the Fleet Commanders was a tremendous factor for the amazingly high morale of the pilots.

—CAPT William L. Erdmann, USN, commander Air-Sea Rescue, Ryukyus (Task Unit 30.5.2), in the unit's command history.[1]

Photo 25-1

Group photograph of Rescue Squadron VH-6 personnel taken 13 May 1945 at Naval Air Station, San Diego, California. The officers are in the first two rows, and enlisted members in the next six rows with chief petty officers at the back.
History of Rescue Squadron Six from 20 September 1944 to 1 September 1945

Air-sea rescue in the Ryukyus existed from 28 March 1945 until the cessation of hostilities with Japan on 15 August, following Emperor Hirohito's announcement of the surrender of Imperial Japan. However, the personnel undertaking these activities did not become a distinct task unit until 14 July 1945, when "Air-Sea Rescue, Ryukyus" (TU 30.5.2, later 95.9.2) was established under the command of CAPT William L. Erdmann, U.S. Navy.[2]

Rescues had, in many cases, been spur-of-the-moment affairs, with individual commanders using such rescue facilities as they had at hand. This system was satisfactory during the earlier stages of the war, when numbers of strikes were relatively few. However, as the magnitude of strikes increased, it became apparent that concurrent rescue operations to retrieve downed flyers by individual commands would lead to confusion and duplication of efforts if allowed to continue. Moreover, there might be some downed flyers overlooked. During the opening phases of the Okinawa campaign, Fleet Air Wing One had one squadron of PBM-5s (VH-3) performing air-sea rescue duties from Kerama Retto, yet there was no dedicated Air-Sea Rescue (ASR) command.[3]

Map 25-1

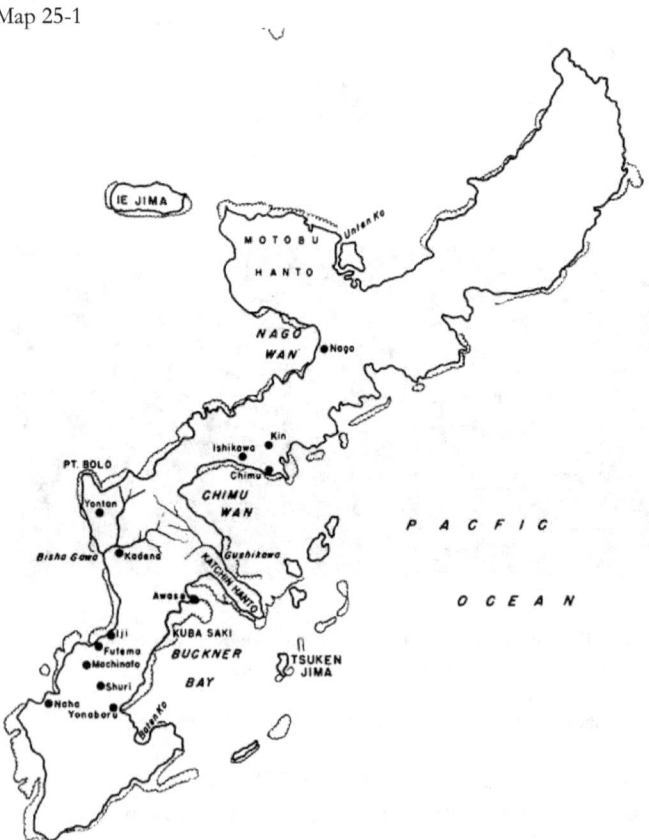

Okinawa (This island, and Kerama Retto, twenty miles to its southwest, are members of the Ryukyu island chain that stretches southwest from Kyushu, Japan, to Taiwan.)
https://www.ibiblio.org/hyperwar/USN/Building_Bases/bases-30.html

With the closing days of the Okinawa campaign, and the greatly enlarged strike program against the Japanese Empire, including B-29 bombers of the US Army Air Force, it became increasingly necessary to have a separate ASR command. Fleet Air Wing One, which had operated from Kerama Retto during the Okinawa campaign, relocated on 14 July to Chimu Wan on the east coast of Okinawa. The move occurred during Third Fleet Operations against Japan from 10-25 July.[4]

Formation of the Air-Sea Rescue Task Unit occurred in conjunction with this move. Comprising the Unit were four squadrons of PBMs—VH-1, VH-3, VH-4, and VH-6—seaplane tenders to service the squadrons, and a motor torpedo boat (PT) squadron to operate in conjunction with the task unit. Surface vessels such as destroyers would aid the task unit, and contact was to be maintained with commander, Submarine Force, U.S. Pacific Fleet, for the use of lifeguard submarines.[5]

COMPOSITION OF AIR-SEA RESCUE, RYUKYUS

When initially formed on 10 July 1945, Air-Sea Rescue, Ryukyus under Fleet Air Wing One (Task Unit 30.5.2) comprised six seaplane tenders, two rescue squadrons, and one motor torpedo boat squadron. (Not all of these units were then present at Chimu Wan.)

Ship/Squadron	Commanding Officer
Pine Island (AV-12)	CDR Henry Hodgkins, USN
Bering Strait (AVP-34)	CDR Walter D. Innis USN
Floyds Bay (AVP-40)	CDR James Robison Ogden, USN
Gardiners Bay (AVP-39)	CDR Carlton C. Lucas, USN
Mackinac (AVP-13)	CDR Paul L. Stahl, USN
Suisun (AVP-53)	CDR James J. Vaughan, USN
VH-3	LCDR William D. Bonvillian, USN
VH-4	LCDR Forrest H. Norvell Jr., USN
MTB Squadron 31	LT John M. Searles, USNR

Separate from the (Navy's Fleet Air Wing One) Air-Sea Rescue, Ryukyus, Task Unit, the U.S. Army Air Force continued the use of its own ASR unit, consisting of PBYs for Dumbo operations, B-17s with droppable rigid boats, and 63- and 85-foot crash boats.[6]

AIR-SEA RESCUE OPERATIONS FROM CHIMU WAN

Flight operations at Chimu Wan to support bomber and fighter strikes against Japanese targets commenced upon arrival of Rescue Squadrons VH-4 and VH-3. In previous operations these units had achieved remarkable success in air-sea rescue. (A total of 151 survivors were rescued throughout the Ryukyu Islands during the assault and occupation of Okinawa and neighboring islands of this chain.) Dumbos at Chimu Wan were stationed each day, based on requests made the previous night by both U.S. Navy and U.S. Army Air Force commanders (the Far Eastern Air Force; Marine Air Wing Two; commanding general, Twentieth Air Force; and commander, Third Fleet). Fighter cover and land-based B-17 and B-25 Dumbo aircraft were provided as necessary at the designated orbit points. Supplementing these specific requests for services, lifeguard submarines were stationed as requested by the task unit at favorable positions along the strike routes; Army PBYs orbited strategic spots along the routes; and four PT boats were on continuous patrol near Okinawa.[7]

Pilots in distress attempted to contact one of the rescue units on the radio distress frequency and, if necessary, tried to ditch near a rescue unit. Information was relayed to all rescue units and where necessary, standby units were dispatched. An example of cooperation provided by Navy ships during a rescue effort occurred on 25 July during bombing strikes on the Empire. LTJG G. B. Smith of VH-4 was orbiting in his PBM-5 Mariner near the fleet when, in late afternoon, he received word of two downed flyers at Jizo Saki off the north coast of Honshu. Smith did not have enough fuel to return to base after a rescue attempt, but was offered fighter cover if he elected to undertake the mission.[8]

Escorted by F6F Hellcats from Night Fighter Squadron VF(N)-91 aboard the carrier USS *Bon Homme Richard* (CV-31), Smith started to climb to 12,000 feet to escape anti-aircraft fire while en route to the scene. While doing so, his right engine cut out, but it started again when altitude decreased. While crossing Shikoku and Honshu to Jizo Saki, AA was encountered by his aircraft and escorts at Miho, Honshu, and the entire flight was amid adverse weather conditions. During the return flight to his carrier, Hellcat pilot ENS K. J. Baldwin shot down one Willow (Yokosuka K5Y1 biplane trainer) and damaged another. Following the mission, a VF(N)-91 war diary entry noted, "This is believed the longest rescue flight ever made by carrier aircraft, 250 miles each way."[9]

Arriving off northern Honshu, the downed pilots—ENS John H. Moore, USNR (VF-85) of the *Shangri-La*, and LT Howard Harrison, USNR (VF-88) of the *Yorktown*—were easily located as they were being orbited by "buddy" fighters. The PBM picked up the aviators ten miles off the coast without mishap, and Smith returned to the fleet position on the south side of the island, by once again climbing above the AA bursts. With insufficient fuel remaining, he was ordered to make a water landing and go aboard a destroyer with his crew and rescued passengers. When he arrived, the seas at the fleet's location were too rough to allow a landing, however the battleship *Missouri* (BB-63) discharged oil to create a slick and calm the surface. This enabled Smith to land, but exigencies of the situation did not permit refueling the PBM. Accordingly, after crew and survivors were taken aboard a destroyer, the aircraft was sunk by naval gunfire.[10]

Lifeguard submarines were always keen to assist in rescue operations, and made several outstanding recoveries, in some cases where rescue by other units was impossible. A daring rescue was made on 18 July 1945, when the USS *Quillback* (SS-424) picked up an Army flyer at the edge of a minefield and within ten miles of the coast of Kyushu.[11]

Motor Torpedo Boat Squadron 31 did excellent rescue work with their PT boats near Okinawa. On one occasion, two Marine pilots bailed out after a mid-air collision thirty-eight miles north of Zanpa Misaki, a cape on Okinawa's western coast. Within three hours of the alert, the survivors had been picked up by *PT-465*, commanded by LT William Grant, and returned safely ashore.[12]

DARING RESCUE IN THE INLAND SEA OFF KOBE

On 24 July 1945, eighteen FG-1D Corsairs of Bombing and Fighter Squadron VBF-88 were launched from USS *Yorktown* to strike airfields at Miho, Yonago, and Himeji on southern Honshu. (When the famed Vought F4U Corsair, referred to as "whistling death" by the Japanese, was subcontracted to be built by Goodyear, the resultant fighter aircraft carried the designation FG.) The VBF-88 Corsairs sank one "Sugar Dog"—Allied codename for a small Japanese merchant ship—in Osaka Harbor and badly damaged seven twin-engine enemy bombers.[13]

Photo 25-2

FG-1D Corsair of VBF-88 aboard the carrier USS *Yorktown* (CV-10). The code letters "RR" on the planes identify the *Yorktown*'s air group. U.S. Navy photograph

This was the first time that carrier-based aircraft had attacked targets in that part of the Inland Sea and, while there was little air opposition, anti-aircraft fire was intense. During the attack on Himeji, ENS Edwin A. Heck, USNR, had to ditch after his F4U Corsair experienced engine failure while making a strafing run on a Sugar Dog. Heck was able to clear Osaka Harbor, but landed about three miles west of Awaji Island, within sight of Kobe (see map). On landing, he was unable to release the life raft, because his harness fell over the release catch, obstructing it, and the plane sank within twenty-five seconds.[14]

In what Alan Emory described in the *Watertown Daily Times* on 4 April 1999 as "perhaps the most daring and the most spectacular of all Pacific air-sea rescues," LTJG Robert MacGill of VH-3 retrieved the downed pilot within sight of the shore. As Heck floated in his "Mae West" life jacket, the pilot's flight leader heard his VHF call, and notified his base. In response to a query, "Is there a Dumbo in the area?" MacGill and his crew, orbiting in a PBM-5 on station near the fleet, answered affirmatively. They were provided a fighter escort to cover the attempted rescue and, but for the actions of LT D. C. Steele and his seven VBF-88 Corsairs, would have likely been shot down.[15]

Upon reaching the Japanese coast from their carrier, Steele left part of his cover with the Dumbo, while he took the remainder of his division to Awaji Island to reconnoiter. En route, LT Gerald C. Hennesy shot down a lurking Judy (Japanese Navy carrier aircraft).

Steele and the other aircraft with him strafed a Sugar Dog headed for Heck and then, with all clear, advised MacGill that it was safe to carry out the rescue.[16]

Photo 25-3

Japanese "Judy" Navy dive bomber.
Source: Japanese Operational Aircraft "Know Your Enemy!"
CinCPac - CinCPOA Bulletin 105-45

MacGill reminisced, in November 1998, about flying into heavily defended Osaka Bay. He recalled that, in 1945 as the PBM Mariner passed over the docks of Kobe at an altitude of about 400 feet, people below stood there watching his aircraft. A Japanese fighter made a run at the seaplane and shore batteries opened up with anti-aircraft fire. The Corsair escort knocked down the Tojo (Nakajima Ki-44 Shoki single-engine fighter aircraft), but then had to depart the area because of low fuel.[17]

Photo 25-4

Tojo 2 Army fighter.
Source: Japanese Operational Aircraft "Know Your Enemy!"
CinCPac - CinCPOA Bulletin 105-45

Map 25-2

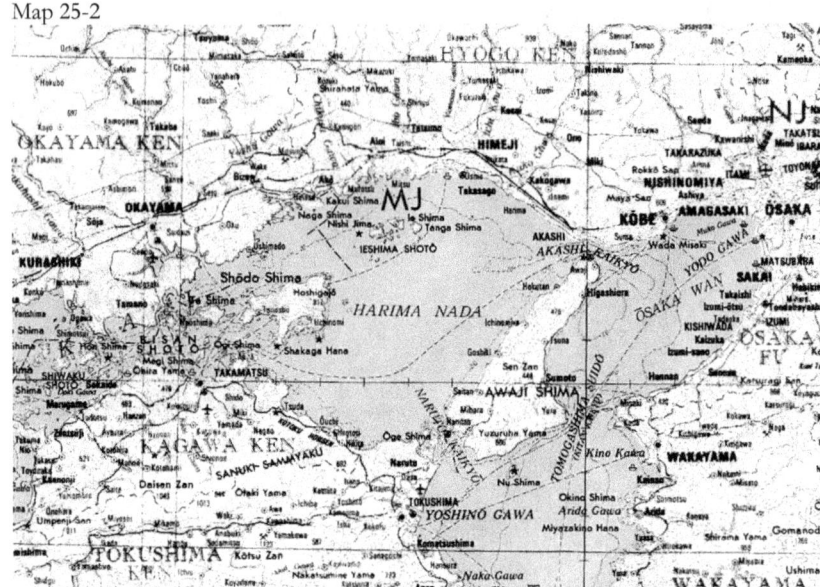

Inland Sea of Japan
http://legacy.lib.utexas.edu/maps/middle_east_and_asia/himeji_topo_1977.jpg

MacGill landed and took aboard the grateful pilot, who had spent five and one-half hours in the water. Only one minute and seventeen seconds elapsed between the time when the PBM's engines were cut, and the Dumbo was aloft again. The next challenge was how to get back to Okinawa with scant fuel. The flight engineer "leaned out" the fuel mixture (increased the quantity of air combined with fuel before combustion) and MacGill reduced airspeed to extend their range.[18]

The PBM hugged the Japanese coast on the return flight to reduce distance and with fighter escort gone, flew at 300 feet to avoid enemy fighters higher up. As a result of proximity to coastal defenses and low altitude, the aircraft was peppered by anti-aircraft fire from shore batteries, but made Chimu Wan—having been given up for lost.[19]

BATTLESHIP *HYUGA* SUNK BY CARRIER AIRCRAFT

That same day of MacGill's spectacular rescue, 24 July, planes from the aircraft carrier USS *Randolph* (CV-15) attacked Hofu, Usa, and Tsuiki airfields on southern Honshu, as well as ships of varying tonnage in the Inland Sea area. Not related to air-sea rescue, but of interest, two strikes concentrated on the *Ise*-class battleship-aircraft carrier *Hyuga*. Photographs taken at the time showed extensive and severe damage, and photo reconnaissance on the 28th revealed that she was resting on the bottom, decks awash.[20]

Photo 25-5

Japanese Battleship *Hyuga* off Kure, 9 October 1945.
National Archives photograph #80-G-351364

A VF-16 war diary entry concerning the attack on *Hyuga*, referred to her as "the bastard monstrosity." Following the loss of four large aircraft carriers during the Battle of Midway, the Japanese Navy had rebuilt the old *Ise*-class battleships *Ise* and *Hyuga* to enable them to operate an air group of floatplanes (a Japanese version of a seaplane tender). A flight deck replaced the rear gun turrets, and together with cranes for recovering aircraft, was designed to carry eleven planes. With an additional two floatplanes on catapults, and nine in the hangar, a maximum of twenty-two aircraft could be embarked.[21]

A SECOND PILOT RESCUED IN THE INLAND SEA

The following day, 25 July, LTJG Samuel Adams Davis of VH-3 rescued three downed aviators, one from the Inland Sea. The first survivor retrieved was ENS Calvin Bert Yoder, USNR, a member of Fighting Squadron VF-16 from the *Randolph*. With carrier aircraft having hit Hofu the preceding day, enemy gun crews were prepared and ready for additional strikes. Yoder's F6F Hellcat was hit by flak while making a rocket attack on a merchant ship and received sufficient damage to require ditching. He landed in the Inland Sea, twenty miles southeast of Hofu Airfield in Bungo Suido, a strait separating the islands of Kyushu and Shikoku and opening into the Pacific.[22]

Yoder set down at 1300 without injury, and was able to extract his life raft from the plane before it sank. There were fishing boats in the area, but he escaped their attention by covering himself and the raft with his poncho, blue side up. Upon sighting the Dumbo approaching in early evening, Yoder put out dye marker and fired his Very pistol (flare gun). Davis arrived at his location via Bungo Suido, encountering AA fire en route. After picking up Yoder, a search was begun five miles farther east, where a second survivor was reported to be located.[23]

This search did not locate the flyer. The Dumbo landed once to investigate an object in the dim light that might have been a survivor, but which proved to be nothing but a floating stump. There were many fishing boats in the vicinity and any one could have picked up the downed aviator. Meanwhile, four F6F Hellcats of Night Fighter Squadron VF(N)-91 from the carrier *Bon Homme Richard* (CV-31), which had escorted Davis' Dumbo to the Inland Sea, had been forced to withdraw because of fuel shortage. During the search, the Dumbo heard from the fighters that they had shot down two Oscars (Nakajima Ki-43 Hayabusa Army Type 1 Fighters) over Beppu Wan (Bay) on the northeast coast of Kyushu on their way to their carrier.[24]

During the Dumbo's unescorted return flight, Davis encountered two of these Hellcats of the fighter escort. As they had insufficient gas to make it back to their carrier, he requested that they ditch two miles south of Ashizuri-saki, the southernmost point of Shikoku. The swells were from ten to fifteen feet high, and there was a 10-knot wind from the southeast but the fighter pilots were able to ditch safely. Nonplussed, Davis made a moonlight landing and picked up LT Warren L. Smith and ENS John Gilbert Selway. It appears that he delivered them to a fleet unit other than the *Bon Homme Richard*. A VF(N)-91 war diary entry notes, "They did not rejoin the squadron until 20 August after journeying around the western Pacific."[25]

Davis was awarded the Navy Cross for heroic actions in carrying out the rescue of Yoder under fire from enemy shore batteries, as described in his medal citation:

> The President of the United States of America takes pleasure in presenting the Navy Cross to Lieutenant, Junior Grade Samuel Adams Davis, United States Naval Reserve, for extraordinary heroism in operations against the enemy while serving as Commander of a Navy Rescue Plane in Rescue Squadron THREE during operations against enemy Japanese forces in the vicinity of the Japanese Homeland on 25 July 1945. Persisting in a search for two downed aviators in the Inland Sea of Japan after his escort was forced to leave him, Lieutenant, Junior Grade, Davis located and

rescued a fighter pilot in the face of anti-aircraft fire from enemy shore batteries. While returning to his base, he sighted two more downed pilots and, executing a skillful landing by moonlight in a rough sea, took them aboard and returned the three airmen safe to base. His superb airmanship and courageous devotion to duty reflect the highest credit upon Lieutenant Davis and the United States Naval Service.26

RESCUE SQUADRON VH-6 REPORTS FOR DUTY

VH-6 patch

Rescue Squadron Six arrived at Chimu Wan, Okinawa, in the Ryukyu Islands, on the morning of 27 July 1945. Her commanding officer, LCDR Leland O. Ebey, reported for duty to commander, Fleet Air Wing One, aboard USS *Norton Sound* (AV-11). The squadron was assigned to the Rescue Task Unit of Fleet Air Wing One (TU 95.9.2), and initially berthed aboard the flagship of the task unit, USS *Pine Island* (AV-12). Within two weeks, several berthing changes were made and, at various times and in various numbers, squadron members were quartered aboard USS *Gardiners Bay* (AVP-39), *Suisun* (AVP-53), *Mackinac* (AVP-13), and *Floyds Bay* (AVP-40). By 14 August, berthing arrangements had stabilized, and crews 2, 3, and 4 were aboard *Pine Island*, and crews 1, 5, 6, 7, 8, and 9 were on *Floyds Bay*.27

LAST RESCUE SQUADRON, VH-6, COMMISSIONED

VH-6 had been commissioned on 20 September 1944 at Naval Air Station, San Diego, California, as the sixth and last Rescue Squadron in the Air Force of the United States Navy. There was no associated ceremony. On 23 May 1945, thirteen officers and forty-one men of VH-6 departed San Diego aboard the escort carrier USS *Cape Gloucester* (CVE-109), bound for Pearl Harbor and NAS Kaneohe Bay, Oahu. The squadron's six PBM-5 Mariners made the so-called "Trans-Pac" flight from the mainland, joining the VH-6 personnel at Kaneohe Bay.28

In early July 1945, orders from commander Pacific Fleet, Air Force, directed the squadron to move forward by increments. These orders indicated that VH-6 would operate in the Ryukyus, but required the squadron commander to report to commander, Fleet Air Wing One, when its first plane arrived at Tanapag, Saipan. The flight route to Tanapag was to be via Johnston Island (located 717 nautical miles west-southwest of Hawaii) and Parry Island, Eniwetok Atoll in the Marshalls. As before, a portion of squadron personnel travelled by ship—this time aboard *Floyds Bay*. The seaplane tender sailed from Pearl Harbor on 3 July, and the planes took flight from Kaneohe on 10 and 11 July.[29]

The squadron's first three planes arrived at Tanapag on 15 July, and the remaining three one day later. No orders were received by VH-6 until 25 July. These directed the squadron to proceed to Chimu Wan, as soon as possible. Following the arrival of VH-6 there on 27 July, its planes were initially flown by crews of Rescue Squadrons Three and Four. One patrol plane commander of VH-6 flew on each mission for indoctrination purposes. These missions were chiefly in support of Third Fleet operations. A failed open sea landing damaged one plane sufficiently that it was unflyable for a considerable period thereafter.[30]

Three crews of VH-6 were ordered to return to Tanapag to ferry replacement aircraft to Chimu Wan for VH-3 and VH-4 Squadrons. The remaining four VH-6 crews available for flying, flew seven air-sea rescue missions from 29-31 July 1944. These consisted of Dumbo flights for the Third Fleet, and for the U.S. Far East Air Forces (comprised of the Fifth, Seventh, and Thirteenth Air Forces), and one search mission for a missing pilot. There were no open sea landings and no recovery of survivors. One survivor was located off the eastern coast of Kyushu on 31 July, but sea conditions did not permit a landing to be made. An accompanying U.S. Army Air Corps B-17 Dumbo dropped the survivor a rigid boat, and instructions about the course to steer. The survivor was seen in the boat, but not located during a subsequent search.[31]

VH-6 SOLO AND COMBINED AIR-SEA RESCUES

Despite arriving in theater at war's end, VH-6 Mariners and aircrews contributed to the rescue of downed airmen prior to the announcement of the surrender of Imperial Japan by Emperor Hirohito on 15 August. After this date, there were few requests for the services of Dumbos.[32]

In a mission flown by LTJG John H. Dierkes, USNR, on 8 August, seven survivors were rescued at two different locations. The first rescue was of a single pilot: 1st Lt. D. L. Schmehl of the USAAF 341st Fighter Squadron. He had ditched his P-51 Mustang off Yaku Shima on 6

August as a result of engine failure. He was "buddied" (monitored by other aircraft) from the time of ditching until two B-17s and three PBYs arrived two hours later. One B-17 attempted to drop a boat, which hung in the bomb bay and had to be chopped loose. Once free, it fell too far away for retrieval by Schmehl. The other B-17 then dropped a boat, which Schmehl easily reached. All five planes circled until dark, then departed.[33]

The morning of 8 August, an Army PBY-5A was sent to the scene. The flying boat located Schmehl ten miles east of Yaku, but decided to land near shore, in the lee of the island where the water was calmer, and then proceed out to him. However, in making the landing, the Army plane was so badly damaged it sank in ten minutes. Later, Dierkes was able to land and rescued the P-51 survivor (Schmehl), then taxied ten miles, retrieved the Army Dumbo crew, and used its Jet Assisted Take Off (JATO) to take-off without mishap.[34]

USAAF 6th Emergency Rescue Squadron emblem, and USAAF 341st Fighter Squadron patch

USAAF 6th Emergency Rescue Squadron (PBY-5A crew)

2nd Lt. E. R. Briggs, USAAF	Sgt. S. D. Bateman, USAAF
2nd Lt. W. D. Riley Jr., USAAF	Sgt. C. P. Pryor, USAAF
Sgt. K. D. Eichner, USAAF	Sgt. J. C. Thompson, USAAF

USAAF 341st Fighter Squadron (P-51 pilot)
1st Lt. D. L. Schmehl, USAAF[35]

On 10 August, five survivors of a B-25 that had ditched off the western coast of Kyushu two days earlier, on 8 August, were transferred from USS *Pomfret* (SS-391), the lifeguard submarine which had originally rescued them, to a Mariner of VH-6 piloted by LTJG Carl L. Richwine, USNR. The bomber had been hit by small arms fire during a run on a Sugar Dog southwest of Fusan, and its starboard engine quit, forcing it down. The survivors had been picked up by the submarine the previous day. Two of the survivors were injured, one seriously, and the transfer was necessary to allow access to better medical care at the VH-6 base than was possible aboard the submarine with its limited facilities and personnel.

USAAF 500th Bombing Squadron B-25J Crewmembers	
1st Lt. P. E. Kent, USAAF	Sgt. G. E. Robertson, USAAF
1st Lt. B. W. Goodson, USAAF	Cpl. M. Gorham, USAAF
1st Lt. H. S. Cushed, USAAF [36]	

Photo 25-6

Open sea transfer of survivors from USS *Pomfret* to VH-6 plane on 10 August 1945. History of Rescue Squadron Six from 20 September 1944 to 1 September 1945

STRAFING OF ENEMY SUB/DOWNING OF ZEROS

The destinations and routes flown by Rescue Squadron Six aircraft during the first half of August varied considerably. Some involved missions to retrieve known survivors, other flights involved standing by in the Inland Sea of Japan during strikes on Kyushu, Honshu, and Shikoku. During one of these, a Mariner (aboard which LCDR Leland Ebey was the plane commander) and an accompanying B-17 and its four-plane P-51 Mustang fighter escort, strafed a small Japanese submarine in the Inland Sea. Although the results of the attack could not be accurately assessed, the submarine was forced to submerge and appeared to be damaged.[37]

On 14 August, a "Zeke" (from a group of three) made a pass on a Dumbo of VH-6 piloted by Lieutenant Dierkes, which was orbiting in the Inland Sea covering an Army Air Force strike. No gunfire was observed from the Mitsubishi A6M "Zero" fighter aircraft, but the Dumbo requested assistance from the P-38 fighter escort. In the ensuing air battle, a P-38 shot down the Zeke that had made the pass, another P-38 claimed a probable kill of another Japanese fighter. One of the four P-38s was hit and spun into the sea with its pilot, who did not survive. The third Zeke, which had scored the hit, escaped unharmed.[38]

SQUADRON VH-1 ARRIVES AT CHIMU WAN

Rescue Squadron VH-1 commanded by LT M. G. Crawford arrived at Chimu Wan on 7 August 1945. Administrative offices were established aboard the *Pine Island* and the squadron's aircrews were berthed aboard the *Suisun* and the *Bering Strait*. The squadron had previously been based ashore at Naval Air Base, Tanapag Harbor, Saipan, under the command of Air-Sea Rescue Unit, Central Marianas.[39]

AIR-SEA RESCUE OFF KAMIKAZE STAGING BASE

On 12 August, LTJG E. L. McClure landed his PBM-5 under fire of shore batteries, one mile off Tomitaka Air Field in western Kyushu, and picked up badly burned 2nd Lt. Donn P. Drake, USAAF of the 333rd Fighter Squadron. An explosive shell had hit the belly of Drake's P-47 Thunderbolt fighter and started a fire, and he'd been forced to bail out before being able to communicate with anyone. On recovery it was noted that his parachute and Mae West functioned well, but his raft was lost, and the dye marker pocket of the life jacket was empty. In the water, Drake used a signal mirror to attract fighter aircraft and they called the Dumbo.[40]

Drake, who was seriously burned, missed catching two passes with a line before he was rescued. The rescue would have been impossible without the aid of the fighters, which made two strafing runs on the shore batteries, and stayed with the Dumbo until it was away from the reach of the guns. Despite fighter cover, a hole was shot in the port rudder of the PBM. For his actions in rescuing the downed pilot under fire, Lieutenant (Junior Grade) McClure was awarded the Distinguished Flying Cross.[41]

Beginning in March 1945, the airfield at Tomitaka had served as a support base for kamikaze operations. Kamikaze squadrons from air bases at Omura, Kyushu; Tsukuba and Honshu, on Honshu; and Yatabe Genzan (Wonsan), Korea, stopped at Tomitaka before proceeding to Kanoya in the southernmost mainland prefecture of Kagoshima, where they made their final sorties toward Okinawa.[42]

A stone monument to the Tomitaka Kamikaze Special Attack Corps Sortie Site was errected in August 1977 on the grounds of the Kyowa Hospital (Hyuga City, Miyazaki Prefecture). The monument has inscriptions in both Japanese and English. The English one reads:

> This place used to be the air base of the Kami-Kaze pilots. And this monument should be dedicated to the peaceful rest of the souls of the Kami-Kaze and the U.S. pilots who were both killed around here for their own mother countries.[43]

273 AIRMEN RESCUED BY RYUKYUS TASK UNIT

The day after Drake's rescue, 13 August 1945, VH-3 was detached from Air-Sea Rescue, Ryukyus, and relocated to Saipan. Between 28 March and 15 August 1945, aircrews of VH-1, VH-3, VH-4 and VH-6 rescued a total of 273 airmen (U.S. Army: 83, U.S. Navy: 150, U.S. Marine Corps: 37, British: 3). Each dot on the diagram on page 248 represents one rescue. These same four rescue squadrons received a battle star for

Third Fleet Operations against Japan. The information about VH-2 and VH-5 in the table is provided for continuity.[44]

Unit	Battle Star	Commanding Officer	Where Based
VH-1	10-25 Jul 45	LCDR Russell R. Barrett Jr., USNR/LT H.G. Crawford, USNR	NAB, Tanapag Harbor, Saipan, under command of Air-Sea Rescue Unit, Central Marianas
VH-2		LCDR Harold A. Wells, USN	NAB, Tanapag Harbor, Saipan, under command of Air-Sea Rescue Unit, Central Marianas
VH-3	10-25 Jul 45	LCDR William D. Bonvillian, USN	*Bering Strait* (AVP-34) *Pine Island* (AV-12)
VH-4	10-25 Jul 45	LCDR Forrest H. Norvell Jr., USN	*Suisun* (AVP-53) *Gardiners Bay* (AVP-39) *Pine Island* (AV-12)
VH-5		LCDR M. E. Brown, USN	NAB Ebeye, Kwajalein; detachments at NAB Majuro, Majuro Atoll, and NAB Parry, Eniwetok Atoll
VH-6	10-25 Jul 45	LCDR Leland O. Ebey, USN	*Pine Island* (AV-12) *Floyds Bay* (AV-40)

248   Chapter 25

Diagram 25-3

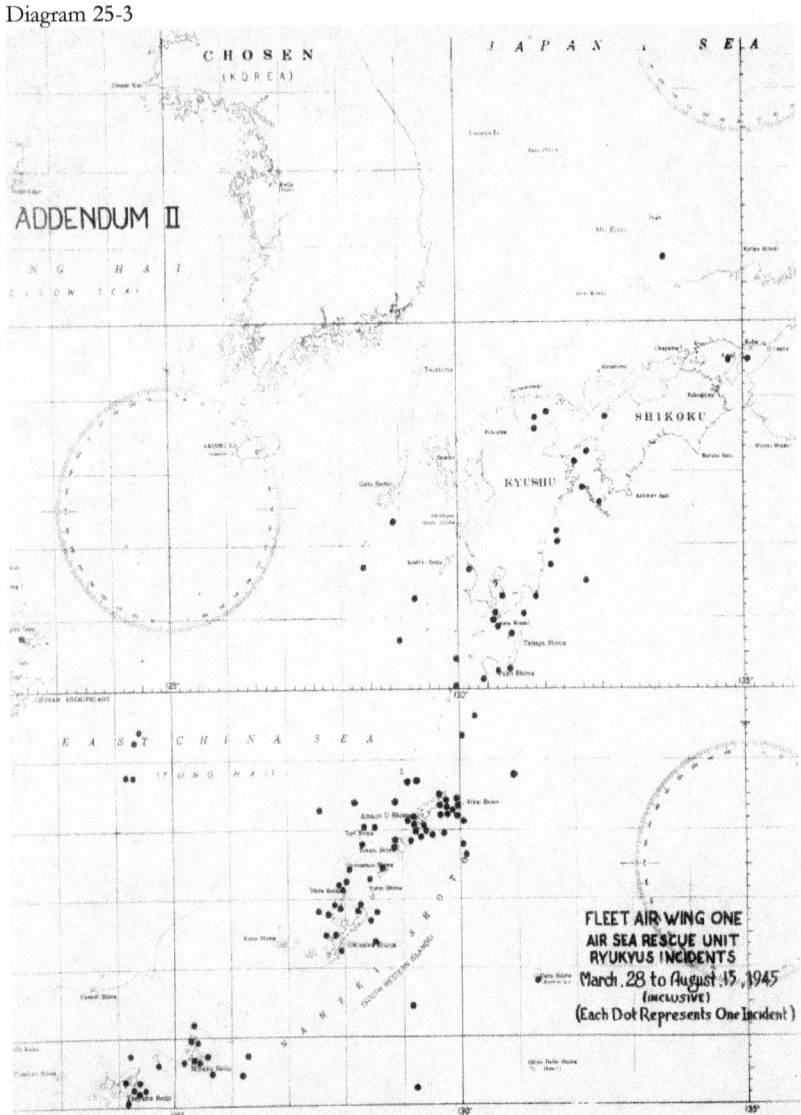

Locations of rescues of 273 airmen, 28 March to 15 August 1945
Commander Air-Sea Rescue, Ryukyus, History, 26 August 1945

26

Mission to Rekata Bay, and War's End

Japanese anti-aircraft gunners at Rekata Bay, who seemed to be the best in the business.
—Observation by Walter Lord in his book,
Lonely Vigil: Coastwatchers of the Solomons.[1]

Photo 26-1

From left: Flight Officer Neil Carr-Smith, Flight Officer Alastair Scott, and Sergeant Andy Bettie at Rekata Bay, 5 August 1945.
Courtesy of Jenny Scott (from Graham Goss Collection)

On the morning of 5 August 1945, a RNZAF Catalina took flight from Halavo Bay at 0700, bound for Rekata Bay on the northeastern coast of Santa Isabel Island. The island lay north-northwest of Guadalcanal in the central Solomons. The mission of those on board the aircraft were to search for a missing Hudson IIIA light bomber of RNZAF No. 3 Squadron (shot down by AA fire on 18 December 1942) and the remains of its aircrew. The importance of this undertaking was reflected in the seniority of the flying boat aircrew, with Group Captain Ronald Joseph Cohen, RNZAF, along as well:

RNZAF Catalina (NZ4051) Flight Crew

Wg Cr K. G. Smith (plane captain)	Wg Cr Avery
FO Neil Carr-Smith (2nd pilot)	FO Alastair M. Scott (diver)
FO Graham Goss (navigator)	Sgt L. Laird (engineer)[2]
GC Ronald Joseph Cohen	

Note: Sergeant Andy Bettie is identified in the preceding photograph, but is not identified as a member of the flight crew in the squadron diary entry from which the above list of individuals was obtained. There appears to be either an omission in the flight manifest or the individual identified as Bettie in the photograph is, in fact, Laird.

Photo 26-2

A RAAF Hudson bomber landing on an airfield after a raid against Japanese bases in the Pacific (the same type of aircraft as the lost RNZAF bomber). Australian War Memorial photograph 012412

The wreckage of the aircraft was located by Alastair M. Scott and others in his party, two miles from Suavana Point bordering the bay. Earlier in the war, the Japanese Navy had used Rekata Bay as a forward base for seaplanes and flying boat operations. During the Guadalcanal campaign, Japanese floatplanes carried out many night-harassment missions from Rekata. American servicemen dubbed the aircraft engaged in bombing them, "Washing Machine Charlie," owing to the sound of their unsynchronized engines.[3]

Rekata Bay offered a place where damaged Japanese aircraft could make emergency landings or ditch, during return to Bougainville or Rabaul following missions against Tulagi or Guadalcanal. Allied bombers and fighters began striking the Rekata Bay area, which was defended by anti-aircraft positions, on 7 August 1942. These missions continued until the end of August 1943.[4]

Sadly, no remains of the crew of the RNZAF bomber (NZ2054) were found. Killed in action were: Flight Sergeant Norman Newall, Sergeant Albert Mahony, Sergeant Henry Downard, Flight Sergeant David Newlands, and Sergeant Richard Andrew.[5]

RETURN HOME FOLLOWING THE END OF THE WAR

I have given serious thought to the situation prevailing at home and abroad and have concluded that continuing the war can only mean destruction for the nation and prolongation of bloodshed and cruelty in the world. I cannot bear to see my innocent people suffer any longer....

I was told by those advocating a continuation of hostilities that by June new divisions would be in place in fortified positions [at Kujukuri Beach, east of Tokyo] ready for the invader when he sought to land. It is now August and the fortifications still have not been completed....

There are those who say the key to national survival lies in a decisive battle in the homeland. The experiences of the past, however, show that there has always been a discrepancy between plans and performance. I do not believe that the discrepancy in the case of Kujukuri can be rectified. Since this is also the shape of things, how can we repel the invaders?...

It goes without saying that it is unbearable for me to see the brave and loyal fighting men of Japan disarmed. It is equally unbearable that others who have rendered me devoted service should now be punished as instigators of the war. Nevertheless, the time has come to bear the unbearable....

I swallow my tears and give my sanction to the proposal to accept the Allied proclamation on the basis outlined by the Foreign Minister....

—Emperor Hirohito addressing his top wartime advisers on 14 August 1945, at a conference in his underground air raid shelter on the grounds of the Imperial Palace in Tokyo.[6]

Following Emperor Hirohito's announcement, on 15 August 1945, of the surrender of Imperial Japan, and signing of peace documents on 2 September, the stimulus of war was gone, and servicemen wanted to go home. In allocating priorities for repatriation of RNZAF personnel, a system of points was worked out, based on length of service overseas, marital status, number of children, and other factors affecting eligibility for release from military service. Men who were due for repatriation in any case were given first priority, followed by others in their turn. For those low on the priority list, time passed slowly. To keep morale up, and help time pass more quickly, sporting activities were increased. Swimming and yachting were popular, and most New Zealand units had cricket teams which played a series of keenly contested matches.[7]

However, by using every available transport aircraft, repatriation efforts proceeded fairly rapidly. There had been over 7,000 RNZAF personnel in the Pacific, including those in Fiji, at the end of August. By the end of 1945, this number had fallen to just over seven hundred. The only men left in the Southwest Pacific Area were small rear parties on Los Negros and Bougainville overseeing stores and equipment awaiting shipping. In the South Pacific, a detached flight of No. 5 Squadron (not covered in this book) at Segond Channel, and rear parties were closing down the stations at Guadalcanal and Espiritu Santo.[8]

RNZAF SERVICE OF ALASTAIR M. SCOTT

One of the men who had to bide his time, awaiting return to New Zealand was Flight Officer Alastair Scott, who had taken on diver duties during the mission to locate the bomber and crew. He, like many citizens of Commonwealth countries, had travelled to Britain for training early in the war. Scott arrived in the UK from Canada by way of a North Atlantic crossing during the height of the U-boat campaign. Returning to New Zealand in February 1943, his journey began with a PanAm Clipper flight to New York City, with stops at Lisbon, Portugal; West Africa; and South America en route. (The narrows of the Atlantic between West Africa and South America, offered the safest crossing by air, in terms of geography and minimal threat posed by the enemy.)[9]

Mission to Rekata Bay, and War's End   253

Photo 26-3

A Boeing 314 "Clipper" (British Overseas Airways Corporation) in flight.
Library of Congress photograph cph.3b37576

Photo 26-4

Anzac Club in New York City, an American hospitality center for Australian and New Zealand forces. Identified personnel are: founder and president of the club, Nola Luxford (seated at far left) and progressing around the back of the table, RNZAF flyer (standing); Lon B. Worth (seated); Larry Timewell RNZAF (standing at far right) and Lieutenant Jack Slee, Canadian Navy (seated in front. The woman wearing the beret, man in uniform standing next to her, and woman seated in front are not identified. Australian War Memorial photograph 106117

Alastair Scott did his pilot training in New Zealand in DH-82 Tiger Moths, flew solo on 21 July 1944, and qualified in multi-engine aircraft before joining No. 6 Squadron.[10]

Photo 26-5

RAAF De Havilland DH82 Tiger Moth A17-285, circa 1940-45. Australian War Memorial photograph P01817.043

During his last nine months of service with the squadron, Scott served as adjutant to the commanding officer, Wing Commander K. G. Smith. The flying boat squadron was disbanded on 9 September 1945. Scott became the de facto commanding officer of Halavo Seaplane Base during his final two days there, because everyone else had departed and the base still officially existed—at least in name. Graham Goss, the navigator on the last PBY to leave, recounted to Jenny Scott (the author of *DUMBO DIARY Royal New Zealand Air Force No.6 (Flying Boat) Squadron 1943-1945*) decades later, that he remembered her father standing on the beach at Halavo waving them good bye. Alastair M. Scott retired with the rank of Wing Commander in 1971 and died in 1992.[11]

27

Post-war RAAF ASR Operations

Photo 27-1

PBY Catalina aircraft moored at Royal Australian Air Force Base Rathmines, New South Wales, Australia. The base was established in 1939 and was the main flying boat base for the RAAF during World War II and the 1950s.
Australian War Memorial photograph P11290.001

Following the cessation of hostilities on 15 August 1945, the PBY-5A Catalinas of No. 113 Air-Sea Rescue Flight operating from Morotai and, after 29 September, at Labuan Island, undertook special missions. These included the evacuation of Australian and Allied prisoners of war, transport of supplies to prisoners of war awaiting repatriation, and other tasks to support Australian units in Borneo. Flight Lieutenant Wally Mills took over as the commanding officer of No 113 Flight on 17 September 1945 as the Flight was relocating to its next base. By the end of September, relocation of the Flight to Labuan (off the northwest coast of Borneo, north of Brunei), which had commenced in July, was complete. The flight returned to Australia in January 1946, and was disbanded at RAAF Base Rathmines at month's end.[1]

WALLACE MILLS RECEIVES THE AIR FORCE CROSS

Photo 27-2

Distinguished Flying Cross

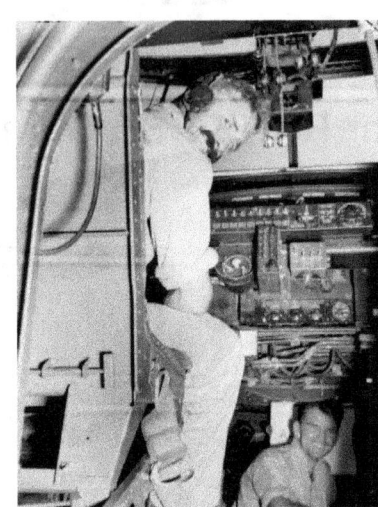

Air Force Cross

FL Wally Mills in flight from Morotai, Malmahera Islands on 20 April 1945. He performed some spectacular air-sea rescues during the war, resulting in the saved lives of several flyers of the First Tactical Air Force (RAAF).
Australian War Memorial photograph OG2443

After being posted to 2 Personnel Depot on 17 December 1945, FL Wallace Mills was discharged from the RAAF on 21 February 1946. He received an Air Force Cross on 25 June 1946. His citation noted in part:

> During his service he has created an unprecedented record in sea rescue work as a captain of Catalina aircraft. On numerous occasions and with a complete disregard for his personal safety Flight Lieutenant Mills has landed his aircraft under dangerous weather conditions and carried out rescues in most difficult circumstances. His fortitudinous spirit has been an inspiration to his fellow pilots and has greatly contributed in making his unit a significant factor in operations. He has completed a total of 162 sorties.[2]

SQUADRON LEADER RONALD RANKIN, AFC RAAF

In every military service, as members are discharged or retire, others take their place continuing the legacy of those who went before them. One such member was Ronald John Rankin who climbed the ranks from wireless (radio) operator to squadron leader over the course of his career, which included the command of several units. Born in Maitland in the southern Hunter Valley area of New South Wales (NSW) on 8

June 1917, he enlisted in the RAAF as a wireless operator in June 1936. After three years' experience as such (which included 400 hours flying) he was selected for the Airman Pilot Course and, after successfully completing the requirement, received his wings in March 1940. He then completed a flying instructor's course and served as a flying instructor with No 5 Service Flying Training School at RAAF Base Uranquinty, NSW. Having risen to the rank of Warrant Officer, he was commissioned as a Pilot Officer in October 1942.[3]

Photo 27-3

Squadron Leader Ronald J. Rankin AFC RAAF (at left, wearing khaki uniform and combination cap) and Col. Louis Lovo Castelar (UN Observer El Salvador) with Korean Military Advisory Group officers at Yongchon, Korea, circa 1950.
Australian War Memorial photograph P00716.053

In April 1943, Rankin was promoted to Flying Officer and in May 1944 he qualified to fly Catalinas. He was posted to No 43 Squadron in August 1944 and promoted to Flight Lieutenant in October. No. 43 squadron was based in Darwin as part of No 76 Wing RAAF, which also included Nos 20 and 42 Squadrons. Rankin participated in twenty-three operational aerial mining missions in the Southwest Pacific. The Catalina's flew up to 25 hours on individual mining tasks, and blockaded areas such as the Hainan, Surabaya, and Laoet Straits, and harbours at Hong Kong, Amoy, Seroea, and Pascadores islands. It was a task which required flying and navigational skills, courage, and a lot of endurance and patience.[4]

Photo 27-4

Darwin, circa 1945. A decorative mural of pin-up girls on the beach, adorning the wall of the metal-roofed mess of either No. 42 or No. 43 Squadron RAAF. Australian War Memorial photograph NWA0847

Rankin transferred to No. 42 Squadron in August 1945 during which time the Squadron conducted long range offensive patrols attacking Japanese aerodromes in southwest Celebes and Japanese shipping in the Flores and Banda Seas. With this squadron, he flew on fourteen operational missions.[5]

In November 1945, he took command of No 112 Air-Sea Rescue Flight, based in Darwin. In addition to search operations for missing aircraft, the Flight was engaged in evacuation of prisoners of war from Labuan, Morotai, and Balikpapan and the transfer of sick and injured prisoners to medical care. It also provided a regular courier and supply service to Australian outposts in East Timor and the Northern Territory. Rankin flew 700 hours' operational flying involving 43 sorties and 200 hours' transport flying on this posting.[6]

Rankin next took command of No 114 Air-Sea Rescue Flight in Townsville, in February 1947, and in June was engaged in search operations for missing aircraft in the Manokwari area on the north coast of West Papua. On 12 June 1947, Rankin was awarded the Air Force Cross (AFC) for exceptional courage and devotion to duty in the air. (An AFC is granted for 'an act or acts of exemplary gallantry while flying, though not in active operations against the enemy.')[7]

Photo 27-5

Manokwari, Dutch New Guinea, circa September 1945.
Australian War Memorial photograph 133807

In October 1947 No 114 Flight was disbanded and reformed under the Search and Rescue Wing, RAAF Base Rathmines. The Wing provided search and rescue capability in and around Australia and took over and consolidated the previously independent Flights in Darwin, Townsville, and Port Moresby. (These Flights had previously been autonomous units, and not part of a parent squadron or wing). In July 1948, Rankin transferred to No. 11 Squadron which was re-formed at Rathmines as a General Reconnaissance Squadron drawing on personnel and aircraft from the former Search and Rescue Wing. He commanded the Port Moresby detachment before joining Headquarters Eastern Area, Penrith, NSW, in March 1949 as Senior Air Operations Room Controller.[8]

Promoted to Squadron leader in March 1950, Rankin was sent to South Korea in May as a UN observer, together with Major Stuart Peach. For two weeks before the initial invasion by the North Koreans, the two Australian officers toured the 38th parallel to gather information about the respective North and South defenses following up with a report which they submitted to the UN. The report noted that the South Koreans were in defensive positions but the North Koreans seemed to be offensively deployed. Within days of their return to Seoul, North Korean forces crossed the 38th Parallel and invaded the South. Armed with Peach and Rankin's report and the blatant action of North Korea, the UN deduced that the North was clearly the aggressor.

The Security Council then passed a resolution to provide forces to oppose North Korea's invasion. Having played a pivotal role in the Korean conflict, Peach and Rankin continued their UN duties until later that year. The Peach-Rankin report became perhaps the most important written by Australian UN observers in the seventy years of Australian peacekeeping.⁹

Photo 27-6

Korea, June 1950. UN Commission of Korea observers, Mr. Charles Coates, Squadron Leader Robert J. Rankin AFC RAAF, and Maj. Stuart Peach, touring the 38th Parallel. Australian War Memorial photograph P00716.018

Rankin was given a permanent commission in January 1951 and took command of No. 30 Squadron at RAAF Base Richmond, a Target Towing and Special Duties Squadron. In March 1952, he undertook the Senior Weapons Officer's course and, in December, joined the Headquarters Eastern Area as the Weapons Officer. His career then followed a series of staff postings until he reached retiring age and left the service in June 1964.¹⁰

Postscript

On 15 August [1945] President Truman was able to announce the unconditional surrender of the Japanese Imperial Government. Orders were sent out for the occupation of Japan and Korea.... Overnight [there was] a change in the attitude of the men in the combat zone from one of war to one of peace. Suddenly everyone wanted to go home. No one was interested in becoming a member of the occupational forces.

—Excerpts from a description by VADM Daniel E. Barbey, USN (Ret.) in *MacArthur's Navy* about the perspective of Sailors and Marines in the Pacific following the end of World War II, which could also describe the attitude of servicemen in Europe after the cessation of fighting there.[1]

Photo Postscript-1

U.S. Navy sailors celebrating VJ (Victory over Japan) Day at sea, 15 August 1945. Naval History and Heritage Command photograph #UA 570.61

Photo Postscript-2

Empty beer bottles strewn along Mindil Beach, Darwin, 16 August 1945, the day after Army soldiers celebrated Victory over Japan (VJ) Day.
Australian War Memorial photograph P04287.006

On 15 August 1945, ADM William F. Halsey, commander Third Fleet, received orders from Fleet Admiral Chester W. Nimitz, commander in chief, Pacific Fleet, to cease all operations against Japan. The cease fire order was received too late to stop the first of the day's air strikes against Japan, but the second strike which had been launched, was recalled in time. Two weeks later, Nimitz arrived in Tokyo Bay aboard a PB2Y Coronado seaplane, in preparation for Japan's formal surrender aboard the battleship *Missouri* (BB-63).

SURRENDER CEREMONY IN TOKYO BAY

It is my earnest hope, and indeed the hope of all mankind, that from this solemn occasion a better world shall emerge out of the blood and carnage of the past—a world dedicated to the dignity of man and the fulfillment of his most cherished wish for freedom, tolerance and justice.

—Closing remark made by Gen. Douglas MacArthur during a speech at the Surrender Ceremony of Japan aboard the battleship USS *Missouri* (BB-63) on 2 September 1945.[2]

Photo Postscript-3

Fleet Admiral Chester W. Nimitz, USN, arrives at Tokyo Bay in a PB2Y Coronado seaplane, 29 August 1945. USS *Missouri* (BB-63) is in the center background.
Naval History and Heritage Command photograph #NH 96809

Present in Tokyo Bay, on 2 September, were 255 Allied naval ships and merchant vessels. Five were the seaplane tenders *Cumberland Sound* (AV-17), *Hamlin* (AV-15), *Gardiners Bay* (AVP-39), *Mackinac* (AVP-13), and *Suisun* (AVP-53).[3]

Photo Postscript-4

Fleet Admiral Chester W. Nimitz, commander in chief, Pacific, and Pacific Ocean Areas, signs the Instrument of Surrender as United States representative, 2 September 1945. Naval History and Heritage Command photograph #NH 58082

Photo Postscript-5

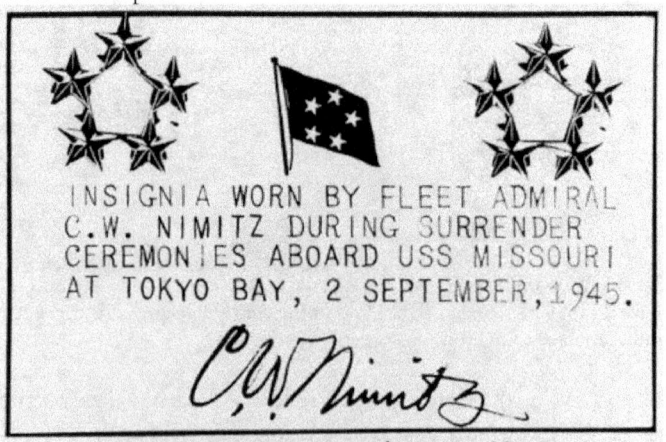

Insignia Worn by Fleet Admiral Chester Nimitz during the ceremony. Naval History and Heritage Command photograph #NH 62680

OPERATION MAGIC CARPET

Photo Postscript-6

Painting *Coming into Golden Gate Bridge* by Franklin Boggs, 1945.
Naval History and Heritage Command Accession #88-159-BR

Following victory in Europe and the surrender of Japan, there remained the enormous task of returning home huge numbers of American and other Allied military personnel scattered across the globe. Operation MAGIC CARPET was the name given to the post-war sealift of military and civilian personnel from Europe, the Pacific, Asia, Alaska, the Caribbean, and South America to the United States.[4]

On 9 September 1945, RADM Henry S. Kendall, USN (commander, Carrier Division 24) was designated commander, Task Group 16.12 and, as such, was in charge of MAGIC CARPET. During its peak operating period in December 1945, 305 ships, including 222 troopships, 12 hospital ships, 6 battleships, 18 cruisers, and 57 aircraft carriers, on all the oceans of the world, were transporting U.S. servicemen home. So many were being separated from military service that Gen. George C. Marshall told historian Samuel Eliot Morison "…that demobilization had become a rout."[5]

Australia, Canada, New Zealand, and other Allied countries implemented similar measures. However, they, like the U.S., did not have enough lift to bring home all the troops as fast as they wanted to come home, or as fast as the public demanded. Priority was rightly given to newly freed prisoners of war, other wounded personnel, and servicemen who had accumulated the necessary points to be discharged.⁶

USS *COOS BAY* IS DECOMMISIONED, EKHOLM AND OTHERS ARE DISCHARGED, THEIR SERVICE DONE

The seaplane tender *Coos Bay*, sent stateside to San Pedro, California, for repair of damage resulting from a collision on 31 March 1945 with the tug *Matagorda*, had sailed from there on 16 August, westward bound. She departed Midway Island on 6 September for Ominato Ko, located on the north coast of Honshu. Following arrival on 13 September, she tended aircraft of patrol bombing squadron VPB-21 and others in transit. In October, she tended aircraft conducting surveillance, search, and courier flights in support of commander, Task Force 56 operations. Finally, in mid-November, *Coos Bay* shifted to Negishi Wan, Tokyo Bay, for occupation duty. Her duty ended on 2 December, when she sailed for Pearl Harbor, to receive further routing orders to take her home to the United States.⁷

Coos Bay arrived at her final destination: Orange, the easternmost city in Texas (located on the Sabine River at the border with Louisiana), on 11 January 1946, for inactivation. Her executive officer, LT Fritiof ("Fritz") Ekholm, USNR, had been the second in command since at least 1 July 1945, as evidenced by his commencement of signing Quarterly Muster Rolls on that date. *Coos Bay* was decommissioned at Orange on 30 April, and laid up in the Atlantic Reserve Fleet, Texas Group, which was located there. Ekholm received an honorable discharge on 9 October 1946. He and his fellow naval reservists had been what the Royal Navy termed "hostilities only" officers during the war and, with massive post-war downsizing in progress, their services were no longer required.⁸

Turning to other officers, not part of *Coos Bay*'s complement, but mentioned earlier in this book, some were able to continue their military careers. These included an Australian and a New Zealand flyer. Squadron Leader Ronald John Rankin, AFC RAAF, who came up through the ranks, served until June 1964. New Zealander, Wing Commander, Alastair M. Scott, RNZAF, retired in 1971. His daughter, Jenny Scott, the author of *Dumbo Diary: Royal New Zealand Air Force No.6 (Flying Boat) Squadron 1943-1945*, generously allowed the use of photographs and other materials from her collection, herein.

Appendix A: RAAF and RNZAF Honours and Awards

Royal Australian Air Force

No. 112 Air-Sea Rescue Flight
Distinguished Flying Cross (DFC)

FL Charles Ralph Bulman	SL Kenneth Arthur Crisp
FL Robin Morton Corrie	

Distinguished Flying Medal

FO Arthur Joseph Scholes

Mention in Despatches

WO William Sydney Andrews	WO Robert Patrick Quirk
A/Sgt Leslie Arthur Hughes	

No. 113 Air-Sea Rescue Flight
Distinguished Flying Cross (DFC)

FL Norman William Hastie	FL Walter Raymond Mills

Air Force Cross (AFC)

FL Walter Raymond Mills

Mention in Despatches

A/Sgt Robert Charles Ballingall	FO Vernon John Boyd Morris
FO Ronald William Thompson Francis	LAC Richard James O'Meara
T/WO Reginald Charles Grilley	T/F/Sgt Arthur Frederick Taylor
Sgt Arvie Maberry	
LAC: Leading Aircraftman	T/WO: Temporary Warrant Officer
T/F/Sgt: Temporary Flight Sergeant	

Royal New Zealand Air Force

No. 6 (Flying Boat) Squadron
Distinguished Flying Cross (DFC)

FL Donald Stanley Beauchamp	FO Christopher David John Regan
FL Winston B. Mackley DFC & Bar	

Distinguished Flying Medal (DFM)

F/S James Benjamin Monk

Air Force Cross (AFC)

SL Ronald Bruce Leslie MacGregor

Flight Lieutenant Donald Stanley Beauchamp received his DFC for rescuing five survivors from the crew of a B-24 from 26th Bombardment Squadron, 7th Air Force, after six days in a raft, as described in his award citation:

By his skill and courage in landing a heavily-laden flying-boat in rough water in the open sea, this officer rescued five United States servicemen on 4th February, 1944. The men, who had been adrift for six days, were transferred to the aircraft, and in spite of rough water and a heavy swell, a take-off was made and the aircraft flown back to base. Flying Officer Beauchamp was captain of the New Zealand 'Catalina' flying-boat concerned in the rescue. On sighting the rafts adrift in the Pacific, he was unable to contact base for instructions, so he decided on his own initiative to attempt the rescue, well knowing that, should he fail, there would be little prospect of rescue for himself and his crew. Although the aircraft was slightly damaged in the open water landing and the subsequent take-off was cross-wind, Flying Officer Beauchamp displayed a high degree of skill and determination, and succeeded in a difficult and hazardous enterprise.

Flying Officer Christopher David John Regan received a DFC for the hazardous rescue of a fighter pilot off New Britain:

NZ4039 0800 – 1500 FGOFF C. Regan (Capt.) PLTOFF Oliver (Nav.) Jacquinot Bay – DUMBO Returning to base after completing an escort mission Jacquinot Bay tower directed aircraft to Ataliklikun Bay where 2 F4U Corsairs were circling off Kabanga River. NZ436816 FLTSGT C.E. Turner, pilot of F4U-1 NZ5468, No.20 Squadron, was located in his dinghy. After circling NZ4039 landed in a heavy sea and FLTSGT Turner was retrieved suffering from shock and exposure. The Navigator administered first aid and the Catalina returned to Jacquinot Bay. FLTSGT Turner's aircraft had been hit by Anti-Aircraft fire.

Flight Lieutenant Bill Mackley got a Bar to his DFC, the first was earned while flying bombers in Europe. The second was for the rescue of ten survivors of a USAAF B-24 on 26 January 1945.

Flight Sergeant James Benjamin Monk was awarded the Distinguished Flying Medal for his heroic actions while serving as the captain of a Catalina, as detailed in Chapter 12.

The first member of No. 6 (Flying Boat Squadron) to be honored for valor had been Squadron Leader Ronald MacGregor, awarded the Air Force Cross following the rescue of eight from the SS *Phoebe A. Hearst* on 2 May 1943. This event is also described in Chapter 12.

Appendix B: RNZAF No. 6 Squadron Rescue Missions

Courtesy of Jenny Scott. A more readable version of this information follows on the next page.

Appendix B

RNZAF 6 Squadron rescues:

DATE	CAPTAIN	TYPE	POSITION	No.	TOTAL
2 May '43	SQLDR McGregor	S.S. Phoebe A. Hearst	120 Miles S of Suva	8	8
26 Jan '44	FGOFF W. Mackley	B-24	S.E. Ontong Java	10	18
4 Feb '44	FGOFF Beauchamp	B-24	S. Nauru	5	23
11 Feb '44	FGOFF Hendry	B-25	40 Miles W. of Buka	6	29
13 Feb '44	FLTLT J. McGrane	P-39	S.E. Cape St George	1	30
16 Feb '44	FGOFF Hendry	P-40 NZ3190	N. Cape Henpan	1	31
18 Feb '44	FGOFF G. Hitchcock	B-25	Buka Island	2	33
18 Feb '44	FLTLT J. McGrane	F4U	St. George's Ch.	1	34
21 Feb '44	FGOFF Hendry	F4U	W. Torokina	1	35
21 Feb '44	FGOFF Hendry	F4U	Cape St. George	1	36
22 Feb '44	FGOFF G. Hitchcock	SBD	W. Buka Passage	2	38
5 Mar '44	FLTSGT J. Monk	B-24	Kabanga Bay	3	41
9 Mar '44	FGOFF Beauchamp	B-25	St. George's Ch.	7	48
16 Mar '44	FLTLT Francis	F4U	Feni Island	1	49
31 Mar '44	FLTLT Scott	TBF	St. George's Ch.	3	52
11 Apr '44	FLTLT Scott	P-38	Santa Isabel	1	53
1 May '44	FLTLT A. Manz	LST	Cape Zelee	1	54
10 May '44	FLTLT R. McHARDY	SC163	San Cristobal	1	55
30 Jun '44	FLTLT R. McHARDY	SBD	S. Guadalcanal	2	57
7 Sept '44	W/O I. Donaldson	FP-147	Malelei Island	9	66
29 Sept '44	SQLDR J.R. Butcher	TBM	Bellona Island	7	73
4 Oct '44	FGOFF S.R. Carleton	APC	S.E. Guadalcanal	1	74
5 Oct '44	WGCDR J. AGAR	PV1 NZ4561	N.W. Florida Is.	1*	74
8 Nov '44	FLTLT Martin	SBD	Beaufort Bay	3	77
16 Nov '44	FGOFF D. Sheehan	SS Waco Victory	S. Cape Henslow	1	78
9 Aug '45	FGOFF C. Regan	F4U NZ5468	Ataliklikum Bay	1	79

* Dumbo recovered body of NZ4213642 PLTOFF Thomas Helier Vautier (23)

Appendix C: *Barnegat*-class Seaplane Tender Battle Stars

Shown below are the commissioning dates of the *Barnegat*-class small seaplane tenders; the number of battle stars ships were awarded, and the associated theater of operations, and identities of the ships initial commanding officers.

A: American Theater
E: Europe-Africa-Middle East Theater
P: Asiatic-Pacific Theater

Puget Sound Navy Yard, Bremerton, Washington

Ship	Comm.	Battle Stars	First Commanding Officer
Barnegat (AVP-10)	3 Jul 41	E *	CDR Felix Locke Baker
Biscayne (AVP-11)	3 Jul 41	E ****	LCDR Carleton Cole Champion Jr.
Casco (AVP-12)	27 Dec 41	P ***	CDR Thomas Selby Combs
Mackinac (AVP-13)	24 Jan 42	P ******	CDR Norman Ridgeway Hitchcock

Boston Navy Yard, Boston, Massachusetts

Ship	Comm.	Battle Stars	First Commanding Officer
Humboldt (AVP-21)	7 Oct 41	A & E	CDR William G. Tomlinson
Matagorda (AVP-22)	16 Dec 41	A & E	CDR Stanley J. Michael

Lake Washington Shipyard, Houghton, Washington

Ship	Comm.	Battle Stars	First Commanding Officer
Absecon (AVP-23)	28 Jan 43	A	CDR Robert Seldon Purvis
Chincoteague (AVP-24)	12 Apr 43	P ******	CDR Ira Earl Hobbs
Coos Bay (AVP-25)	16 May 43	P **	CDR William Miller Jr.
Half Moon (AVP-26)	15 Jun 43	P **	CDR William O. Gallery
Barataria (AVP-33)	13 Aug 44	P *	CDR Garrett S. Coleman
Bering Strait (AVP-34)	19 Jul 44	P ***	CDR Walter Deans Innis
Castle Rock (AVP-35)	8 Oct 44	Pacific	CDR George S. James Jr.
Cook Inlet (AVP-36)	5 Nov 44	P *	CDR William P. Woods
Corson (AVP-37)	3 Dec 44	Pacific	CDR Samuel P. Weller
Duxbury Bay (AVP-38)	31 Dec 44	P **	CDR Frank N. House
Gardiners Bay (AVP-39)	11 Feb 45	P **	CDR Carlton C. Lucas
Floyds Bay (AVP-40)	26 Mar 45	P *	CDR James R. Ogden
Onslow (AVP-48)	22 Dec 43	P ****	CDR Alden D. Schwarz
Orca (AVP-49)	23 Jan 44	P ****	CDR Morton K. Fleming Jr.
Rehoboth (AVP-50)	23 Feb 44	E & P	CDR Robert C. Warrack
San Carlos (AVP-51)	21 Mar 44	P ***	LCDR DeLong Mills
Shelikof (AVP-52)	17 Apr 44	P ***	LCDR Reuben E. Stanley
Suisun (AVP-53)	10 Sep 43	P **	CDR James J. Vaughn

Associated Ship Building Inc., Seattle, Washington

Rockaway (AVP-29)	6 Jan 43	E *	CDR Henry C. Doan
San Pablo (AVP-30)	15 Mar 43	P ****	CDR Roy R. Darron
Unimak (AVP-31)	31 Dec 43	A, E, P	CDR Hilfort C. Owen
Yakutat (AVP-32)	31 Mar 44	P ****	CDR George K. Fraser

Appendix D: David S. McCampbell Medal of Honor, and Roy Warrick Rushing Navy Cross Citations

CITATION:

The President of the United States of America, in the name of Congress, takes pleasure in presenting the Medal of Honor to Commander David S. McCampbell, United States Navy, for conspicuous gallantry and intrepidity at the risk of his life above and beyond the call of duty as Commander, Air Group FIFTEEN (AG-15), attached to the U.S.S. *ESSEX* (CV-9), during combat against enemy Japanese aerial forces in the first and second battles of the Philippine Sea. An inspiring leader, fighting boldly in the face of terrific odds, Commander McCampbell led his fighter planes against a force of 80 Japanese carrier-based aircraft bearing down on our fleet on 19 June 1944. Striking fiercely in valiant defense of our surface force, he personally destroyed seven hostile planes during this single engagement in which the outnumbering attack force was utterly routed and virtually annihilated. During a major fleet engagement with the enemy on 24 October, Commander McCampbell, assisted by but one plane, intercepted and daringly attacked a formation of 60 hostile land-based craft approaching our forces. Fighting desperately but with superb skill against such overwhelming airpower, he shot down nine Japanese planes and, completely disorganizing the enemy group, forced the remainder to abandon the attack before a single aircraft could reach the fleet. His great personal valor and indomitable spirit of aggression under extremely perilous combat conditions reflect the highest credit upon Commander McCampbell and the United States Naval Service.

CITATION:

The President of the United States of America takes pleasure in presenting the Navy Cross to Lieutenant, Junior Grade Roy Warrick Rushing, United States Navy, for extraordinary heroism in operations against the enemy while serving as Pilot of a carrier-based Navy Fighter Plane in Fighting Squadron FIFTEEN (VF-15, attached to the U.S.S. *ESSEX* (CV-9), in action against enemy Japanese surface forces over the Sibuyan Sea during the Battle for Leyte Gulf in the Philippine Islands on 24 October 1944. With the support of his Section Leader, Lieutenant, Junior Grade, Rushing attacked a formation of 40 enemy fighter planes and shot down six of the enemy planes in flames, probably destroying two others. His actions effectively broke up the enemy formation and prevented an attack upon our surface forces. Lieutenant, Junior Grade, Rushing's outstanding courage and determined skill were at all times inspiring and in keeping with the highest traditions of the United States Naval Service.

Appendix E: Solicitation of Goldfish Club Members

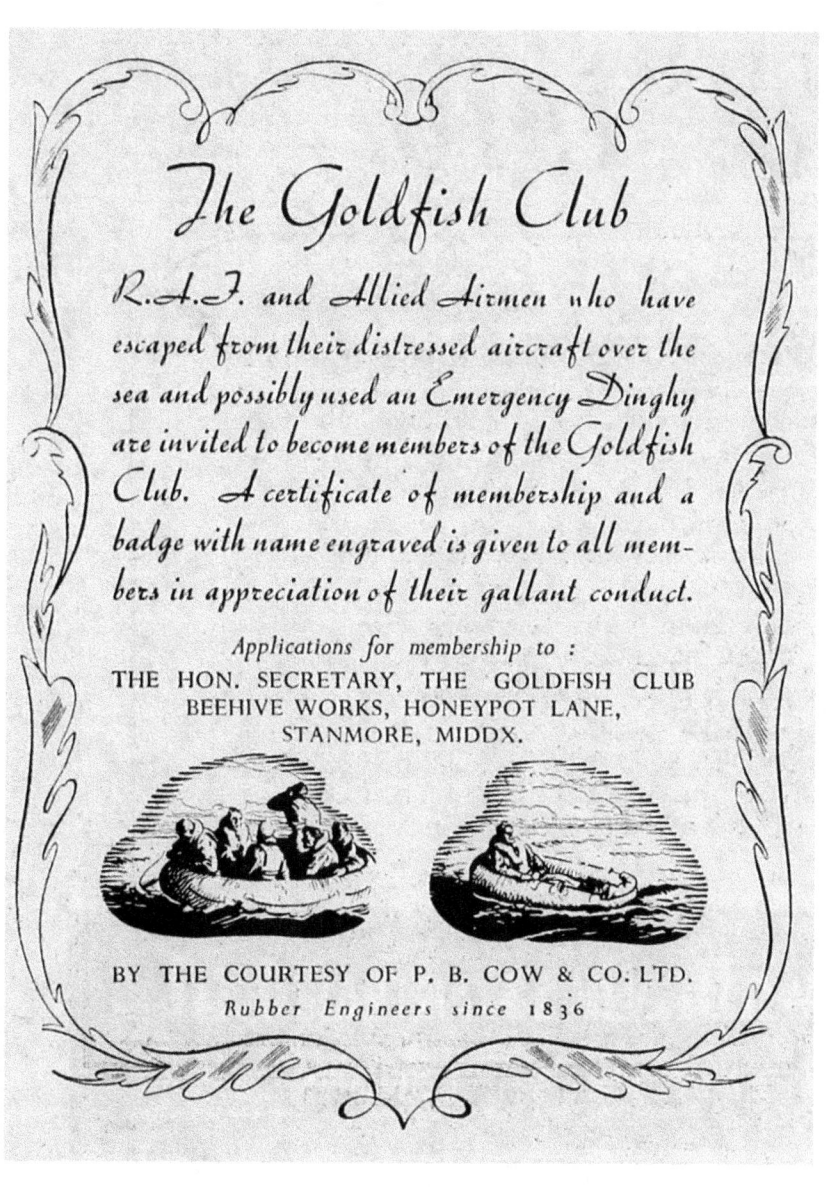

Appendix F: VH-3 Instructions to Downed Airmen

**UNITED STATES PACIFIC FLEET
AIR FORCE
RESCUE SQUADRON THREE**

IMPORTANT

YOU HAVE BEEN SIGHTED BY RESCUE SQUADRON THREE. FOLLOW THESE INSTRUCTIONS AND YOUR RESCUE WILL BE EXPEDITED THROUGH YOUR **COOPERATION**.

1. **MAINTAIN IDENTIFICATION:** If you will trail dye markers, show smoke, use mirror, it will enable rescue craft to keep track of your position.
2. **SIGNAL IF INJURED BADLY:** By crossing arms across body. If uninjured, hold arms straight out at side. Rescue plane will acknowledge by dipping wings.
3. **DO NOT USE EMERGENCY EQUIPMENT NOT REQUIRED:** It may not be possible to pick you up at once so conserve your gear, particularly pyrotechnics, which are of little value by day.
4. **PLAN FOR RESCUE:**
   You will be picked up by:

   CIRCLE ONE   This Airplane   Submarine

                      Seaplane   Surface Craft

   In approximately _____ Minutes. / Hours.   Approximate bearing _____ °.

Detailed instructions regarding cooperation expected of you for each type of rescue following. In each type of rescue remember: **REMOVE HELMET AS RESCUE CRAFT APPROACHES AND SECURE ALL FIREARMS.**

RESCUE

1. **FLYING BOAT.**
   A. Favorable Sea.
      1. Airplane will land; remember it is hard to keep you in sight when plane is on the water so **CONTINUE TO SIGNAL** by mirror or smoke.
      2. **STAY CLEAR OF PROPELLERS** as a swell may wash you into them. It is recommended that you remain outboard of the wing tip floats. If one of the engines is cut, approach the plane between the wing tip float and the hull on the side of the cut engine.
      3. A life preserver with a line will be thrown to you, or if you are incapacitated a man in a life raft with a line attached will guide you to the plane.
   B. Unfavorable Sea.
      1. This plane will report your position and arrange for you to be picked up by a submarine or surface craft.
      2. This plane will drop provisions as required for your rescue. The equipment or supplies dropped will depend on the circumstances. Most of this equipment will be dropped

(over)

with lines having floats to assist you in recovery. Check out in the use of all of your equipment.
   3. Rescue Squadron THREE will stand by as necessary to direct your rescue by leading pick up craft to your position.
2. **SEAPLANE** (Cruiser or Battleship type)
   A. **CONTINUE TO SIGNAL** when plane is on water.
   B. If **EXHAUSTED**, swim or paddle outboard of wing tip float if forward of wing as swell may wash you into propellers. Crew will try to get you to the main float. **DO NOT HOLD ONTO WING TIP FLOAT—DANGER OF CAPSIZING PLANE.**
   C. If capable, swim or paddle to main float aft of wing.
   D. Plane may or may not have two (2) man crew. If only pilot aboard, you must be prepared to get in plane with little help.
   E. **DO NOT ATTEMPT TO CLIMB FROM FLOATS TO WING UNTIL** pilot or radioman instructs you or you may capsize plane. Crew will give you verbal or hand signals.
   F. If seaplane is equipped with small life preserver and line, crew will show this so stand by for line throw.
3. **SUBMARINE.** (Day)
   1. **CONTINUE TO SIGNAL** preferably with mirror or smoke. Also dye marker can be seen.
   2. If **SUBMARINE SURFACES** look for life preserver and line to be thrown.
   3. If **PERISCOPE TOW** is desirable due to proximity of shore batteries.
      (a) Submarine will make pass at raft with periscope showing.
      (b) Have line ready to throw around periscope (remember if you miss on first pass, submarine cannot reverse course submerged and must make another circle). Best tow length 20 feet aft.
      (c) Do not secure line to periscope and raft unless it can be slipped quickly in case submarine dives.
      (d) Maintain surface and aircraft watch for submarine and signal to periscope if unfriendly aircraft or surface craft in vicinity.
      (e) Submarine will tow until safe to surface; **MAINTAIN CONTACT THROUGH PERISCOPE.**
   SUBMARINE. (Night)
   1. **WATCH** for flares from submarine, answer them with flares, flashlight; **LISTEN** for whistle, answer with whistle, flares or flashlight.
   2. **SHIELD FLASHLIGHT** to maintain contact with approaching submarines without giving away operation to enemy.
4. **SURFACE SHIP.**
   A. **DAY:** Best signal is smoke, then mirror, less dye marker.
   B. **NIGHT:** Best, signal flares, then flashlight.

Appendix G: Mills and Hastie DFC Award Citations

Flight Lieutenant Walter (Wally) Raymond Mills, RAAF
Distinguished Flying Cross Award Citation

ROYAL AUSTRALIAN AIR FORCE

HONOURS AND AWARDS

DISTINGUISHED FLYING CROSS

FLIGHT LIEUTENANT WALTER RAYMOND MILLS (412170)

CITATION:

Flight Lieutenant MILLS, whilst on routine air sea rescue patrol, was diverted to attempt the rescue of a crew forced down within range of enemy ground fire. Whilst manoeuvring for his landing approach in a slow and extremely vulnerable flying boat, Flight Lieutenant MILLS was under continuous A.A. fire. Nevertheless he executed a landing in an open sea swell and, while on the water, was under fire from at least 3 machine gun posts and a concentration of rifle fire from enemy troops on the shore 300 yards away.

To save risk of death or injury to the crew in the dinghy whilst being brought alongside the aircraft, Flight Lieutenant MILLS stopped his engines, lowered his wheels, and remained stationary on the water during the rescue operations. The take off was executed with great skill, through A.A. fire, and the aircraft returned to base with the rescued crew.

Flying Officer Norman William Hastie, RAAF
Distinguished Flying Cross Award Citation

On the morning of 31st March, a Catalina aircraft, captained by Flight Lieutenant W.R. Mills with Flight Lieutenant N.W. Hastie as a member of the crew, was diverted from a routine air-sea rescue patrol to attempt the rescue of a crew forced down in the sea within vulnerable range of enemy ground fire. The Catalina landed under continuous A.A. fire and whilst on the water was subjected to the same continued rate of fire from at least three machine gun posts and a concentration of rifle fire from enemy troops on the shore not more than 300 yards away.

Flight Lieutenant Hastie was the gunnery leader and fire control officer of the aircraft. He directed the captain on the positions of A.A. guns and their range whilst making their landing approach. During the time the Catalina was on the water, taking the rescued crew aboard, Flight Lieutenant Hastie assumed control of the fire of the aircraft's .5 guns and whilst on this duty received a severe abdominal wound. Despite this handicap and in great pain he remained at his post until relieved by Sergeant Ballingall, another member of the crew. Flight Lieutenant Hastie continued at his post and directed Sergeant Ballingall's fire on to the target before receiving medical attention.

Bibliography/Notes

Bonvillian, William. *The Forgotten Heroes: The Story of Rescue Squadron VH-3 in World War II*. Ridgewood, NJ: DoGo Publishing, Ezeread Books, 2000.

Bruhn, David D. *Battle Stars for the "Cactus Navy": America's Fishing Vessels and Yachts in World War II*. Berwyn Heights, MD: Heritage Books, 2014.

—*Eyes of the Fleet: The U.S. Navy's Seaplane Tenders and Patrol Aircraft in World War II*. Berwyn Heights, MD: Heritage Books, 2016.

—*MacArthur and Halsey's "Pacific Island Hoppers": The Forgotten Fleet of World War II*. Berwyn Heights, MD: Heritage Books, 2014.

Bruhn, David D., Rob Hoole. *Enemy Waters: Royal Navy, Royal Canadian Navy, Royal Norwegian Navy, U.S. Navy, and Other Allied Mine Forces Battling the Germans and Italians in World War II*. Berwyn Heights, MD: Heritage Books, 2019.

—*Nightraiders: U.S. Navy, Royal Navy, Royal Australian Navy, and Royal Netherlands Navy Mine Forces Battling the Japanese in the Pacific in World War II*. Berwyn Heights, MD: Heritage Books, 2019.

Campbell, Douglas E. *Save Our Souls: Rescues Made By U.S. Submarines During WWII*. Lulu.com, Inc., 2016.

—*VP Navy! USN, USMC, USCG and NATS Patrol Aircraft Lost or Damaged During World War II*. Lulu.com, 2018.

Chesneau, Roger. *Conway's All the World's Fighting Ships 1922–1946*. New York: Mayflower Books, Inc., 1980.

Dyer, George C. *The Amphibians Came to Conquer: The Story of Admiral Richmond Kelley Turner*. Washington, DC: U.S. Government Printing Office, 1969.

Frank, Richard B. *Downfall: The End of the Imperial Japanese Empire*. New York: Penguin, 1999.

Frank, Wesley, James Lea Cate, *Army Air Forces in World War II Vol. VII: Services Around the World*. Chicago, Illinois: University of Chicago Press, 1958.

Galdorisi, George. *Leave No Man Behind - The Saga of Combat Search and Rescue*. Minneapolis, MN: Zenith Press, 2008.

Gandt, Robert. *The Twilight Warriors*. New York: Broadway Press, 2010.

Hulver, Richard A., Peter C. Luebke, *A Grave Misfortune: The USS Indianapolis Tragedy*. Washington DC: Naval History and Heritage Command, 2018.

MacFarlane, John M., Robbie Hughes. *Canada's Naval Aviators*. British Columbia: Maritime Museum of British Columbia. 1994.

—*Canada's Naval Aviators (The Nominal List*. http://www.nauticapedia.ca/Articles/NavalAviator.php.

McCart, Neil. *The Illustrious & Implacable Classes of Aircraft Carrier 1940–1969*. Cheltenham, UK: Fan Publications, 2000.

Morison, Samuel Eliot. *The Two-Ocean War*. Boston: Little, Brown, 1963.

Oliver, David. *Airborne Espionage: International Special Duty Operations in the World Wars*. Thrupp, UK: Sutton Publishing, 2005.

RAAF Historical Section. *Units of the Royal Australian Air Force A Concise History, Volume 4 Maritime and Transport Units*. Canberra: Australian Government Publishing Service, 1995.

Rentz, John N. *Bougainville and the Northern Solomons*. Washington, DC: Headquarters, U.S. Marine Corps, 1946.

Ross, J. M. S. *Royal New Zealand Air Force*. Wellington, New Zealand: Historical Publications Branch, 1955.

Shaw Jr., Henry I., Douglas T. Kane, *History of U.S. Marine Corps Operations in World War II Volume II: Isolation of Rabaul*. Washington, DC: Headquarters, U.S. Marine Corps, 1963.

Thompson, H. L. *New Zealanders with the Royal Air Force Vol. II*. Wellington, Historical Publications Branch, 1956.

Turner, Mike, Hector Donohue. *Australian Minesweepers at War*. Canberra: Sea Power Centre – Australia, 2018.

U.S. Government. *Building the Navy's Bases in World War II: History of the Bureau of Yards and Docks and the Civil Engineer Corps 1940-1946 Volume II*. Washington DC: U.S. Government Printing Office, 1947.

U.S. Government. *U.S. Naval Aviation in the Pacific*. Washington, DC: Office of the Chief of Naval Operations, 1947.

FOREWORD NOTES:

[1] The Army Air Force in World War 2, vol. 7, 493.
[2] *U.S. Naval Aviation in the Pacific* (Washington, DC: Chief of Naval Operations, 1947).
[3] Capt. Earl Telford Jr., "*Seenotdienst*: The Development of Air-Sea Rescue," Air Services Review, Nov.-Dec 1976, 53.

PREFACE NOTES:

[1] *U.S. Naval Aviation in the Pacific*, 9.
[2] Commander Third Fleet, Action Report, Okinawa Operation, 28 May to 10 August, 13 September 1945.
[3] *U.S. Naval Aviation in the Pacific*, 11.
[4] Ibid.
[5] David D. Bruhn, *Battle Stars for the "Cactus Navy"*, 101.
[6] "Here Come the 'Cats'!" *Popular Mechanics*, February 1943 (https://books.google.com/books?id=xdYDAAAAMBAJ&pg=PA19#v=onepage&q&f=false); "PBY-5A Catalina" (https://www.navalaviationmuseum.org/aircraft/pby-5a-catalina/): both accessed 5 December 2019.
[7] Ut supra.
[8] "Cat Tales: Consolidated's PBY Flying Boat" by Stephan Wilkinson (https://www.historynet.com/cat-tales-consolidateds-pby-flying-boat.htm); PBY Specifications The Catalina Preservation Society" (http://pbycatalina.com/specifications/): both accessed 5 December 2019.
[9] "Here Come the 'Cats'!" *Popular Mechanics*, February 1943; "PBY Specifications The Catalina Preservation Society."

[10] "PBY Specifications The Catalina Preservation Society."
[11] Ibid.
[12] Bruhn, *Battle Stars for the "Cactus Navy,"* 11-12.
[13] "Martin PBM Mariner Patrol Bomber"
(http://worldwar2headquarters.com/HTML/aircraft/americanAircraft/mariner.html: accessed 2 November 2019).
[14] Ibid.
[15] "The Savior of Ceylon: The Pilot who Stopped a Repeat of Pearl Harbor" (https://www.warhistoryonline.com/instant-articles/air-commodore-leonard-birchall.html: accessed 30 November 2019).
[16] Ibid.
[17] *U.S. Naval Aviation in the Pacific*, 12-13.
[18] Historical Officer, Rescue Squadron Three, Historical Manual, 1 January 1945.
[18] Historical Officer, Rescue Squadron Three, Historical Manual, 1 January 1945; Commanding Officer, Rescue Squadron Five, Squadron History, 23 January 1945.
[20] "United States Pacific Fleet Organization 1 May 1945"
(https://www.history.navy.mil/content/history/nhhc/research/library/online-reading-room/title-list-alphabetically/u/us-pacific-fleet-organization-1may1945.html: accessed 4 November 2019).
[21] *U.S. Naval Aviation in the Pacific*, 13.
[22] Ibid.
[23] J. M. S. Ross, *Royal New Zealand Air Force* (Wellington, New Zealand: Historical Publications Branch, 1955), 1, 28.
[24] Ibid, 32-33.
[25] Ibid.
[26] Ibid.
[27] Ibid, 70-71.
[28] Ross, *Royal New Zealand Air Force*, 125; Bruhn, *Battle Stars for the "Cactus Navy,"* 9.
[29] Ross, *Royal New Zealand Air Force*, 125.
[30] Ibid, 224-225.
[31] Ibid, 314, 323-324
[32] Ibid, 225-226.
[33] Ibid.
[34] Ibid, 225-226, 258, 310.
[35] *U.S. Naval Aviation in the Pacific*, 14.
[36] "RAAF Experience in Combat Sar," *Pathfinder*, Air Power Development Centre Bulletin, Issue 87, April 2008.
[37] "Supermarine Seagull V (Walrus)"
(https://www.navy.gov.au/aircraft/supermarine-seagull-v-walrus: accessed 15 November 2019).
[38] "Air-Sea Rescue Units"
https://www.airforce.gov.au/sites/default/files/minisite/static/7522/RAAF museum/research/units/rescue.htm: accessed 15 November 2019).

[39] RAAF Historical Section, *Units of the Royal Australian Air Force A Concise History, Volume 4 Maritime and Transport Units* (Canberra: Australian Government Publishing Service), 156.
[40] "Dayaks" (http://factsanddetails.com/indonesia/Minorities_and_Regions/sub6_3f/entry-4016.html: accessed 15 November 2019).
[41] David D. Bruhn, *Eyes of the Fleet*, 17, 26.
[42] Ibid, 17.
[43] Ibid, 18.
[44] Ibid, 20.
[45] Ibid,
[46] Ibid, 22.
[47] Ibid.
[48] Ibid, 5, 26.
[49] Ibid, 23.
[50] Ibid, 25.
[51] *U.S. Naval Aviation in the Pacific*, 13-14; "Introducing One of America's Deadliest Submarine" (That Was Sunk by Its Own Torpedo)" (https://nationalinterest.org/blog/buzz/introducing-one-americas-deadliest-submarine-was-sunk-its-own-torpedo-39577: accessed 13 November 2019.
[52] *U.S. Naval Aviation in the Pacific*, 14.
[53] Ibid.

CHAPTER 1 NOTES:
[1] "COMINCH U.S. Fleet, Chart Room Collection of Dispatches Pacific Dispatches, 1 July-15 August 1945 (https://www.history.navy.mil/content/history/nhhc/research/histories/ship-histories/loss-of-uss-indianapolis-ca-35/rescue-operations/rescue-ship-dispatches.html: accessed 30 October 2019).
[2] "7 August 1945 Memorandum for Operations Officer, Sub Area, Extracts from VPB-23 Duty Officer's Log, Supplemented to Give Fuller Picture of Operations" (https://www.history.navy.mil/content/history/nhhc/research/histories/ship-histories/loss-of-uss-indianapolis-ca-35/rescue-operations/summations-of-rescue-operations.html: accessed 30 October 2019).
[3] *Building the Navy's Bases in World War II: History of the Bureau of Yards and Docks and the Civil Engineer Corps 1940-1946 Volume II* (Washington DC: U.S. Government Printing Office, 1947), 326.
[4] Ibid.
[5] VP 152, VP-133, VP-23, *Dictionary of American Naval Aviation Squadrons* (*DANAS*).
[6] VP-152, *DANAS*; "Lockheed PV-1 Ventura" (http://www.joebaugher.com/usaf_bombers/b34_4.html: accessed 2 November 2019).
[7] "PBY-5A Catalina" (https://www.navalaviationmuseum.org/aircraft/pby-5a-catalina/: accessed 2 November 2019).

[8] History of the USS *Coos Bay* (AVP-25) 15 May 1943 – 2 September 1945.
[9] Richard A. Hulver and Peter C. Luebke, *A Grave Misfortune: The USS Indianapolis Tragedy* (Washington DC: Naval History and Heritage Command, 2018), 115-116.
[10] Ibid, 116.
[11] Ibid, 116-117.
[12] "The Saga of a Ship and Its Crew" https://www.history.navy.mil/content/history/nhhc/browse-by-topic/disasters-and-phenomena/indianapolis.html: accessed 2 November 2019).
[13] Hulver, Luebke, *A Grave Misfortune*, xxv.
[14] Hulver, Luebke, *A Grave Misfortune*, xxv-xxvi; "The Saga of a Ship and Its Crew."
[15] Hulver, Luebke, *A Grave Misfortune*, xxviii-xxix.
[16] Ibid, 118.
[17] Ibid.
[18] Ibid.
[19] Ibid, 119-120.
[20] Ibid, 120.
[21] Ibid.
[22] Ibid, 120-121.
[23] Ibid, xxx, 123.
[24] "Setting the Record Straight: The Loss of USS *Indianapolis* and the Question of Clarence Donnor" by Richard Hulver and Sara Vladic (https://www.history.navy.mil/content/history/nhhc/browse-by-topic/disasters-and-phenomena/indianapolis/setting-the-record-straight--the-loss-of-uss-indianapolis-and-th.html: accessed 4 November 2019); Hulver, Luebke, *A Grave Misfortune*, xxx-xxxi.
[25] "Adrian Marks, 81, War Pilot Who Led Rescue of 56, Is Dead" by Richard Goldstein, *New York Times*, March 15, 1998.

CHAPTER 2 NOTES:
[1] Bruhn, *Eyes of the Fleet*, 19.
[2] Ibid.
[3] Roger Chesneau, *Conway's All the World's Fighting Ships 1922–1946* (New York: Mayflower Books, Inc., 1980), 157.
[4] Bruhn, *Eyes of the Fleet*, 21.
[5] Ibid.
[6] *Navy and Marine Corps Awards Manual* NAVPERS 15,790 (Rev. 1953), 61.
[7] "Those Other Grads" by Thomas J. Cutler (https://www.usni.org/magazines/naval-history-magazine/1994/april/those-other-grads: accessed 17 November 2019).
[8] Ibid.
[9] Samuel Eliot Morison, *The Two-Ocean War* (Boston: Little, Brown, 1963), 586.

[10] "USS *Coos Bay* (AVP-25)" (http://www.navsource.org/archives/09/43/4325.htm: accessed 17 November 2019).
[11] USS *Coos Bay* War Diary, May 1943.
[12] History of the USS *Coos Bay*, 15 May 1943 - 30 April 1946.
[13] History of the USS *Coos Bay*, 15 May 1943 - 30 April 1946; *Concord, DANFS*.
[14] Ibid.
[15] USS *Coos Bay* War Diary, June 1943.
[16] Ibid.
[17] Ibid.
[18] Ibid.
[19] Ibid.
[20] Ibid.
[21] Ibid.
[22] Ibid.
[23] Ibid.
[24] USS *Coos Bay* War Diary, June and July 1943.
[25] USS *Coos Bay* War Diary, July 1943.
[26] Ibid.
[27] Ibid.
[28] Ibid.
[29] Stephen Ekholm interview materials.
[30] Ibid.
[31] Ibid.
[32] USS *Coos Bay* War Diary, July 1943.
[33] Ibid.
[34] Ibid.
[35] USS *Coos Bay* and *Lansdowne* War Diary, August 1943.
[36] Ibid.
[37] Ibid.

CHAPTER 3 NOTES:
[1] General Orders: Commander Southern Pacific Forces: Serial 020 (January 7, 1943).
[2] Bruhn, *Eyes of the Fleet*, 171.
[3] Ibid, 172.
[4] Ibid, 174.
[5] Ibid, 175.
[6] Bruhn, *Eyes of the Fleet*, 175; Commander Bombing Squadron Twenty-Four, Patrol Bombing Squadron Twenty-four, History of, 31 December 1944.
[7] Commanding Officer VPB-71 War History, forwarding of, 11 July 1945.
[8] Bruhn, *Eyes of the Fleet*, 254-255.
[9] Ibid.
[10] Ibid, 255.
[11] Ibid.

[12] Ibid, 256.
[13] Ibid, 256-257.
[14] Ibid, 257.
[15] Ibid.
[16] Ibid, 258.
[17] Commanding Officer VPB-71 War History, forwarding of, 11 July 1945; Bruhn, *Eyes of the Fleet*, 260.
[18] History of the USS *Coos Bay* (AVP-25) 15 May 1943 – 30 April 1946.
[19] Commander Air Center, Navy 140, History of Air Command Center, Navy 140 – Forwarding of, 13 June 1945.
[20] Ibid.

CHAPTER 4 NOTES:

[1] "No 6 Squadron History" (http://airforce.mil.nz/about-us/who-we-are/squadrons/6-squadron/history.htm); "Wings over Cambridge, Cambridge's Connections with the Wartime Air Force" (http://www.cambridgeairforce.org.nz/RNZAF%20Stations%20North%20Island.htm): both accessed 5 November 2019.
[2] Ross, *Royal New Zealand Air Force*, 225-226; "No 6 Squadron History"; "Air Force Museum of New Zealand May 24, 2018" (https://www.facebook.com/AirForceMuseumofNewZealand/posts/on-this-day-75-years-ago-no-6-squadron-rnzaf-was-formed-at-laucala-bay-fiji-as-a/10155653096121279/); "HyperWar The Official Chronology of the US Navy in World War II" (https://www.ibiblio.org/hyperwar/USN/USN-Chron/USN-Chron-1943.html); "No.6 Flying Boat Squadron (RNZAF): Second World War" by J. Rickard (http://www.historyofwar.org/air/units/RNZAF/No_6_sqn_RNZAF.htmlall): all accessed 4 November 2019.
[3] Ross, *Royal New Zealand Air Force*, 225-226.
[4] Commanding Officer, USS *Curtiss* (AV-4), Ship's History, Resubmission of, 1 March 1946.
[5] Commanding Officer, USS *Curtiss* (AV-4), Ship's History, Resubmission of, 1 March 1946; *Wright*, *Coos Bay* War Diary, August 1943.
[6] "No.6 Flying Boat Squadron (RNZAF): Second World War"; "*Cape San Juan*" (http://www.ssarkansan.com/home/american-hawaiian-in-wwii/cape-san-juan: accessed 6 November 2019); Ross, *Royal New Zealand Air Force*, 225-226.
[7] "No.6 Flying Boat Squadron (RNZAF): Second World War"; Commander Naval Bases, Fiji, S.S. *San Juan*, Sinking of, 19 November 1943.
[8] "Sinking of the *San Juan*" by Bill Leadley (http://rnzaf6squadron.blogspot.com/2009/09/sinking-of-san-juan-by-walter-leadley.html: accessed 7 November 2019).
[9] "*Cape San Juan*"; Commander Naval Bases, Fiji, S.S. *San Juan*, Sinking of, 19 November 1943.
[10] Ut supra.

[11] Commander Naval Bases, Fiji, S.S. *San Juan*, Sinking of, 19 November 1943.
[12] Ibid.
[13] Ibid.
[14] "SS *Edwin T. Meredith*" (http://www.ssarkansan.com/home/american-hawaiian-in-wwii/cape-san-juan/ss-edwin-t-meredith: accessed 8 November 2019).
[15] Commander Naval Bases, Fiji, S.S. *San Juan*, Sinking of, 19 November 1943; "No. 6 Flying Boat Squadron (RNZAF): Second World War"; "*Cape San Juan*." (Different sources cite different numbers of survivors. I have used the one from the second reference.)
[16] "*Cape San Juan*."
[17] Ibid.
[18] Ibid.
[19] Ibid.
[20] "Fleet Admiral William Frederick Halsey, Jr." (https://www.history.navy.mil/research/library/research-guides/modern-biographical-files-ndl/modern-bios-h/halsey-william-f/fleet-admiral-halsey.html: accessed 8 November 2019).
[21] "*Cape San Juan*."
[22] "Sinking of the *San Juan*" by Bill Leadley.
[23] Ibid.
[24] Ibid.
[25] Ibid.
[26] Ibid.
[27] Ibid.
[28] Ibid.
[29] Ibid.
[30] Ibid.
[31] Ibid.
[32] *Dempsey* War Diary, November 1943.
[33] U.S. Naval Forces Fiji Islands War Diary, November 1943; "USS *SC-654*" (http://www.ssarkansan.com/home/american-hawaiian-in-wwii/cape-san-juan/uss-sc-654: accessed 9 November 2019).
[34] "Sinking of the *San Juan*" by Bill Leadley.

CHAPTER 5 NOTES:
[1] Bruhn, *Eyes of the Fleet*, 266.
[2] Bruhn, *Battle Stars for the "Cactus Navy,"* 4.
[3] "Halavo Seaplane Base (Halavo Bay)" (https://www.pacificwrecks.com/airfields/solomons/halavo/index.html); "Florida Island (Nggela Sule, Big Gela)" (https://www.pacificwrecks.com/provinces/solomons_nggela_sule.html): both accessed 22 November 2019.
[4] Ibid, 266-267.

[5] Commanding Officer Patrol Bombing Squadron Seventy-one, War History, forwarding of, 11 July 1945.
[6] History of the *Coos Bay* (AVP-25), 15 May 1943-2 September 1945.
[7] ONI Combat Narratives: Solomon Islands Campaign: XII The Bougainville Landing and the Battle of Empress Bay.
[8] Bruhn, *Eyes of the Fleet*, 263.
[9] Ibid, 263-264.
[10] Bruhn, *Eyes of the Fleet*, 264; "H-024-1: Operation Cherryblossom—The Invasion of Bougainville and Victory in the Solomon Islands" (https://www.history.navy.mil/about-us/leadership/director/directors-corner/h-grams/h-gram-024/h-024-1.html: accessed 23 November 2019).
[11] Ibid, 265.
[12] Ibid, 265-266.
[13] Commanding Officer Patrol, Bombing Squadron Seventy-one, War History, forwarding of, 11 July 1945; VP-71 War Diary, November 1943.
[14] VP-71 War Diary, November 1943.
[15] Commanding Officer, Patrol Bombing Squadron Twenty Three, History of Patrol Bombing Squadron Twenty Three, 23 June 1945; Combat Chronology of the U.S. Army Air Forces, November 1943 (http://paul.rutgers.edu/~mcgrew/wwii/usaf/html/Nov.43.html: accessed 23 November 2019).
[16] Commanding Officer, Patrol Bombing Squadron Twenty Three, History of Patrol Bombing Squadron Twenty Three, 23 June 1945; "North from Guadalcanal" by Richard B. Frank (https://www.usni.org/magazines/naval-history-magazine/2013/july/north-guadalcanal: accessed 23 November 2019).
[17] Ibid.
[18] General Orders: Bureau of Naval Personnel Information Bulletin No. 336 (March 1945).

CHAPTER 6 NOTES:
[1] David D. Bruhn, *MacArthur and Halsey's "Pacific Island Hoppers"*, 123.
[2] Ibid, 125.
[3] Ibid.
[4] Ibid, 125-126.
[5] Ibid, 126.
[6] Ibid, 123-124.
[7] Ibid.
[8] Ibid, 124-125.
[9] Ibid, 125.
[10] Ibid.
[11] Ibid.
[12] Ibid.
[13] Ibid, 126-127.
[14] Ibid, 127.
[15] Ibid.

[16] Ibid.
[17] Ibid, 127-128.
[18] Ibid, 128.
[19] Ibid.
[20] Ibid.

CHAPTER 7 NOTES:
[1] History of the *Coos Bay* (AVP-25), 15 May 1943-2 September 1945; *Coos Bay* War Diary, December 1943; Patrol Bombing Squadron Seventy-One War History.
[2] History of the *Coos Bay* (AVP-25), 15 May 1943-2 September 1945.
[3] *Coos Bay* War Diary, November 1943.
[4] *Coos Bay* War Diary, November 1943; Marine Scout Bombing Squadron Two Thirty Six War Diary, December 1943.
[5] Marine Scout Bombing Squadron Two Thirty Six War Diary, December 1943.
[6] Ibid.
[7] Ibid.
[8] *Coos Bay* War Diary, November 1943.
[9] Marine Scout Bombing Squadron Two Thirty Six War Diary, December 1943.
[10] Ibid.
[11] *Coos Bay* War Diary, December 1943.
[12] Morison, *The Two-Ocean War*, 167-168.
[13] *Coos Bay* War Diary, December 1943; Henry I. Shaw Jr. and Douglas T. Kane, *History of U.S. Marine Corps Operations in World War II Volume II: Isolation of Rabaul* (Washington, DC: Headquarters, U.S. Marine Corps, 1963), 491.
[14] "B-25 Mitchell Serial Number?" (https://www.pacificwrecks.com/aircraft/b-25/marx.html: accessed 1 December 2019).

CHAPTER 8 NOTES:
[1] Bruhn, *Eyes of the Fleet*, 267-268.
[2] Ibid, 268.
[3] Ibid.
[4] Ibid.
[5] VP-14 and *Coos Bay* War Diary, January 1944.
[6] *Coos Bay* and VP-14 War Diary, January 1944.
[7] *Coos Bay* and VP-14 War Diary, January 1944; "339th Fighter Squadron (339th FS)" (https://www.pacificwrecks.com/units/usaaf/347fg/339fs.html: accessed 13 December 2019).
[8] VP-14 War Diary, January 1944.
[9] *Coos Bay* War Diary, January 1944.
[10] VP-14 War Diary, February 1944.
[11] VP-71 War Diary, February 1944.
[12] Ibid.

[13] VP-71 War Diary, February 1944; information provided by Jenny Scott.
[14] *Coos Bay* and VP-14 War Diary, February 1944.
[15] VP-14 War Diary, February 1944.
[16] *Coos Bay* War Diary, February 1944; "Vunakanau Airfield"
(https://www.pacificwrecks.com/airfields/png/vunakanau/index.html: accessed 2 December 2019).
[17] *Coos Bay* War Diary, February 1944.
[18] Ibid.
[19] VP-71 War Diary, March 1944.
[20] *Coos Bay*, VP-71 War Diary, February-March 1944.
[21] VP-91 War Diary, April 1944.
[22] Ibid.
[23] *Coos Bay* and VB-148 War Diary, April 1944.
[24] VB-148 War Diary, April 1944.
[25] Ibid.
[26] Ibid.
[27] Ibid.
[28] Ibid.
[29] Ibid.
[30] VP-91 War Diary, April 1944.
[31] *Coos Bay* War Diary, May 1944.

CHAPTER 9 NOTES:
[1] Bruhn, *Eyes of the Fleet*, 268-269.
[2] John N. Rentz, *Bougainville and the Northern Solomons* (Washington, DC: Headquarters, U.S. Marine Corps, 1946), 114
(https://www.ibiblio.org/hyperwar/USMC/USMC-M-NSols/USMC-M-NSol-4.html: accessed 8 December 2019).
[3] Ibid, 114-115.
[4] Ibid, 115.
[5] Ibid, 116.
[6] Rentz, *Bougainville and the Northern Solomons*; Bruhn, *Eyes of the Fleet*, 268-269.
[7] "Top of the Ladder: Marine Operations in the Northern Solomons" by John C. Chapin
(https://www.nps.gov/parkhistory/online_books/npswapa/extContent/usmc/pcn-190-003141-00/sec11.htm: accessed 8 December 2019)

CHAPTER 10 NOTES:
[1] "Interview with Nathan Gordon [No date] by an unidentified interviewer" (https://memory.loc.gov/diglib/vhp-stories/loc.natlib.afc2001001.89690/transcript?ID=mv0001: accessed 15 December 2019).
[2] Commanding Officer, VPB-34, The History of Patrol Bombing Squadron Thirty Four, 22 January 1945,
[3] Ibid.
[4] Ibid.

[5] "345th Bombardment Group" (http://www.historyofwar.org/air/units/USAAF/345th_Bombardment_Group.html); "Flight Out of Hell: The Harrowing Bombing Mission to Kavieng" by Steven D. Smith (https://warfarehistorynetwork.com/2019/03/20/flight-out-of-hell-the-harrowing-bombing-mission-to-kavieng/): both accessed 16 December 2019.
[6] "Flight Out of Hell" by Steven D. Smith;
[7] "Interview with Nathan Gordon"; "Flight Out of Hell" by Steven D. Smith; USS *San Pablo* War Diary, February 1944.
[8] "Interview with Nathan Gordon"; "Flight Out of Hell" by Steven D. Smith.
[9] "Interview with Nathan Gordon."
[10] Ibid.
[11] "Flight Out of Hell" by Steven D. Smith.
[12] "Interview with Nathan Gordon."
[13] Ibid.
[14] "Interview with Nathan Gordon"; "Flight Out of Hell" by Steven D. Smith.
[15] "Flight Out of Hell" by Steven D. Smith.
[16] "Interview with Nathan Gordon."
[17] "Flight Out of Hell" by Steven D. Smith; "The End of Gremlin's Holiday" (https://airwarworldwar2.wordpress.com/2014/05/15/the-end-of-gremlins-holiday/: accessed 17 December 2019).
[18] Ut supra.
[19] "Interview with Nathan Gordon."
[20] "Interview with Nathan Gordon"; USS *San Pablo* War Diary, February 1944.
[21] "Flight Out of Hell" by Steven D. Smith.
[22] USS *San Pablo* War Diary, February 1944.
[23] "Nathan Green Gordon" (https://www.legacy.com/obituaries/name/nathan-gordon-obituary?pid=178584427: accessed 17 December 2019).
[24] "Nathan Green Gordon" (https://valor.militarytimes.com/hero/2429: accessed 5 December 2019).

CHAPTER 11 NOTES:
[1] Bruhn, *Eyes of the Fleet*, 269-270.
[2] Ibid, 270-271
[3] Ibid, 271.
[4] Ibid.
[5] Ibid, 271-272.
[6] Ibid, 272-273.
[7] Ibid, 269.
[8] *Coos Bay*, VP-91 War Diaries, May 1944.
[9] Ut supra.
[10] VP-91 War Diary, May 1944.

[11] VMSB-241 War Diary, May 1944.
[12] VMSB-244 War Diary, May 1944.
[13] VMSB-235 War Diary, June 1944.
[14] *Coos Bay* War Diary, May-June 1944.
[15] *Coos Bay* War Diary, June 1944; History of the USS *Coos Bay* (AVP-25) 15 May 1943 – 2 September 1945.

CHAPTER 12 NOTES:

[1] Ross, *Royal New Zealand Air Force*, 256.
[2] Ibid.
[3] Ibid, 280.
[4] Ross, *Royal New Zealand Air Force*, 225-226, 256; "Obituary: Winston Brooke (Bill) Mackley," 15 April 2005 (https://www.nzherald.co.nz/nz/news/article.cfm?c_id=1&objectid=10120697: accessed 11 December 2019).
[5] "Combat Chronology of the US Army Air Forces January 1944" (http://paul.rutgers.edu/~mcgrew/wwii/usaf/html/Jan.44.html: accessed 11 December 2019); Ross, *Royal New Zealand Air Force*, 225-226, 256.
[6] Information provided by Jenny Scott.
[7] "Obituary: Winston Brooke (Bill) Mackley," 15 April 2005.
[8] Ibid.
[9] "My Dad's Catalina" (https://m.sunlive.co.nz/news/107698-my-dads-catalina.html: accessed 11 December 2019).
[10] Ross, *Royal New Zealand Air Force*, 256-257; "My Dad's Catalina."
[11] Ut supra.
[12] Information provided by Jenny Scott.
[13] "Kabanga Bay East New Britain Province Papua New Guinea (PNG)" (https://www.pacificwrecks.com/provinces/png_kabanga_bay.html: accessed 12 December 2019).
[14] WWII Distinguished Flying Medals for British Soldiers (https://www.fold3.com/title/980/wwii-distinguished-flying-medals-for-british-soldiers: accessed 13 November 2019).
[15] Jenny Scott correspondence of 13 November 2019.
[16] Ibid.
[17] Ibid.
[18] "NZDF-Serials Australian & New Zealand Military Aircraft Serials & History" (http://www.adf-serials.com.au/nz-serials/nzhudson.htm: accessed 13 December 2019).
[19] Lt (jg) Paul R. Campagna 14 November 1943 interview, Office of Naval Records and Library.
[20] U.S. Naval Forces Fiji Islands War Diary, May 1943; Jenny Scott correspondence of 13 November 2019.
[21] "IJN Submarine *I-19*: Tabular Record of Movement" (http://www.combinedfleet.com/I-19.htm); "SS *Phoebe A. Hearst* [+1943]" (https://www.wrecksite.eu/wreck.aspx?177958): both accessed 15 December 2019.

[22] Lt (jg) Paul R. Campagna 14 November 1943 interview, Office of Naval Records and Library.
[23] USS *Dash* War Diary, May 1943.
[24] Ibid.
[25] U.S. Naval Forces Fiji Islands War Diary, May 1943.
[26] "New Zealand Disasters and Tragedies R.N.Z.A.F. Catalina Flying Boat Crash FIJI Saturday 05 June 1943"
(https://www.sooty.nz/fijiaircrash1943.html: accessed 15 December 2019).

CHAPTER 13 NOTES:
[1] History of the *Coos Bay* (AVP-25), 15 May 1943-2 September 1945.
[2] Ibid.
[3] Ibid.
[4] History of the *Coos Bay* (AVP-25), 15 May 1943-2 September 1945; Bruhn, *Eyes of the Fleet*, 193.
[5] Bruhn, *Eyes of the Fleet*, 193.
[6] History of the *Coos Bay* (AVP-25), 15 May 1943-2 September 1945.
[7] Ibid.
[8] VPB-18 War History.

CHAPTER 14 NOTES:
[1] Commander, Patrol Bombing Squadron Eighteen, History of Patrol Bombing Squadron Eighteen – Submission of, 20 December 1944.
[2] "Martin PBM-5A Mariner" (https://airandspace.si.edu/collection-objects/martin-pbm-5a-mariner: accessed 30 December 2019).
[3] "Martin PBM-5A Mariner."
[4] "Martin PBM-5A Mariner."; "The Martin Mariner, Mars, & Marlin Flying Boats" (http://www.airvectors.net/avmars.html#m3: accessed 2 January 2020).
[5] "Martin PBM-5A Mariner."
[6] *Dictionary of American Naval Aviation Squadrons, Volume 2*, Appendix 1.
[7] "Martin PBM-5A Mariner."
[8] *Dictionary of American Naval Aviation Squadrons, Volume 2*, Appendix 1.
[9] Ibid.
[10] "Martin PBM-5A Mariner."

CHAPTER 15 NOTES:
[1] History of the *Coos Bay* (AVP-25), 15 May 1943-2 September 1945.
[2] Ibid.
[3] History of the Coos Bay (AVP-25), 15 May 1943-2 September 1945; "Navy Air Transport"
(https://www.globalsecurity.org/military/systems/aircraft/cargo-navy.htm: accessed 31 December 2019).

[4] History of the *Coos Bay* (AVP-25), 15 May 1943-2 September 1945; Douglas E. Campbell, *VP Navy! USN, USMC, USCG and NATS Patrol Aircraft Lost or Damaged During World War II* (Self-published, 2018), 129.
[5] VP-18 War Diary, August 1945; VH-1 War Diary, September 1944.
[6] Bruhn, *Eyes of the Fleet*, 312.
[7] Ibid, 312-313.
[8] VP-18 War Diary, August 1945.
[9] Ibid.
[10] Ibid.
[11] Ibid.
[12] VH-1 War Diary, November 1945.
[13] Ibid.
[14] Morison, *The Two-Ocean War*, 428.
[15] VH-1 and VP-18 War Diary, November 1945; "USS *Ellet* (DD-398)" (http://www.navsource.org/archives/05/398.htm: accessed 1 January 2020).
[16] VH-1 War Diary, November 1945.
[17] History of the *Coos Bay* (AVP-25), 15 May 1943-2 September 1945.
[18] Ibid.
[19] History of the *Coos Bay* (AVP-25), 15 May 1943-2 September 1945; *Coos Bay* War Diary, March 1945.
[20] *Coos Bay* War Diary, March 1945.
[21] *Coos Bay* War Diary, March 1945; "*Matagorda*" (http://www.tugboatinformation.com/tug.cfm?id=6509: accessed 1 January 2020).
[22] *Coos Bay* War Diary, April and May 1945.
[23] History of the *Coos Bay* (AVP-25), 15 May 1943-2 September 1945.

CHAPTER 16 NOTES:

[1] Commanding Officer, Squadron History, VT-51, 22 Sept. 1943 to 30 Nov. 1944, 20 December 1944; USS *San Jacinto* War Diary, September 1944.
[2] Ut supra.
[3] Timothy J. Christmann, "Vice President Bush Calls World War II Experience 'Sobering'," *Naval Aviation News 67* (March-April 1985) (https://www.history.navy.mil/research/histories/biographies-list/bios-b/bush-george-h-w/ltjg-george-bush-in-world-war-ii.html: accessed 17 December 2016).
[4] Commanding Officer, Squadron History, VT-51, 22 Sept. 1943 to 30 Nov. 1944, 20 December 1944; Morison, *The Two-Ocean War*, 424-425.
[5] Christmann, "Vice President Bush Calls World War II Experience 'Sobering'"; VF-51 Aircraft Action Report No. 32-44, 2 September 1944; Commanding Officer, USS *Finback* (SS-230), USS *Finback* – Report of War Patrol Number Nine, 21 July 1944.
[6] Christmann, "Vice President Bush Calls World War II Experience 'Sobering'"; VF-51 Aircraft Action Report No. 32-44, 2 September 1944.
[7] Christmann, "Vice President Bush Calls World War II Experience 'Sobering'"; "Lt. j.g. George H. W. Bush (USN) and USS *Finback* (SS-230) A

Focused Narration of the Silent Services WWII Lifeguard Operations," by Rolfe L. Hillman, *Undersea Warfare*, Summer 2014 (http://www.public.navy.mil/subfor/underseawarfaremagazine/Issues/Archives/issue_55/Lifeguarding.html: accessed 17 December 2016); Commanding Officer, USS *San Jacinto*, Action Report – Operations against the Bonin Island and Iwo Jima, 7 September 1944.
[8] Commanding Officer, Squadron History, VT-51, 22 Sept. 1943 to 30 Nov. 1944, 20 December 1944; "Lt. j.g. George H. W. Bush (USN) and USS *Finback*" by Hillman.
[9] "Lt. j.g. George H. W. Bush (USN) and USS *Finback*" by Hillman.
[10] "Lt. j.g. George H. W. Bush (USN) and USS *Finback*" by Hillman; Commanding Officer, USS *Finback* (SS-230), USS *Finback* – Report of War Patrol Number Nine, 21 July 1944.
[11] "Lt. j.g. George H. W. Bush (USN) and USS *Finback*" by Hillman.
[12] Ibid.
[13] Ibid.
[14] "Lt. j.g. George H. W. Bush (USN) and USS *Finback*" by Hillman; Commanding Officer, USS *Finback* (SS-230), USS *Finback* – Report of War Patrol Number Nine, 21 July 1944.
[15] General Orders: Commander 2d Carrier Task Force Pacific: Serial 0596 (December 23, 1944).
[16] "Lt. j.g. George H. W. Bush (USN) and USS *Finback*" by Hillman.
[17] *San Jacinto* War Diary, November 1944; Commanding Officer, USS *San Jacinto*, Action Report – Operations against Yap, Ulithi, and Palau Island Group, 19 September 1944; Commanding Officer, USS *San Jacinto*, Action Report, Operations against Okinawa Jima, Formosa, Luzon, P.I., and Visayas, P.I., in support of the occupation of Leyte, P.I., during the period 7 through 21 October 1944, 31 October 1944.
[18] Commanding Officer, Squadron History, VT-51, 22 Sept. 1943 to 30 Nov. 1944, 20 December 1944.
[19] Gerald Rudolph Ford, Jr. 14 July 1913 - 26 December 2006 (https://www.history.navy.mil/research/library/research-guides/modern-biographical-files-ndl/modern-bios-f/ford-gerald.html: accessed 19 December 2016).
[20] Ibid.
[21] Ibid.

CHAPTER 17 NOTES:
[1] Bruhn, *Eyes of the Fleet*, 329.
[2] *U.S. Naval Aviation in the Pacific*, 34; Morison, *The Two-Ocean War*, 424-435.
[3] *U.S. Naval Aviation in the Pacific*, 34.
[4] *U.S. Naval Aviation in the Pacific*, 34-35; Morison, *The Two-Ocean War*, 425-428.
[5] *U.S. Naval Aviation in the Pacific*, 35; Morison, *The Two-Ocean War*, 428; Bruhn, *Eyes of the Fleet*, 325.

[6] Bruhn, *Eyes of the Fleet*, 326; Commander Fleet Air Wing One, Action Report, 10 September to 15 October 1944, 8 December 1944.
[7] *U.S. Naval Aviation in the Pacific*, 35.
[8] Bruhn, *Eyes of the Fleet*, 330.
[9] Ibid.
[10] Ibid, 331.
[11] Ibid, 331-332.
[12] "H-038-1: Battle of Leyte Gulf—A Synopsis" (https://www.history.navy.mil/content/history/nhhc/about-us/leadership/director/directors-corner/h-grams/h-gram-038/h-038-1.html: accessed 6 January 2020.
[13] "H-038-1: Battle of Leyte Gulf—A Synopsis"; "The Battle of Leyte Gulf 23–26 October 1944" (https://www.history.navy.mil/browse-by-topic/wars-conflicts-and-operations/world-war-ii/1944/battle-of-leyte-gulf.html: accessed 5 January 2020).
[14] "H-038-1: Battle of Leyte Gulf—A Synopsis"; "The Battle of Leyte Gulf 23–26 October 1944"; "Battle of Leyte Gulf World War II" (https://www.britannica.com/event/Battle-of-Leyte-Gulf#ref186855: accessed 5 January 2020).
[15] "The Battle of Leyte Gulf 23–26 October 1944"; "Battle of Leyte Gulf World War II."
[16] "H-038-1: Battle of Leyte Gulf—A Synopsis."
[17] "David S. McCampbell" (https://valor.militarytimes.com/hero/630: accessed 7 January 2020).
[18] "H-038-1: Battle of Leyte Gulf—A Synopsis."
[19] "Cape Engaño" (http://pwencycl.kgbudge.com/C/a/Cape_Engano.htm); "Leyte Naval Battles" (http://combinedfleet.com/battles/Leyte_Campaign: accessed 6 January 2020).
[20] "The Battle of Leyte Gulf 23–26 October 1944."
[21] "Battle of the Sibuyan Sea" (http://pwencycl.kgbudge.com/S/i/Sibuyan_Sea.htm); "Battle off Samar" (http://pwencycl.kgbudge.com/S/a/Samar.htm): both accessed 7 January 2020.
[22] "The Battle of Leyte Gulf 23–26 October 1944"; Morison, *The Two-Ocean War*, 451-455.
[23] "H-038-1: Battle of Leyte Gulf—A Synopsis."
[24] "The Battle of Leyte Gulf 23–26 October 1944"; "USS *Portland* (CA-33) Battle Pennant from the Battle of Leyte Gulf" (https://www.history.navy.mil/content/history/nhhc/our-collections/artifacts/ship-and-shore/Flags/Pennants/uss-portland--ca-33--battle-pennant-battle-of-leyte-gulf.html: accessed 7 January 2020).
[25] "Battle of Surigao Strait" (https://www.history.navy.mil/content/history/museums/nmusn/explore/photography/wwii/wwii-pacific/leyte/surigao-strait.html: accessed 5 January 2020).

[26] "H-038-1: Battle of Leyte Gulf—A Synopsis."
[27] "Battle of Surigao Straits" (http://pwencycl.kgbudge.com/S/u/Surigao.htm); "The Japanese Southern Force" (http://www.angelfire.com/fm/odyssey/LEYTE_GULF_IJN_Southern_Force_.htm): both accessed 6 January 2020.
[28] "The Battle of Leyte Gulf 23–26 October 1944."
[29] "H-038-1: Battle of Leyte Gulf—A Synopsis"; "The Battle for Leyte Gulf" (http://www.angelfire.com/fm/odyssey/LEYTE_GULF_JAPANESE_FORCES_.htm): accessed 9 January 2020.

CHAPTER 18 NOTES
[1] Bruhn, *Eyes of the Fleet*, 335.
[2] Wesley Frank and James Lea Cate, *Army Air Forces in World War II Vol. VII: Services Around the World* (Chicago, Illinois: University of Chicago Press, 1958), 479-480.
[3] Ibid, 480-481.
[4] Ibid, 498-499.
[5] Ibid, 499.
[6] Ibid, 499-500; Enclosure A (of unidentified document) titled Aviation History of the Forward Area, Central Pacific and the Marianas Area.
[7] Bruhn, *Eyes of the Fleet*, 334.
[8] Ibid, 334-335.
[9] Ibid, 335.
[10] Ibid.
[11] Ibid, 335-336.
[12] Ibid, 336-337.
[13] Ibid, 337.
[14] Ibid.
[15] Ibid, 337-338.
[16] Commanding Officer Patrol Bombing Squadron Thirty Four, The History of Patrol Bombing Squadron Thirty Four, 22 January 1945.
[17] War Diary of Patrol Bombing Squadron Thirty Four, Fleet Air Wing Ten, Aircraft Seventh Fleet, for November 1944, December 1944, and January 1945.
[18] Ibid.
[19] *Hollandia, DANFS*; War Diary of Patrol Bombing Squadron Thirty Four, Fleet Air Wing Ten, Aircraft Seventh Fleet, for November 1944, December 1944, and January 1945.
[20] The Commander Patrol Bombing Squadron Thirty-Three, History of Patrol Bombing Squadron Thirty-Three, Addenda to, transmittal of, 24 March 1945.
[21] The Commander Patrol Bombing Squadron Thirty-Three, History of Patrol Bombing Squadron Thirty-Three, Addenda to, transmittal of, 24

March 1945; "Vanderpump, Mortimer Tuke, 1920-1955" (https://natlib.govt.nz/records/22417674); "Mortimer Tuke Vanderpump" (https://www.aucklandmuseum.com/war-memorial/online-cenotaph/record/183555): both accessed 11 January 2020.
[22] "Vanderpump, Mortimer Tuke, 1920-1955"; "Mortimer Tuke Vanderpump."
[23] The Commander Patrol Bombing Squadron Thirty-Three, History of Patrol Bombing Squadron Thirty-Three, Addenda to, transmittal of, 24 March 1945
[24] H. L. Thompson, *New Zealanders with the Royal Air Force Vol. II* (Wellington, Historical Publications Branch, 1956), 183.
[25] The Commander Patrol Bombing Squadron Thirty-Three, History of Patrol Bombing Squadron Thirty-Three, Addenda to, transmittal of, 24 March 1945.
[26] *Tangier* War Diary, January 1945.
[27] Ibid.
[28] Ibid.
[29] VPB-104 War Diary, January 1945.
[30] *Currituck* and *Tangier* War Diary, January 1945.
[31] *Tangier* War Diary, January 1945.
[32] Ibid.
[33] *Tangier* War Diary, January 1945; The War History of Patrol Bombing Squadron Twenty.
[34] *Tangier* War Diary, January 1945.
[35] *Tangier* War Diary, January 1945; "American Prisoners of War: Massacre at Palawan" (https://www.historynet.com/american-prisoners-of-war-massacre-at-palawan.htm: accessed 12 January 2020).
[36] "American Prisoners of War: Massacre at Palawan."
[37] Ibid.
[38] Ibid.
[39] Ibid.
[40] Ibid.
[41] Ibid.
[42] Bruhn, *Eyes of the Fleet*, 341-342.
[43] *U.S. Naval Aviation in the Pacific*, 37; Bruhn, *Eyes of the Fleet*, 342.
[44] *U.S. Naval Aviation in the Pacific*, 39.

CHAPTER 19 NOTES:
[1] "S 2 Headquarters" (http://clik.dva.gov.au/history-library/part-3-order-battle/ch-2-order-battle-air-force/s-2-headquarters): accessed 25 November 2019; "Organising for War: The RAAF Air Campaigns in the Pacific," Pathfinder, Air Power Development Centre Bulletin, Issue 121, October 2009.
[2] "Near Wewak, New Guinea, 1944-11-10. A RAAF Catalina Rescuing Five American Airmen from a 5th..."
(https://www.awm.gov.au/collection/C40600: accessed 7 December 2019).

[3] No. 111 Air-Sea Rescue Flight Operations Record Book entries for December 1944, and January 1945.
[4] No. 111 Air-Sea Rescue Flight Operations Record Book for January 1945.
[5] No. 111 Air-Sea Rescue Flight Operations Record Book for January-April 1945.
[6] No. 111 Air-Sea Rescue Flight Operations Record Book for April 1945.
[7] No. 111 Air-Sea Rescue Flight Operations Record Book for May 1945.
[8] No. 111 Air-Sea Rescue Flight Operations Record Book for May-July 1945.
[9] No. 111 Air-Sea Rescue Flight Operations Record Book for July 1945.
[10] "P-38 Lightning Serial Number?" (https://www.pacificwrecks.com/aircraft/p-38/carter.html: accessed 25 November 2019)
[11] No. 111 Air-Sea Rescue Flight Operations Record Book for July 1945; "P-38 Lightning Serial Number?"
[12] Ut supra.
[13] "P-38 Lightning Serial Number?"
[14] "Tadji Airfield (Aitape, Korako)" (https://www.pacificwrecks.com/airfields/png/tadji/index.html: accessed 26 November 2019); "P-38 Lightning Serial Number?"

CHAPTER 20 NOTES:
[1] "IJN Light Cruiser ISUZU: Tabular Record of Movement" by Bob Hackett and Sander Kingsepp (http://www.combinedfleet.com/isuzu_t.htm: accessed 24 November 2019).
[2] Ibid.
[3] Ibid.
[4] No. 24 Squadron Operations Record Book entry for 6 April 1945.
[5] No. 24 Squadron Operations Record Book entry for 6 April 1945; 112 Air Sea Rescue Flight Darwin Operations Record Book entry for 6 April 1945.
[6] Ut supra.
[7] David Oliver, *Airborne Espionage: International Special Duty Operations in the World Wars* (Thrupp, UK: Sutton Publishing, 2005); 112 Air Sea Rescue Flight Darwin Operations Record Book entry for 6 April 1945.
[8] Oliver, *Airborne Espionage: International Special Duty Operations in the World Wars.*
[9] "IJN Light Cruiser ISUZU: Tabular Record of Movement."

CHAPTER 21 NOTES:
[1] Hector Donohue correspondence of 25 November 2019.
[2] Ibid.
[3] Ibid.
[4] Ibid.
[5] Ibid.
[6] No. 113 Air Sea Rescue Flight Operations Record Book, February 1945.
[7] Ibid.

[8] No. 113 Air Sea Rescue Flight Operations Record Book, March 1945.
[9] Ibid.
[10] Ibid.
[11] "VALE - Norman William (Bill) Hastie DFC (1921-2014) Flight Lieutenant N.W. Hastie DFC W/Airgunner 20 Squadron" (https://www.catalinaflying.org.au/documents/CatalinaMagazineJan-Mar2014.pdf: accessed 29 November 2019).

CHAPTER 22 NOTES:
[1] Bruhn, *Eyes of the Fleet*, 353.
[2] Ibid.
[3] *U.S. Naval Aviation in the Pacific*, 39.
[4] David D. Bruhn, Rob Hoole, *Nightraiders*, 241
[5] Ibid, 242.
[6] Amphibious Operations - Capture of Iwo Jima - 16 February to 16 March 1945 (https://www.history.navy.mil/research/library/online-reading-room/title-list-alphabetically/a/amphibious-operations-capture-iwo-jima.html: accessed 1 December 2019).
[7] Commander Fleet Air Wing One, Fleet Air Wing One – History of (1 January – 2 September 1945), no date on cover letter.
[8] Commander Fleet Air Wing One, Fleet Air Wing One – History of (1 January – 2 September 1945); Bruhn, *Eyes of the Fleet*, 354.
[9] Bruhn, *Eyes of the Fleet*, 354-355.
[10] Ibid, 355.
[11] VPB-19 War Diary, 1 February-17 March 1945; Bruhn, *Eyes of the Fleet*, 356.
[12] VPB-19 War Diary, 1 February-17 March 1945.
[13] VH-2 and VPB-23 War Diary, February and March 1945.
[14] Bruhn, *Eyes of the Fleet*, 356; VPB-19 War Diary, February-March 1945.
[15] Ibid, 359.

CHAPTER 23 NOTES:
[1] Bruhn, Hoole, *Nightraiders*, 249.
[2] Ibid, 249-250.
[3] Commander Fleet Air Wing One, Fleet Air Wing One – History of (1 January – 2 September 1945).
[4] Ibid.
[5] VPB-21 Squadron History (Period 1 March-1 June 1945) 1 June 1945; *U.S. Naval Aviation in the Pacific*, 40.
[6] USS *Morrison* War Diary, March 1945.
[7] VPB-21 War History, March 1945.
[8] Commanding Officer, USS *Stockton*, Action Report – Capture, Occupation, and Defense of Okinawa Gunto, 13 March 1945-28 May 1945, 1 June 1945.
[9] Ibid.
[10] Ibid.

[11] Ibid.
[12] Ibid.
[13] Ibid.
[14] Commanding Officer, USS *Stockton*, Action Report – Capture, Occupation, and Defense of Okinawa Gunto, 13 March 1945-28 May 1945, 1 June 1945; "IJN Submarine I-8: Tabular Record of Movement" (http://www.combinedfleet.com/I-8.htm: accessed 20 January 2020).
[15] Commander Fleet Air Wing One, Fleet Air Wing One – History of (1 January – 2 September 1945).
[16] "Death of the Battleship: Sinking the *Yamato* and *Musashi*" by Herb Kugel (https://warfarehistorynetwork.com/2018/12/30/death-of-the-battleship-sinking-the-yamato-and-musashi/: accessed 21 January 2020); *U.S. Naval Aviation in the Pacific*, 40.
[17] Commander Fleet Air Wing One, Fleet Air Wing One – History of (1 January – 2 September 1945).
[18] Commander Fleet Air Wing One, Fleet Air Wing One – History of (1 January – 2 September 1945); "Death of the Battleship: Sinking the *Yamato* and *Musashi*"; *U.S. Naval Aviation in the Pacific*, 40.
[19] Commander Fleet Air Wing One, Fleet Air Wing One – History of (1 January – 2 September 1945); "Death of the Battleship: Sinking the *Yamato* and *Musashi*."
[20] "Death of the Battleship: Sinking the *Yamato* and *Musashi*."
[21] Ibid.
[22] Ibid.
[23] Ibid.
[24] Ibid.
[25] Ibid.
[26] Ibid.
[27] "Death of the Battleship: Sinking the *Yamato* and *Musashi*"; Robert Gandt, *The Twilight Warriors* (New York: Broadway Press, 2010), 199.
[28] "Death of the Battleship: Sinking the *Yamato* and *Musashi*"; *U.S. Naval Aviation in the Pacific*, 40.
[29] Commander Fleet Air Wing One, Fleet Air Wing One – History of (1 January – 2 September 1945); VPB-21 War Diary, April 1945.
[30] VPB-21 War Diary, April 1945.
[31] Commander Fleet Air Wing One, Fleet Air Wing One – History of (1 January – 2 September 1945).
[32] Ibid.
[33] Ibid.
[34] Commander Amphibious Forces, U.S. Pacific Fleet, General Action Report, Capture of Okinawa Gunto, Phases I and II, 17 February 1945 to 17 May 1945 – Submission of, 25 July 1945; George C. Dyer, *The Amphibians Came to Conquer: The Story of Admiral Richmond Kelly Turner* (Washington: DC: U.S. Government Printing Office, 1991), 1058.
[35] Ibid.
[36] Ibid.

[37] Ibid.
[38] Ibid.
[39] Ibid.
[40] Ibid.
[41] VH-3 War Diary, April, May, June, July, August 1945.
[42] Ibid.
[43] VPB-21 Squadron History (Period 1 March – 1 June 1945), 1 June 1945.
[44] Forty-five Minute Running Battle between Two PBM-5 Mariners and Six or Seven Jacks and Tojos, Addendum (A) VPB-21 History, 1 June 1945, Excerpt from VPB-21 ACA-1 Report, No. 16-45.
[45] Ibid.
[46] Ibid.
[47] Ibid.
[48] Ibid.
[49] Ibid.
[50] Ibid.
[51] Ibid.
[52] Ibid.
[53] Ibid.
[54] Ibid.
[55] VPB-21 Squadron History (Period 1 March – 1 June 1945), 1 June 1945; Forty-five Minute Running Battle between Two PBM-5 Mariners and Six or Seven Jacks and Tojos.
[56] Forty-five Minute Running Battle between Two PBM-5 Mariners and Six or Seven Jacks and Tojos.
[57] *U.S. Naval Aviation in the Pacific*, 40.
[58] *U.S. Naval Aviation in the Pacific*, 41-42; "Adm Kantaro Suzuki" (https://www.findagrave.com/memorial/54269211/kantaro-suzuki: accessed 22 January 2020).
[59] *U.S. Naval Aviation in the Pacific*, 41-42; "Adm Kantaro Suzuki"; "Suzuki Kantaro (1867-1948)" (http://pwencycl.kgbudge.com/S/u/Suzuki_Kantaro.htm): both accessed 22 January 2020.

CHAPTER 24 NOTES:
[1] "Commander Dickie Reynolds" (https://www.telegraph.co.uk/news/obituaries/1346476/Commander-Dickie-Reynolds.html: accessed 17 March 2018).
[2] "The British Pacific Fleet Task Force 57 Politics & Logistics: Sakishima Gunto, Okinawa Campaign, 1945" (http://www.armouredcarriers.com/task-force-57-british-pacific-fleet: accessed 13 January 2020)
[3] Mike Turner and Hector Donohue, *Australian Minesweepers at War* (Canberra: Sea Power Centre – Australia, 2018), 134-139; *U.S. Naval Aviation in the Pacific*, 41; "Operation "Meridian" – Palembang Oil Refineries, 1945" (https://www.navyhistory.org.au/operation-meridian-palembang-oil-refineries-1945/6/: accessed 24 January 2020).

[4] Turner, Donohue, *Australian Minesweepers at War, 134-139*; *U.S. Naval Aviation in the Pacific*, 41.
[5] Turner, Donohue, *Australian Minesweepers at War, 134-139*; "The British Pacific Fleet Task Force 57 Politics & Logistics: Sakishima Gunto, Okinawa Campaign, 1945" (http://www.armouredcarriers.com/task-force-57-british-pacific-fleet: accessed 13 January 2020).
[6] "The Royal Navy's Pacific Strike Force" (https://www.usni.org/magazines/naval-history-magazine/2013/january/royal-navys-pacific-strike-force: accessed 13 January 2020).
[7] "The British Pacific Fleet Task Force 57 Politics & Logistics: Sakishima Gunto, Okinawa Campaign, 1945" (http://www.armouredcarriers.com/task-force-57-british-pacific-fleet: accessed 13 January 2020); Neil McCart, *The Illustrious & Implacable Classes of Aircraft Carrier 1940–1969* (Cheltenham, UK: Fan Publications, 2000), p. 173.
[8] "Canada's last Victoria Cross winner – Lieutenant (N) Robert Hampton Gray, VC, DSC" (https://militarybruce.com/canadas-last-victoria-cross-winner-lieutenant-n-robert-hampton-gray-vc-dsc/: accessed 13 January 2020).
[9] "HMS King George V" (https://www.naval-history.net/xGM-Chrono-01BB-KGV.htm#iceberg: accessed 24 January 2020).
[10] "Canada's last Victoria Cross winner – Lieutenant (N) Robert Hampton Gray, VC, DSC"; "Task Force 57: The British Pacific Fleet, Iceberg 1 Redux Sakishima Gunto, Okinawa Campaign - April 16 - 20, 1945" (http://www.armouredcarriers.com/task-force-57-iceberg-2-ii-british-pacific-fleet: accessed 17 January 2020).
[11] "An Aussie "Loaner" with the British Pacific Fleet" by aussieradar (https://www.bbc.co.uk/history/ww2peopleswar/stories/54/a6132854.shtml: accessed 19 January 2020).
[12] Ibid.
[13] "Task Force 57: The British Pacific Fleet, Iceberg 1 Redux Sakishima Gunto, Okinawa Campaign - April 16 - 20, 1945."
[14] "An Aussie "Loaner" with the British Pacific Fleet."
[15] "Mission: Lifeguard American Submarines in the Pacific Recovered Downed Pilots" by Nathaniel S. Patch (https://www.archives.gov/files/publications/prologue/2014/fall/lifeguard.pdf: accessed 19 January 2020); *Kingfish, DANFS*.
[16] Douglas E. Campbell, *Save Our Souls: Rescues Made By U.S. Submarines During WWII* (Self Published, 2016), 73-74, 287-289.
[17] Ibid, 287-289.
[18] "Goldfish Club History" (https://web.archive.org/web/20130911033017/http://www.thegoldfishclub.co.uk/tgfc/history.html: accessed 24 January 2020).
[19] Campbell, *Save Our Souls: Rescues Made By U.S. Submarines During WWII*, 73-74.
[20] "Canada's last Victoria Cross winner – Lieutenant (N) Robert Hampton Gray, VC, DSC."

[21] "Robert Hampton Gray VC, DSC" *The Crowsnest* Vol. 2 no. 12. Ottawa: Queen's Printer. October 1950. p. 16-7; "A Brilliant Flying Spirit" Lieutenant Hampton Gray, VC, DSC RCNVR by Stuart E. Soward (https://web.archive.org/web/20120415204124/http://www.navalandmilitarymuseum.org/resource_pages/heroes/gray.html: accessed 13 January 2020).
[22] "Brilliant Flying Spirit" Lieutenant Hampton Gray, VC, DSC RCNVR."
[23] Ibid.
[24] "Brilliant Flying Spirit" Lieutenant Hampton Gray, VC, DSC RCNVR"; "Canada's last Victoria Cross winner – Lieutenant (N) Robert Hampton Gray, VC, DSC."
[25] "Canada's last Victoria Cross winner – Lieutenant (N) Robert Hampton Gray, VC, DSC."
[26] "Our Victoria Cross Recipients" (https://www.fleetairarm.com/victoria-cross-medals.aspx: accessed 24 January 2020); David D. Bruhn, Rob Hoole, *Enemy Waters*, 141-145.
[27] "Sutton, Arthur William" (http://www.nauticapedia.ca/dbase/Query/Biolist3.php?&name=Sutton%2C%20Arthur%20William&id=21674&Page=40&input=1): accessed 24 January 2020).
[28] "Just One Cup of Fuel" by Dave O'Malley (http://www.vintagewings.ca/VintageNews/Stories/tabid/116/articleType/ArticleView/articleId/552/The-First-and-the-Last.aspx: accessed 16 January 2020).
[29] Ibid.
[30] Ibid.
[31] Ibid.
[32] "Cdr. William 'Bill' Atkinson" (https://navalandmilitarymuseum.org/archives/articles/high-achievers/cdr-william-bill-atkinson/; "Index of /documents/Royal Canadian Navy Citations" (https://www.blatherwick.net/documents/Royal%20Canadian%20Navy%20Citations/): both accessed 16 January 2020.
[33] "Cdr. William 'Bill' Atkinson."
[34] "HMS Formidable Roll of Honour" (http://www.maritimequest.com/warship_directory/great_britain/pages/aircraft_carriers/hms_formidable_67_roll_of_honour.htm: accessed 19 January 2020).
[35] "Archive Report: Allied Forces" (http://www.aircrewremembered.com/haberfield-john-kerle-tipaho.html: accessed 24 January 2020).
[36] Ibid.

CHAPTER 25 NOTES:

[1] Commander Air Sea Rescue, Ryukyus, Air Sea Rescue, Ryukyus – History of, 26 August 1945.
[2] Ibid.

[3] Ibid.
[4] Ibid.
[5] Ibid.
[6] Ibid.
[7] Ibid.
[8] Commander Air Sea Rescue, Ryukyus, Air Sea Rescue, Ryukyus – History of, 26 August 1945; VF(N) 91 War Diary, July 1945.
[9] Ut supra.
[10] Commander Air Sea Rescue, Ryukyus, Air Sea Rescue, Ryukyus – History of, 26 August 1945.
[11] Ibid.
[12] Ibid.
[13] Bomber Fighting Squadron VBF-88 Unit History, 1 June 1945 to 15 September 1945; "FG-1D Corsair" (https://www.navalaviationmuseum.org/aircraft/fg-1d-corsair/: accessed 25 December 2019).
[14] USS *Yorktown* War Diary, July 1945; Bomber Fighting Squadron VBF-88 Unit History, 1 June 1945 to 15 September 1945; Fleet Air Wing One Air Sea Rescue File No. Cons Inc #25, 26 July 1945.
[15] "Seaplane Crew's Battle for Recognition," Congressional Record, June 18, 1999 106th Congress, 1st Session Issue: Vol. 145, No. 87 — Daily Edition (https://www.congress.gov/congressional-record/1999/6/18/senate-section/article/S7275-1: accessed 25 December 2019); Fleet Air Wing One Air Sea Rescue File No. Cons Inc #25, 26 July 1945.
[16] "Seaplane Crew's Battle for Recognition"; Bomber Fighting Squadron VBF-88 Unit History, 1 June 1945 to 15 September 1945; Fleet Air Wing One Air Sea Rescue File No. Cons Inc #25, 26 July 1945.
[17] Commander Air Sea Rescue, Ryukyus, Air Sea Rescue, Ryukyus – History of, 26 August 1945; Fleet Air Wing One Air Sea Rescue File No. Cons Inc #25, 26 July 1945; "Seaplane Crew's Battle for Recognition"; Donald H. Sweet, Lee Roy Way, William Bonvillian, *The Forgotten Heroes: The Story of Rescue Squadron VH-3 in World War II* (Ridgewood, New Jersey: DoGo Publishing, Ezeread Books, 2000); George Galdorisi, *Leave No Man Behind - The Saga of Combat Search and Rescue* (Minneapolis, MN: Zenith Press, 2008), 87-96.
[18] Ut supra.
[19] Ut supra.
[20] USS *Randolph* War History.
[21] "Hyuga" (https://ww2db.com/ship_spec.php?ship_id=29: accessed 27 December 2019).
[22] USS *Randolph* and VF-16 War History; Commander Air Sea Rescue, Ryukyus, Air Sea Rescue, Ryukyus – History of, 26 August 1945.
[23] Fleet Air Wing One Air Sea Rescue File No. Cons Inc #29, 26 July 1945.
[24] Fleet Air Wing One Air Sea Rescue File No. Cons Inc #29, 26 July 1945; USS *Bon Homme Richard* and VF(N) 91 War Diary, July 1945.
[25] Ut supra.

[26] "Samuel Adams Davis" (https://valor.militarytimes.com/hero/19933: accessed 27 December 2019).
[27] Commanding Officer, Rescue Squadron Six, History of Rescue Squadron Six from 20 September 1944 to 1 September 1945, 3 September 1945.
[28] Ibid.
[29] Ibid.
[30] Ibid.
[31] Ibid.
[32] Ibid.
[33] Commanding Officer, Rescue Squadron Six, History of Rescue Squadron Six from 20 September 1944 to 1 September 1945, 3 September 1945; Commander Air Sea Rescue, Ryukyus, Air Sea Rescue, Ryukyus – History of, 26 August 1945.
[34] Ut supra.
[35] Ut supra.
[36] Commanding Officer, Rescue Squadron Six, History of Rescue Squadron Six from 20 September 1944 to 1 September 1945, 3 September 1945; Commander Air Sea Rescue, Ryukyus, Air Sea Rescue, Ryukyus – History of, 26 August 1945; Fleet Air Wing One Air Sea Rescue File No. Cons Inc #72, 11 August 1945.
[37] Commanding Officer, Rescue Squadron Six, History of Rescue Squadron Six from 20 September 1944 to 1 September 1945, 3 September 1945.
[38] Ibid.
[39] VH-1 War Diary, August 1945.
[40] VH-1 War Diary, August 1945; Fleet Air Wing One Air Sea Rescue File No. Cons Inc #86A, 15 August 1945.
[41] Ut supra.
[42] Campbell, *VP Navy! USN, USMC, USCG and NATS Patrol Aircraft Lost or Damaged During World War II*, 130.
[43] "Tomitaka Kamikaze Special Attack Corps Sortie Site Monument Hyūga City, Miyazaki Prefecture"
(http://www.kamikazeimages.net/monuments/tomitaka/index.htm: accessed 28 December 2019).
[44] Commander Air Sea Rescue, Ryukyus, Air Sea Rescue, Ryukyus – History of, 26 August 1945.

CHAPTER 26 NOTES:
[1] "Rekata Bay Seaplane Base"
(https://www.pacificwrecks.com/airfields/solomons/rekata/index.html: accessed 13 December 2019).
[2] Jenny Scott correspondence of 17 November 2019.
[3] "Rekata Bay Seaplane Base."
[4] Ibid.
[5] "NZDF-Serials Australian & New Zealand Military Aircraft Serials & History" (http://www.adf-serials.com.au/nz-serials/nzhudson.htm: accessed 13 December 2019).

[6] Richard B. Frank, *Downfall: The End of the Imperial Japanese Empire* (New York: Penguin, 1999), 295-296.
[7] Ross, *Royal New Zealand Air Force*, 314.
[8] Ibid.
[9] Jenny Scott correspondence of 17 November 2019.
[10] Ibid.
[11] Ibid.

CHAPTER 27 NOTES:
Authored by Commodore Hector Donohue, AM RAN (Retired).

POSTSCRIPT NOTES:
[1] Bruhn, *Eyes of the Fleet*, 397.
[2] Ibid, 393.
[3] Ibid, 395.
[4] Stewart B. Milstein L-7205, Operation Magic Carpet, Universal Ship Cancellation Society Data Sheet #31, April 2008.
[5] Ibid.
[6] Ibid.
[7] *Coos Bay* War Diary, August-December 1945.
[8] *Coos Bay, DANFS*.

Index

Ada, Bruce Laird (PO, RAAF), 223
Adam, J. D., xxxviii
Adie, Gordon Marsh (F/S, RAAF), 114
Agar, John R. S. (Wg Cr, RNZAF), xli
Agnew, Neil M. E. (FO, RAAF), 177-178
Akers, R. G., 77
Atkinson, William Henry Isaac (Comdr., DSC, CD, RCN), 229
Alagna, A. F., 76
Alexander, A. G., 96
Allen, Robert W., 212
Allison, E. M. (FL, RAAF), xliv
Allsbury, John W., 107
Anderson, Gerald Arthur (Lt., RCNVR), 228-230
Appelbaum, Alex, 39
Armour, Lawrence B., 212
Arnett, Fred, 95
Arthur, R. E., 211
Asbridge, William Bell (Lt., RCNVR), 229-230
Atteberry, George C., 5-8
Australia/Australian
  Army
    2nd Infantry Battalion, 116
    7th Infantry Battalion, 53
  Bowen, 186
  Cairns, xviii, xlii, xliv, xlv, 177, 185-186,
  Darwin, 170, 179-184, 257-262
  Garbutt, xlv
  Lindfield, 186
  Maitland, 256
  Penrith, 259
  Royal Australian Air Force
    No. 1 Rescue and Communications Squadron (later No. 8
      Communication Unit), xviii, xli, xliii, 173
    No. 9 Operational Group (later Northern Command), xvii-xviii
    No. 10 Operational Group (later First Tactical Air Force), xviii
    Air-Sea Rescue Flight
      No. 111, xviii, xliii, xliv, lvii, 170-175
      No. 112, xviii, xlii, xliv, lvii, 179, 184, 267
      No. 113, xviii, xlii, xliv, xlv, lvii, 185, 255, 267
      No. 114, xviii, xlii, xliv, xlv
      No. 115, xviii, xlii, xliv, xlv
    Elementary Flying School, No. 8 (at Narrandera),186

RAAF Base
  Nowra/HMS Nabbington/HMAS Albatross, 223
  Rathmines (Lake Macquarie), xix, xliv, 255, 259
Service Flying Training School
  No. 5, RAAF Base Uranquinty, 257
  No. 7, RAAF Station Deniliquin, 186
Squadrons
  No. 11, 259
  No. 20, 111
  No. 21, 180
  No. 24, 180-181
  No. 30, 260
  No. 31, 187, 189
  No. 42, No. 43, 257-258
  No. 87, 180
Supermarine Walrus amphibian aircraft, 105-106, 176-178, 221-223
Sydney, 115-118, 186, 220, 223
Townsville, 38, 185, 258-259
Baker, Felix Locke, 271
Baker, James E., 55
Baldwin, K. J., 234
Ball, Joe Frederick, 155-161
Ballingall, Robert Charles (Sgt, RAAF), 189-190, 267
Barbey, Daniel E., 261
Barrett Jr., Russell R., xxxviii, 247
Barrowclough, Harold Eric (Maj Gen, NZA), 83
Barta, Joseph, 169
Barth, Robert A., 38
Bateman, S. D., 243
Bauwens, John H., 76
Bayer, Richard W., 10
Beauchamp, Donald Stanley (FL, DFC, RNZAF), 76, 107-109, 267-268
Beckman, James, 137
Bell, John F. (Sub. Lt., RNVR), 230
Benim, Thomas A., 73
Benson, Eugene, 93
Bentley, Jack F., 225
Bergman, D. D., 77
Bergmann (FO, RAAF), 177
Biens, Earl F., 212
Bigger, Warner T., 57-60
Birchall, Leonard (SL, RCAF), xxxvi
Black, William, 20
Bond, Bruce Alexander Godley (F/S, RNZAF), 114
Bogue, Douglas, 169
Bonvillian, William D., xxxviii, 211-212, 233, 236, 247

Boutall, John C., 45
Boyington, Gregory, 31, 51
Bradmore, Alan E. W. (FO, RNZAF), 114
Brandley, Frank A., 55
Brately, R. E., 96
Briggs, E. R., 243
Britain/British
  British Pacific Fleet,
    1st Aircraft Carrier Squadron, 220
    Fleet Air Arm Squadrons
      No. 801, 820, 828, 849, 854, 857, 880, 887, 894, 1700, 1771, 1830, 1833, 1834, 1836, 1839, 1842, 1844: 221
      830, 221, 227
      848, 221-223, 229-230
      1841, 221, 225, 230
  RAF
    No. 58 Squadron, 107
    No. 122 Squadron, No. 222 Squadron, No. 504 Squadron, 164
    No. 333 Squadron, xxxiii
Brown, E. D., 211
Brown, M. E., xxxviii, 247
Brubaker, Ray Edward, 107
Bulman, Charles Ralph (FL, RAAF), 181-184, 267
Burger, Alan D. (Lt., RNVR), 230
Burgess, Emery L., 45
Burke, Arleigh, 67
Burke, Thomas J., 212
Burks, Jesse B., 118
Burris, Howard M., 73
Burton, Leo Boyd, 107
Bush, George H. W., 131-140, 191
Butchart, Stanley, 134
Butterworth, Charles Edgar (Lt., DSC, RCNVR), 228-229
Callahan, H. E., 211
Calo, Vincent J., 212
Cameron, L. M. (FL, RAAF), xliv
Camp, Horace W., 80
Campagna, Paul R., 11-112
Canada/Canadian
  13th Infantry Brigade, xxiii
  Royal Rifles/Winnipeg Grenadiers Battalions, xxiii
Carlisle, W. W., 80
Carlton, William C., 83
Carter, John P., 176-178
Castelar, Louis Lovo, 257
Cavin, Edgar R., 94-95

312 Index

Chiaro, Roland D., 80
Cheverton, Milton R., 55-56
Christiansen, Allen E., 64-66
Chute, D. T. (Lt., RNVR), 225
Clark, R. T. (FL, RAAF), xlv
Claytor Jr., Graham, 9
Coates, Charles, 260
Colin, L. L., 211
Congrove, J. E., 80
Conley, Delbert L., 55, 64, 118
Cook, C. H., 80
Cook, W. C., 80
Cooper, A. G., 211
Cornelius, Robert J., 212
Corrie, Robin Morton (FL, RAAF), xliv, 179-184, 267
Cottrell, L., 211, 213
Cougar, R. D., 212
Crawford, H. G., 245, 247
Crawford, William F., 59
Creech, C. P., 76
Crisp, Kenneth Arthur (SL, RAAF), xliv, 267
Crump, H. P. (Sgt, RNZAF), 75
Cushed, H. S., 244
Cuthbertson (FL, RAAF), 175
Czarnecki, Raymond J., 73
D'Amico, E. E., 76
Davis, Samuel Adams, 240-241
Davis Jr., G. A., 66
Decker, Clayton, liii
Delahunty, A. E. (FL, RAAF), xliv
Delaney, John L., 131-134
Delaney, W. E., 205
DeLong, Phillip C., 83
DeMott, Arthur L., 77
Denison, J. M., xxxvii
Denny, Michael Maynard (Adm. Sir. GCB, CBE, DSO, RN), 223
Dierkes, John H., 243, 245
Digulio, J. M., 211, 217
Doheny (LTJG), 128
Doherty, John R., 136
Donlan (FL, RAAF), 174
Dorton (LTJG), 211
Drake, Donn P., 246-247
Dumas, P. A., 68
Dunn (Lt.), 211
Duckworth, D. D., 107

Durkin, Joseph R., 68
Duxbury, Earl, 10
Eaton, C. E., 76
Ebey, Lealand O., xxxviii, 241, 245, 247
Eddy (Lt.), 211-216
Eichner, K. D., 243
Elkey, John Raymond, 78-79
Ekholm, Fritiof L., xxviii, 20, 266
Ellet family, 128-129
Erdmann, William L., 231-232
Essary, Melvin S., 160
Faella, C. D., 76
Fairley, J. V., 76
Farmiloe, Harry (Sgt, RNZAF), 44-46
Farrington, E. L., 55
Ferganis, J. L., 211, 213
Fields, Thomas M., 191
Fitch, Aubrey W., 27, 35, 74
Flodquist, Robert D., 211
Ford, Eric Valentine (FL, RAAF), 181
Ford Jr., Gerald R., 139-140
France, Robert G., 10
Francis, Alfred Cecil (Lt., RNVR), 230
Francis, H. T. (FL, RNZAF), 77
Francis, Robert William Thompson (FO, RAAF), 190, 267
Frankfort (Lt.), 90
Fraser, Bruce (Admiral of the Fleet, GCB, KBE), 220
Fraser, George K., 272
Frazier, E. F., 77, 79
Fredenburg, F. E., 76
Freet, John Francis, 107
Fuchay (CPhoM), 211
Fulmer (ENS), 96
Galilei, D., 211
Gargas (ENS), 195
Garland, Anthony M. (Lt. Comdr., RNVR), 230
Gass, John (Sub Lt.), 223
Germany, Berlin, Bremen, Cologne, Hamburg, 107
Goodson, B. W., 244
Gordon, Nathan Green, 87-97
Gorham, M., 244
Gotchling, Charles A., 66
Grant, William, 234
Graves, W. I., 211, 216
Gray, Robert Hampton (Lt., VC, DSC & bar, RCNVR), xxiii, xxiv, 225-230
Gregerson, Geoffrey Francis (FL, RAAF), xlv

314 Index

Gresham (ENS), 125
Griffiths, Allen Robert (FO, RAAF), 174
Grilley, Reginald Charles (WO, RAAF), 190, 267
Gross, Albert, 93
Gwinn, Wilbur G., 5
Hall (LTJG), 129
Hall, Donald M., 10
Hall, N. M. (PO, DFM RAAF), 173
Halsey Jr., William F., xxix, xlvi, xlviii, liii, lvi, 25, 41, 51, 74, 81-85, 96, 115, 148-150, 231, 262
Hamman, Charles H., xiii
Hannum, Martin H., 75
Harden, W. A., 68
Hargis, Robert N., 20
Harper, Cecil Kelly, 27, 29, 55, 74
Harrison, Howard, 234
Harrison (LTJG), 16
Hart, Glen E., 72
Hashimoto, Mochitsura, 6
Hastie, Norman William (FO, DFC, RAAF), 189-190, 267, 272, 280
Hayden, Nerva Malcolm, 107
Hayes, J. A., 80
Healan, Claude, 93
Heath, Larry (F/S, RNZAF), 44, 46
Heck, Edwin A., 236-237
Heintz, John A., 225
Hemphill, Norman S., 20
Henderson, William Thomas, 77-79
Hendry, J. A. (FO, RNZAF), 75-76
Hennesy, Gerald C., 237
Hensley, Morgan F., 10
Herbst, Lawrence, 95
Higgins (Lt.), 128
Hitchcock, G. C. (FO, RNZAF), 75-76
Hodgkins, Henry, 233, 236
Hoffman, Herbert Allen, 314
Holdeman, Richard B., 107
Hoover, John H., 156, 206
Hornell, Daniel E. (FL, VC, RCAF), xxxiii, xxxvi
Horner, Theodore J., 83
Howard, Dave (FL, RAAF), 223
Howell, John F., 75
Huff, Robert G., 94-95
Hullett, R. W., 211
Humphrey, Thomas Phillip, 77-79
Hyakutake, Harukichi, 53

Ikeda (Captain), 230
Illingworth, W. (Sub. Lt., RNVR), 225
Inada, Hiroshi, 37
Innis, Walter D., 233, 236, 271
Irwin, Charles W., 230
Iungerich, Alexander, 20
James, William F., xxviii, 20
Japan/Japanese
  Airfields
    Ballale (Ballale Island), 56, 65-66, 79
    Hofu (Honshu Island), 239-240
    Honshu (Honshu Island), 246
    Kara (Bougainville Island), 65
    Lakunai (New Britain), 73-80, 106
    Omura (Kyushu Island), Tomitaka (Kyushu Island), Tsukuba (Honshu Island), Yatabe Genzan (Wonsan, Korea), 246
    Tsuiki (Honshu Island), Usa (Honshu Island), 239
  Bungo Suido, 202-203, 240
  Honshu, 130, 156, 194, 226, 234-239, 245-256, 266
    Himeji, Yonago, 236
    Hiroshima, 10, 226
    Kobe, Osaka Harbor, 236-237
    Maisuru, 226
    Miho, 234, 236
    Ominato Ko, 130, 266
    Tokyo, 41, 128-132, 191, 227, 251, 262-266
      Negishi Wan, 266
  Jizo Saki, 234
  Kazaretto Archipelago, Volcano Islands, Iwo Jima, 131, 137, 156, 170, 191-196, 206, 210
  Kyushu, 197, 202-214, 232-246
    Beppu Wan, 240
    Nagasaki, 226
  Matsushima, 226
  Military
    38th Infantry Division, xxiii
    131st Airfield Battalion, 167
    Combined Fleet, 126-127, 147, 203
    Prisoner Camp 10-A (Palawan), 167
  Ogasawara (Bonin) Islands, 132
  Onagawa Bay, 226-229
  Ryukyu Islands, 197, 220, 231-235, 241-247
    Amami Islands, 197, 202, 212
    Okinawa Islands
      Kerama Retto, 197-216, 232-233
      Okinawa, 41, 139, 150, 170, 191, 197-246

316   Index

       Zanpa, 235
    Osumi Islands, 197
       Yaku Shima, 212
    Sakishima Islands, 197, 220-225
       Miyako Island, 221-212, 222-223
    Tokara Islands, 197
    Yaeyama Islands, Ishigaki Island, 223
   Shikoku, 202, 221, 234, 240, 245
    Ashizuri-saki, 240
   Special Sea Attack Force, 201-205
   Tokuyama Bay, 201
Jeraeum, J. Paul, 96
Johnson, Brengle S., 212
Johnson, W. G., 80
Johnson, W. J., 74
Johnston (Lt.), 177-178
Johnston, Arthur H., 73
Johnston, C. L., 77
Joicey, Ian (Lt., RN), 224
Jones, A. E. (F/S, RAAF), 182
Jones, J. F., 74
Jones, Jack W., 164
Jupp, Donald G. (Sub. Lt., RNVR), 230
Jury, Alan Victor (SL, RNZAF), 34
Kataoka, Toshio, 230
Keefer, E. G., 76
Keene, Thomas R., 136-137
Keller, Clarence A., xxxviii
Kelley, Jack, 96
Kendall, Harry S., 24, 74, 265
Kennedy, Henry Marton (Sgt, RNZAF), 114
Kennedy, John F., 57-62
Kent, P. E., 244
Keresey Jr., Richard E., 60-61
Ketcham, Dixwell, 193, 206
Kilian, John Rutherfold Clark (SL, RNZAF), 164
Kinashi, Takakazu, 112
King, Ernest J., 142
King, Henry, 71
Kinkaid, Thomas C., 87, 148, 169
Kirchoff, Warren A., 10
Kirkland, Elmer, 95
Kirkpatrick, Earl D., 75
Knoertzer, Halford A., 45
Kojima, Nagayoshi, 167
Kouns (Lt.), 211-212

Krueger, Walter, 169
Krulak, Victor H., 59
Kuniaki, Koiso, 217
Kurita, Takeo, 141, 148-151
Langelies, Louis C., 75
Laurenson, Ross (PO, RNZAF), 46
Leadley, Walter (Sgt, RNZAF), 37-46
Leahy, William D., xlvi
Lecklider, Russel P., 77
Lee, F. E., 80
Lefkowitz, Irving D., 10
Little, Ralph C., 75
Lofgren, Mattis G., 107
Lohmeyer, Viv (PO, RAAF), 223
Lord, Thurman C., 212
Lucas, Charlton C., 233, 236, 271
Ludlow, George H., xiii
Lund, Martin, 76
Maberry, Arvie, 189-190, 267
MacArthur, Douglas, xvii, xlviii, lvi, 82, 141-148, 167, 261, 263
MacGill, Robert, 237-239
Macgrane, John (FO, RNZAF), 46
MacGregor, Ronald Bruce Leslie (SL, AFC, RNZAF), 33, 111, 114, 267-268
MacInnis, R. J., 212
Mackley, Winston ("Bill") Brook (FL, DFC & Bar, RNZAF), 106-107, 267-268
MacRae, Murdock Daniel, 39
Maitland, Leslie Alan (Sub. Lt., RNVR), 229-230
Makgill, Richard (SL, RNZAF), 46
Malone, William J., 15
Manser, Burton R., 20
Mansueto (Lt.), 211
Maraist, Joseph George, 77
Marcy, Clarence P., 76
Marks, Robert A., 5-10
Marr, William H., 211
Marsh, Richard, 72
Marshall, A. R. (Sub. Lt.), 223
Marshall, George C., 265
Martin, P. A., 68
Martin, W. R., 76
Marx, Albert B., 68
Mason, G. M. (SL, RAAF), xlv
Mateja, H. M., 66
Maxwell (Lt.), 90

May (Corporal), 60
McCain, John S., 27, 35
McCampbell, David, 147-149, 273
McCloud, Kenneth I., 73
McClure, E. L., 246
McClure, James G., 75
McClure, John C., 71
McCormick, Maxwell William (FL, RNZAF), 114
McCready, David B., 95
McDole, Glen, 169
McGrane, J. R. (FL, RNZAF)., 75
McGregor (SL, RNZAF), 34
McGregor, Hector A. G. (LAC, RNZAF), 34, 114
McLaughlin, J. E., 80
McLean, Gordon Alexander, 194, 198, 207
McMaster, George Willam (FO, DFM RAAF), 173
McQuaid, R. R., 73
McVay III, Charles Butler, 5
Melvill, Noel (Sgt, RNZAF), 46
Michael, Fred D., 19-20
Miller, C. W. (FL, DFM RAAF), xliv, 173-177
Miller, William, 15, 19, 20, 24, 30, 55, 64
Mills, A. L., 76
Mills, Walter Raymond (SL, DFC, AFC RAAF), xliv, xlv, 185-190, 255-256, 267, 279
Mitchell, John N., 62
Mitchell, Ralph J., 74
Mitscher, Marc A., 203
Molinaro, Victor T., 225
Monbart, W. A., 66
Monk, James Benjamin (F/S, DFM, RNZAF), 76, 109-110, 267-268
Moore, John H., 234
Moose, K. A., 211
Morehouse, Albert K., 55
Morison, Jack Eric (FO, RNZAF), 114
Morris, Jack G., 75
Morris, Vernon John Boyd (FO, RAAF), 190, 267
Mortimer, R. F., 225
Morton, Derick (RNZAF), 223
Moss Jr., William W., 39-42
Nadeau, Leo W., 133-138
Netherlands/Dutch
  18th Squadron, 180
  East Indies, 179-180, 186-187
    Ambon Island, 187-188
    Celebes Island, 258

Flores Island, 179-181, 258
  Java, 180
    Padjang Island, 187
    Paternoster Island, 180
    Sumatra Island, 220-221, 227
      Palembang, 227
      Pladjoe, Soengei Gerong, 220
    Sumbawa Island, 180
    Surabaya Island, 180, 257
    Timor Island, Kupang, 180
  Rotterdam, 107
New Zealand
  Army
    3rd Division, 52
  New Plymouth, 107
  Paekakariki, 33
  Royal Air Force
    Base Depot Workshop Unit 60, 111
    Hobsonville, 105, 111
    No. 6 (Flying Boat) Squadron, xxvii, xxxviii, xl, xli, xlvi, xlvii, l, lvi, 33-36, 46, 64-76, 104-111, 254, 267-269
    No. 14 Fighter Squadron, No. 16 Fighter Squadron, No. 19 Squadron, No. 485 Squadron, 164
    No. 20 Squadron, 111, 186, 268
    No. 24 Squadron, 164, 180-181
    Milson Aerodrome (later RNZAF Station Palmerston North), 33
  Wellington, xxxviii
  Woodbourne, 107
Newkirk, Edward C., 198-201
Newton, John H., 115
Nicholson, John Robert Mills (FL, RNZAF), 114
Nimitz, Chester W., xvii, xlvi, 10, 21, 262-264
Nishimura, Shoji, 148-153
Nixon, Richard M., 62, 99-102
Norvell Jr., Forrest H., xxxviii, 233, 236, 247
Nottingham, Fred Charles (Lt. Comdr., RNVR), 224-225
Oakley, R. D., 73
O'Brien, W. J., 76
Ogden, James Robison, 233, 236, 271
Oldendorf, Jesse B., 154
Operation
  BLISSFUL, 59
  FORAGER, 125
  KIKUSUI, 203
  CHERRY BLOSSOM, 53-54
  JUBILEE, 164

MAGIC CARPET, 265
MERIDIAN, xxvi, 220, 230
SHO, 148, 154
Ormesby, Abb (F/S, RNZAF), 45-46
Orr Jr., John L., 158
Osmena, Sergio, 146
Oughton, R., 66
Overby (Lt.), 90
Owen, Hilford C., 272
Owen, R. G., 72
Ozawa, Jisaburo, 148, 150
Pacific Islands
  Admiralty Islands, 32, 82, 110, 162-163, 223
    Los Negros, 110, 162, 164, 252
    Manus, 32, 82, 162, 186, 223
  Bismarck Archipelago, 81-89, 97, 102-110, 140, 163
    Emirau, 32, 77, 81-85, 102, 110, 163-164
    Feni Islands, 77, 163
    Lihir Islands, 163
    New Britain, 54, 67, 75-81, 90, 102-110, 163, 171-175, 186, 268
      Adler Bay, Wide Bay, 174
      Ataliklikun Bay, 110, 163, 268
      Kabanga Bay, 103-111, 268
      Jacquinot Bay, 110-111, 173-177, 268
      Rabaul, 51-54, 68, 73-86, 103, 109-110, 164, 251
    New Ireland, 67, 77-91, 110, 163-164
      Kavieng, 77-97, 164
    St. Matthias Islands, 82, 84
    Tanga Islands, 163
    Vitu Islands, Narega, 90, 97
  Borneo, 3, 185, 187, 255
    Balikpapan, 258
    Brunei, 3, 152, 255
    Kapit, xliv-xlv
    Labuan Island, 185, 255, 258
  Caroline Islands, 2, 81, 132, 139-144, 198, 206
    Palau Islands, 1-2, 32, 123-143
      Angaur, 2, 132-133, 143-144
      Babelthuap, 139
      Kossol Passage, 123-127, 143-144
      Ngesebus, 139
      Peleliu, 1-10, 32, 132-139, 143-144, 156
    Truk, xxix, 127-128, 208
    Ulithi, 123-139, 156
    Yap, xxix, 139-144
  Ellice Islands, Funafuti, 162

Index   321

Fijian Islands, xxxviii-xli, 33-46, 105, 111, 114, 252
   Viti Levu, 45, 105, 112
     Laucala Bay, 3
     Suva, 33, 46, 112-113
Johnston Island, 242
Mariana Islands, 5, 32, 81, 118, 123-128, 140, 150-156, 191-206, 245, 247
   Saipan, 32, 118, 123-129, 143, 156, 194-198, 242-247
   Tinian, 5-10, 126, 129
Marshall Islands
   Eniwetok, 118, 130, 162, 196, 242, 247
     Parry Island, 192, 242, 247
   Kwajalein, 108, 125, 162, 247
Molucca (Maluku) Islands, 141, 170, 187-188
   Morotai, xviii, xlii, xliv, xlv, 3, 133, 141-143, 156, 170, 185-189, 255-258
   Halmahera, 141
Nauru Island, 29, 108
New Caledonia, 34, 39, 46, 100, 117, 186
   Noumea, 34, 39, 41-46, 100, 112-118
New Hebrides, 23, 31-35
   Espiritu Santo, 19, 24-36, 46, 104-105, 111-118, 252
Papua New Guinea, 3, 26, 32, 54, 77, 82, 88-95, 140-145, 157-162, 170-178, 186, 258-259
   Alexishaven, 88
   Annanberg, 174
   Biak, 157
   Boram, xliii
   Dobodura, 95
   Finschhafen, 89-90
   Goodenough Island, xviii, xli
   Hansa Bay, 88
   Hollandia, 88, 140,
   Langemak Bay, 88, 90, 95
   Madang, xviii, xlii, xliv, 170-175
   Manokwari, 258-259
   Milne Bay, xvii, xviii, xli, 82
   Mios Woendi, xxix, 157, 162, 164
   Nadzab, xxx
   Nightingale Bay, 175
   Port Moresby, xviii, xliv, 26, 82, 259
   Sansapor, 3, 141
   Wewak, 54, 88, 172-176
Philippines
   Luzon, 88, 139, 143-149, 161-169, 191
     Dagupan, Rabon, San Fabian, 169
     Lingayen, 165-169
     Manila, 139, 144, 147, 165, 169, 186

322   Index

    San Fernando, 165
  Mindanao, 141, 151, 161
    Mangarin Bay, 166
  Visayan Islands, 142, 144
    Bohol, 142, 161
    Cebu, 88, 142, 157, 161, 165
    Jinamoc Island, 155
    Leyte, 6, 88, 133, 139-169, 191, 274
      Baybay, 157
      Dulag, 146
      Ormoc Bay, 88, 155-161
      San Pedro Bay, 88, 155, 160-167
      Tacloban, 146, 165
    Masbate, Dimasalang, 159
    Mindoro, 140, 147, 166-167
    Negros, 110, 162
    Palawan, 150-151, 166-169
    Panay, 142, 165
    Pilar, 157
    Ponson Island, 157-158
    Samar, 142, 144, 150-166
Samoan Islands, 24, 35
Solomon Islands
  Choiseul, 57-74
  Florida Island, Halavo Bay, 25, 30, 32, 47-49, 104-105, 110, 115
  Gaomi ("Palm Island"), 50
  Gavutu, 27, 48-49
  Green (Nissan) Island, 73, 77, 80-89, 99-104, 110, 163
  Guadalcanal, 24-35, 47-49, 74, 100, 118, 249-252
    Kukum, Lunga Point, 27
  New Georgia Islands, 50-51, 55, 61-64, 71, 77, 115
    Bougainville
      Cape Torokina, Empress Augusta Bay, 53-59, 71-72, 81-82, 176-177
      Kahili, 56, 78
      Moisuru Harbor, Mutupina Point, 78
    Buka Island, 66-67
    New Georgia
      Munda, 51, 71-79, 115
    Rendova, 51-55, 63-71
    Vella Lavella, 51, 58-61
  Pauro Island, 78
  Shortland Islands, 51, 56, 64-66, 77-78
    Ballale, 56, 64-66, 77-79
    Faisi, 56
    Maifu, 56, 79
    Tomass, 64, 66

Taiwan (formerly Formosa), 41, 139-140, 148, 150, 197, 220, 232
Tanambogo, 27, 48-49
Treasury Islands,
   Blanche Harbor, 64-73, 102
   Mono, 69
   Stirling, 53, 69-70
Tonga
   Tongatapu, Nukualofa Harbor, 36, 46, 112-113
Palm (Lt.), 211
Paltridge, Norman John, 114
Parker, B. (FL, RAAF), xliv, 102
Patch, Alexander M., 117
Patrick, Walter L., 96
Patterson, O. H., 71-72
Paulson, G. A., 68
Payne, O., 211
Paynter, William K., 20
Peach, Stuart, 259-260
Peacock, John R., 199
Pehl, H. A., 76
Perry, E. M., 80
Peterson, Mell A., 155, 158
Phillips, B. W., 190
Pine, W. (CPO, RN), 224
Plotczyk, John A., 107
Porterfield, M. H., 76
Powell, D. E., 211
Price, John D., 206
Price, W. N., 80
Priest, G. H. (FL, RAAF), xliv
Prince Harry (Duke of Gloucester), 175
Pryor, C. P., 243
Puddle, Eric Samuel (LAC, RNZAF), 114
Purcell, George H., 71
Quinn, Arthur L., 212
Radford, Arthur W., 224
Randall, Richard H., 20
Rankin, E. P., 55
Rankin, Ronald John (FL, RAAF), xliv, 256-266
Rawlinson, Gordon C. (T/A/PO Airman), 230
Rawnsley, Ivan Evan (Wg Cr, RNZAF), 74
Read, R. W., 80
Read, Robert W., 20
Regan, Christopher David John (FO, DFC, RNZAF), 110-111, 267-268
Replogle, M. C., 211
Rice, R. L., 211

324　Index

Riddle, J. R. (PO, DFM, RAAF), 173
Ridgeway, Norman, 271
Richwine, Carl L., 244
Rigger, Ralph (Sgt, RNZAF), 46
Riley Jr., W. D., 243
Robran, Billy, 77
Roberts, R. K., 190
Robertson, C. A., 225
Robertson, G. E., 244
Robey, F. S. (SL, RAAF), xliv
Rollins, W. F., 66
Rosasco, Robert A., 55
Ross, James Finlay (Lt., RCNVR), 229-230
Routon, R. W., 96
Rushing, Hollie, 93
Rushing, Roy Warrick, 149, 273-234
Russell, John Wilfred (LAC, RNZAF), 114
Rutz, W., 211
Sandell, Mandel, 76
Sato, Yoshikazu, 168
Sayer, Walter W. T. (F/S, RAAF), 182
Schaeffer, R. A., 76
Schmehl, D. L., 243-244
Schmidt, Harry, 192
Schmidt, John W., 34
Schnell, Edward J., 60-61
Schoenweiss, Carl W., 55
Schutz, Homer L., 212
Schwartz, Jack Bogart, 71
Scott, Alastair (FO, RNZAF), xxviii, 249-254, 266
Scott, I. A. (Wg Cr, RNZAF), xli,
Scott, Marcellius James, 114
Scott, R. L. (FL, RNZAF), 77
Searles, John M., 233, 236
Seely, Robert W., 80
Selway, John Gilbert, 240
Shanks, E. M., 76
Sheppard, Donald John (Comdr., DSC, RCN), 229
Sheppard, Robert Ross (Lt., RCNVR), 229-230
Shigeowas, Shinohara, 201
Shilling, K. R. (WO, RAAF), 182

Ships and Craft
　Australian *Quiberon*, 182
　British
　　Euryalus, 222
　　Formidable, Implacable, Indefatigable, Indomitable, Victorious, 219-230

Illustrious, xxi, 219-230
King George V, 222, 224
Prince of Wales, xxii
Repulse, xxii
Spark, 180
Canadian, *Uganda*, xxiii
German *Gneisenau, Prince Eugen, Scharnhorst, Tirpitz*, 227
Japanese
  Abukuma, Akebono, Asagumo, Fuso, Hatsuharu, Michishio, Mogami, Nachi, Shigure, Shiranuhi, Ushio, Wakaba, Yamagumo, Yamashiro, 153
  Akebono Maru, xxxi
  Akikaze, Akizuki, Chitose, Chiyoda, Hatsuzuki, Kiri, Kuwa, Maki, Oyodo, Shimotsuki, Sugi, Tama, Wakatsuki, Zuiho, 150
  Akishimo, Ashigara, Atago, Chikuma, Chokai, Fujinami, Haguro, Hamanami, Haruna, Hayashimo, Kishinami, Kiyoshimo, Kongo, Kumano, Maya, Musashi, Naganami, Nagato, Noshiro, Nowaki, Okinami, Shimakaze, Suzuya, Takao, Tone, Urakaze, 151
  Amakusa, Etorofu, 226
  Asashimo, 151, 202-204
  Fuyutsuki, 202, 205
  Hakuun Maru, Hassho Maru, 137
  Hamakaze, 151, 202, 204
  Hatsushimo, 153, 202, 205
  Hyuga, 150, 239
  I-8, 201
  I-19, 33, 112
  I-21, 36-37
  I-58, xxv, lvi, 1, 5-6
  Ise, 150
  Isokaze, Yahagi, Yukikaze, 151, 202, 205
  Isuzu, 150, 180-183
  Kari, W-34, 180
  Kasumi, 153, 202, 205
  Kikuzuki, 47
  Kongo Maru No. 2, 228
  Shokaku, Taiho, 126
  Suzutsuki, 202, 204-205
  W-12, 180, 183
  Yamato, 201-205
  Zuikaku, 149-150
United States
  Merchant Marine/War Shipping Administration
    Cape San Juan, 36-44
    Edwin T. Meredith, 39-43
    Matagorda, 130, 266
    Monroe, 100

Phoebe A. Hearst, 112-113, 268
William K. Vanderbilt, 33
Navy
  aircraft carriers
    Battan, Bennington, Independence, Santee, Suewanee, 212
    Belleau Wood, 205, 212
    Bon Homme Richard, 234, 240
    Cape Gloucester, 242
    Chenango, 212, 224-225
    Essex, 147, 149, 273-274
    Enterprise, Franklin, 136-137, 153
    Gambier Bay, 152, 154
    Hornet, 144, 212
    Lexington, 126
    Marcus Island, xxxvii
    Monterey, 139-140
    Princeton, St. Lo., 154
    Randolph, 212, 239-240
    San Jacinto, 131-140
    Saratoga, 224
    Shangri-La, 212, 234
    Yorktown, 47, 234, 236
  aviation support ships
    seaplane tenders
      Absecon, Barnegat, Biscayne, Castle Rock, Corson, li, 271
      Albemarle, lii
      Avocet, Gannet, Heron, Lapwing, Pelican, Sandpiper, Swan, Teal, Thrush, xlix
      Ballard, xxxii, l
      Barataria, li, 13, 271
      Belknap, Childs, Clemson, George E. Badger, Gillis, Goldsborough, Greene, Hulbert, McFarland, Osmond Ingram, William B. Preston, Williamson, l
      Bering Strait, li, 13, 198, 207, 233, 236, 245, 247, 271
      Casco, Cook Inlet, Duxbury Bay, li, 13, 271
      Chandeleur, lii, 28, 55, 143, 198, 212-213
      Chincoteague, li, 13, 28-30, 52, 55, 70, 80, 102, 193-196, 271
      Coos Bay, xxi, xxviii, xli, l, li, lvi, lvii, 11-24, 28, 30, 32, 47, 50-51, 55, 64, 66-80, 99, 102-104, 115-118, 123-124, 127-130, 266, 271
      Cumberland Sound, lii, 263
      Currituck, lii, 166
      Curtiss, lii, 27-35
      Floyds Bay, li, 13, 233-247, 271
      Gardiners Bay, li, 13, 233-247, 263, 271
      Greenwich Bay, li, 13
      Half Moon, li, 13, 104, 271
      Hamlin, lii, 193-198, 207, 263

Humboldt, Matagorda, Rehoboth, Rockaway, Unimak, li, 271-272
Kenneth Whiting, lii, 123-127
Langley, li-lii
Mackinac, li, 13, 28, 143, 233, 236, 241, 263, 271
Norton Sound, lii, 207, 241
Onslow, li, 13, 143, 198, 207, 271
Orca, li, 13, 155-161, 271
Pine Island, lii, 233-247
Pocomoke, lii, 55, 143
Salisbury Sound, xlvii, lii
San Carlos, li, 13, 164-165, 271
San Pablo, li, 13, 96, 272
Shelikof, li, 13, 127, 198, 207, 271
St. George, lii, 198
Suisun, li, 13, 207, 233-247, 263, 271
Tangier, lii, 117, 156, 165-167
Thornton, xxxii, l, 30
Timbalier, Valcour, li, 13
Wright, xli, li, lii, 24, 35-36, 55, 66-74
Yakutat, li, 13, 143, 198, 207, 272
amphibious/high speed transports
  American Legion, 54
  Eldorado, 207
  Crosby, Kilty, McKean, Ward, 59
  LST-446, 81
auxiliaries
  Allioth, 130
  Kewaydin, 18
  Las Vegas Victory, Tomahawk, 199
  Mataco, 17
  Mindanao, 115
  Nantahala, 118
combatants
  battleship
    Arizona, 15-16
    Illinois, 14
    Missouri, 235, 262-263
    North Carolina, 208
  cruisers
    Concord, 15
    Indianapolis, xxv, lvi, 1-10, 158
    Louisville, 154
  destroyers
    Allen M. Sumner, 157-158
    Baldwin, 17-18
    Braine, 210, 212

Buchanan, 143
Cecil J. Doyle, xxv, lvi, 9
Cooper, 155-160
Dempsey, 45-46
Ellet, 128-129
Eversole, 154
Gillespie, 199
Halford, Wadsworth, Waller, 74
Hoel, Johnston, Samuel B. Roberts, 152, 154
Lansdowne, 24
LeHardy, 18
McCalla, 44-46
Moale, 157-158
Morrison, 199-201
Stockton, 199-201
Whipple, lii
submarine chaser
SC-654, 44-46
motor torpedo boats
PT-59, PT-236, 57-62
PT-109, 58
PT-164, 63
PT-465, 235
yard patrol craft *YP-72, YP-284, YP-290, YP-345, YP-350*, xxxii
mine warfare
minesweepers
Dash, 111-113
Pursuit, 18
YMS-89, 113-114
YMS-241, 44-46
submarines
Besugo, Charr, Gabilan, 180-183
Bluefish, Kingfish, 224-225
Finback, 131-137
Hackleback, Threadfin, 202
Quillback, 235
Peto, 223
Pomfret, 244
Tang, liii, 208

Simms, Richard L., 205-216
Smiley, Curtis S., 55
Smith (Rear. Adm.), 127
Smith, B. (Lt.), 71-73
Smith, Donald W., 80
Smith, G. B., 234-235
Smith, Eugene Victor, 72

Smith, K. G. (Wg Cr, RNZAF), xli, 250, 254
Smith, Warren L., 240
Smith, William J., 93, 95
Solomon (Lt.), 211-212
Sowers, R. A., 80
Spratlin, Dean, 137
Spruance, Raymond A., 141, 231
Stahl, Paul L., 233
Staniforth, T. H. (Sub. Lt., RNVR), 225
Stead, George Gatonby (Wg Cr, RNZAF), xli, 34
Steele, D. C., 237
Steenis, Dallas Earle, 107
Stovall, James T., 136
Stradwick, Walter Thomas (Sub. Lt., RNVR), 230
Strong, Walter Mervyn, 37
Stuart, William (Lt. Comdr., RNVR), 224
Studley, E. M., 72
Sturdevent, H. W., 212
Suaggi, Eugene V., 200
Sutherland, Woodford W., 165
Sutton, Arthur William (Lt., RCNVR), 227
Sutton, Frank C., 55
Suzuki, Kantaro, 217
Swart, Bouke, K. (Sub. Lt., RNNAS), 221, 230
Takamasa, Mukai, 201
Taylor, Arthur Frederick (T/F/Sgt, RAAF), 190, 267
Teetsel, Ernest W., 212
Terauchi, Hisaichi, 167
Thacker, J. R., 211
Thomas, Gordon, 174
Thomas, William H., 73
Thomason, Ernest Y., 212
Thompson, J. C., 243
Thompson, R. C., 76
Thorn, F. B., 71
Thorvaldson, J., xxx
Thurman, Jack L., 163
Toguri, Iva (Tokyo Rose), 71
Toyoda, Soemu, 141, 147, 203
Trojnar, Thaddeus J., 83
Truman, Harry S., 261
Tucker, A. F. (FO, RNZAF), 80
Turner, C. E. (or L.), 111
United States
  Army
    Sixth Army, 145, 148, 169

Tenth Army, 197-198
27th Infantry Division, 125, 197
31st Infantry Division, 143
81st Infantry Division, 128, 143
Air Corps
  1st Fighter Control Squadron, 38
  Fifth Air Force, xvii
  Thirteenth Air Force, 106, 156, 242
  Twentieth Air Force, 217, 234
  6th Emergency Rescue Squadron, 243
  26th Bombardment Squadron, 108, 267
  31st Bombardment Squadron, 76
  69th Bombardment Squadron, 68, 80
  70th Bombardment Squadron, 71, 75
  71st Bombardment Squadron, 93
  75th Bombardment Squadron, 76
  106th Reconnaissance Squadron, 77
  253rd Ordnance Company, 38
  333rd Fighter Squadron, 246
  339th Fighter Squadron, 72-73
  341st Fighter Squadron, 243-244
  345th Bombardment Group, 89
  498th Squadron, 94-95
  500th Bombardment Squadron, 95, 165, 244
  855th Engineers, 38
  Air-Sea Rescue Group 5276 (5th Air Force), 165
  Far Eastern Air Force, 234
  Nadzab Airfield (New Guinea), 177
  Tadji Airfield (New Guinea), 178
Marine Corps
  1st Marine Division, 27, 49, 143, 197
  2nd Marine Division, 21
    2nd Marine Parachute Battalion, 59
  3rd Marine Division, 53, 59, 85, 192
  4th Marine Division, 84, 125
  5th Marine Division, 191-192
  6th Marine Division, 197
  Air Wing Two, 234
  VMF-114, 80
  VMF-123, VMF-312, 212
  VMF-212, 83
  VMF-214, 31, 51, 80
  VMF-215, 72
  VMF-216, 76, 83-84
  VMF-218, VMF-222, 76
  VSMB-235, 80, 103

VMSB-236, 65
VSMB-241, 103
VMSB-244, 76, 103
VMTB-134, 80
VMTB-143, 72-73
VMTB-242, 80
Navy/Naval
  Aviation
    Air Base/Air Station/Air Field
      Corpus Christi (Texas), 132
      Henderson (Guadalcanal Island), 27, 74
      Honolulu (Hawaii), 124
      Kaneohe Bay (Oahu), 162, 242
      Majuro (Majuro Atoll), 247
      Ondonga (New Georgia Island), 55-56
      Ottumwa Naval Reserve Aviation Base (Iowa),
      Parry (Eniwetok Atoll), 100
      Pearl Harbor (Hawaii), 15, 23, 32, 137
      Quonset Point (Rhode Island), 100
      Seattle (Washington), 16
      Tanapag (Saipan), 124, 195, 245, 247
    Air-Sea Rescue Unit
      Marianas/Central Marianas, 245, 247
      Ryukyus, 231-248
    Air Transport Service (NATS), xxxvii, 39, 124
    Bombing Squadron VB-148, 77-78
    Bombing and Fighter Squadron
      VBF-12, 212
      VBF-85, 211-212
      VBF-88, 236-237
    Composite Squadron VC-40, 77
    Fighting Squadron
      VF-16, 126
      VF-17, 72-73, 212
      VF-24, VF-25, VF-30, VF-46, VF-47, 211-212
      VF-51, 131, 133
      VF-85, 211-212, 234
    Night Fighter Squadron VF(N)-91, 234, 240
    Patrol Aircraft Service Unit PATSU 17-3-1, 127, 165
    Patrol Squadron/Patrol Bombing Squadron
      VPB-11, 27, 156-157
      VP-14, 71-73
      VPB-16, VPB-202, VPB-216, 127, 143
      VP-17/VPB-17, 123-127
      VP-18/VPB-18, 118-128
      VPB-19, 194-196

332   Index

   VPB-20, 166
   VPB-21, 198-213, 265
   VP-23/VPB-23, 1, 3, 5, 10, 27-28, 53-55, 63-66, 195
   VPB-25, 165
   VPB-27, 198, 207
   VP-33/VPB-33, 122, 162-165
   VP-34/VPB-34, 87-97, 155-162
   VP-52, xxix
   VPB-54, 55, 162, 165
   VP-71/VPB-71, 27-32, 48-55, 63-64, 69, 73-77, 165
   VP-91, 19, 27, 69, 77-80, 102-103
   VP-102, 123
   VPB-104, 165
   VPB-109, VPB-118, 206
   VPB-133, 4
   VPB-152, 1-7
  Rescue Squadron
   VH-1, xxxvi, xxxviii, 124-128, 143-144, 198, 233, 245, 247
   VH-2, 194-195, 233, 247
   VH-3, 207-213, 232-247
   VH-4, xxxvi, xxxviii, 233-247
   VH-5, 233, 247
   VH-6, 231-247
  Scouting Squadrons One and Three, 15
  Torpedo Squadron
   VT-25, 212, 225
   VT-40, 212
   VT-51, 131-139
 Destroyer Division 120, 158
 Hospitals No. 20 Peleliu and No. 114 Samar, 10
 Indian Island Bomb Depot/Torpedo Station, Keyport, Washington, 16
 Motor Torpedo Boat Squadron
  19, 58
  31, 235
 Naval Construction ("Seabee") Battalion, 52, 101
 Submarine Chaser Training Center, Miami, Florida, 62
 West Coast Sound School, 19
Valentin, A. C., 80
Van Pelten, R. E., 76
Vandchaick, John H., 75
Vanderpump, Mortimer Tuke (SL, RNZAF), 163-164
Vaughan, James J., 233, 236
Vian, Philip (Admiral of the Fleet, GCB, KBE, DSO & Two Bars), 220, 226
Vickers, Colin George (WO, RAAF), 181-182
Vieweg, Walter V. R., 55
Villalbos, Vernon H., 73

Vraciu, Alexander, 126
Yoder, Calvin Bert, 240
Voges, C. A. (SL, RAAF), xliv
Vorachek, C. W., 73
Wagner, Frank D., 16
Wakeford, Jack (Sgt, RNZAF), 46
Waldie, Ivan James (LAC, RNZAF), 114
Warner, John Edward, 71
Watts, P. H., 80
Weimer, Edward L. B., 55
Welch, L. D., 211
Wells, Harold A., xxxviii, 195, 247
West, Douglas, 135
West, Robert, 20
White, William G. (FL, RAAF), xliv, 131-134
Whitehead, Douglas R. (Sub. Lt., RNVR), 230
Whitehead, Richard F., 207
Whiteley (Lieutenant), 66
Wilkinson, Theodore S., 83, 128, 132, 281
Williams Jr., Robert R., 136
Wolf, R. W., 211
Wood, Douglas Edmund (PO, RNZAF), 78, 114
Wood, Ian James Lock (FL, DFC, RAAF), xliii, xliv, 172-173
Wood, Norman Edwin, 78
Wood, T. S., 90-91
Woolery, R. C., 80
Zollinhofer, Joseph D., 77

About the Author

Stephen Allen Ekholm was born on June 24, 1949 at Swedish Hospital in Seattle, Washington, the second son of Fritz and Dorothea Elholm. His father was in Alaska at the time, working at his factory off the Bering Sea. Upon his return to Seattle, it was learned that both father and son had polio. The former succumbed to the disease two months later.

Stephen attended Lakeside Academy in Seattle through high school and then the University of Washington. He earned a Master's degree in Divinity at Andover Newton Theological School in Newton, Massachusetts, and was ordained as a Congregational Minister. He served local churches in New England and later earned his D.Min. at Andover Newton, as well.

He met and married his wife, Lorraine, in New Hampshire. They went on to serve a church in Massachusetts, where their two sons were born. The family later moved to Dallas, Texas, where he was accepted at the Pastoral Counseling and Education Center. He also served as a pastor at a church in Plano, Texas.

After earning his credentials for counseling, Stephen joined a small practice in Oklahoma City, Oklahoma. He worked there for several years until the early 1990s when it became necessary to relocate back to the northwest to care for his ill mother. He and his family continue to reside in the Seattle area.

About the Author

Commander David D. Bruhn, U.S. Navy (Retired) served twenty-two years on active duty and two in the Naval Reserve, as both an enlisted man and as an officer, between 1977 and 2001.

After completion of basic training, he served as a sonar technician aboard USS *Miller* (FF-1091) and USS *Leftwich* (DD-984). He was commissioned in 1983 following graduation from California State University at Chico. His initial assignment was to USS *Excel* (MSO-439), serving as supply officer, damage control assistant, and chief engineer. He then served in USS *Thach* (FFG-43) as chief engineer and Destroyer Squadron Thirteen as material officer.

After graduation from the Naval Postgraduate School, Commander Bruhn was assigned to Secretary of the Navy and Chief of Naval Operations staffs as a budget analyst and resources planner before attending the Naval War College in 1996, following which he commanded the mine countermeasures ships USS *Gladiator* (MCM-11) and USS *Dextrous* (MCM-13) in the Persian Gulf.

Commander Bruhn's final assignment was executive assistant to a senior (SES 4) government service executive at the Ballistic Missile Defense Organization in Washington, D.C.

Following military service, he was a high school teacher and track coach for ten years, and is now a USA Track & Field official. He lives in northern California with his wife Nancy and has two grown sons, David and Michael.